Identification of Plant Metabolites: Characterization and Biological Activities

Identification of Plant Metabolites: Characterization and Biological Activities

Guest Editors

Ramona Paltinean
Irina Ielciu

MDPI

Basel • Beijing • Wuhan • Barcelona • Belgrade • Novi Sad • Cluj • Manchester

Guest Editors
Ramona Paltinean
Department of
Pharmaceutical Botany
"Iuliu Hațieganu" University
of Medicine and Pharmacy
Cluj-Napoca
Romania

Irina Ielciu
Department of
Pharmaceutical Botany
"Iuliu Hațieganu" University
of Medicine and Pharmacy
Cluj-Napoca
Romania

Editorial Office
MDPI AG
Grosspeteranlage 5
4052 Basel, Switzerland

This is a reprint of the Special Issue, published open access by the journal *Metabolites* (ISSN 2218-1989), freely accessible at: www.mdpi.com/journal/metabolites/special_issues/Plant_Metabolites_Characterization.

For citation purposes, cite each article independently as indicated on the article page online and using the guide below:

Lastname, A.A.; Lastname, B.B. Article Title. *Journal Name* **Year**, *Volume Number*, Page Range.

ISBN 978-3-7258-3518-8 (Hbk)
ISBN 978-3-7258-3517-1 (PDF)
https://doi.org/10.3390/books978-3-7258-3517-1

Contents

metabolites

Review

Nutraceutical Properties of *Medicago sativa* L., *Agave* spp., *Zea mays* L. and *Avena sativa* L.: A Review of Metabolites and Mechanisms

Tannia A. Quiñones-Muñoz [1,*], Socorro J. Villanueva-Rodríguez [2] and Juan G. Torruco-Uco [3]

1 Consejo Nacional de Ciencia y Tecnología (CONACYT)—Centro de Investigación y Asistencia en Tecnología y Diseño del Estado de Jalisco (CIATEJ), Av. Normalistas 800, Colinas de la Normal, Guadalajara C.P. 44270, Mexico
2 Centro de Investigación y Asistencia en Tecnología y Diseño del Estado de Jalisco (CIATEJ), Av. Normalistas 800, Colinas de la Normal, Guadalajara C.P. 44270, Mexico
3 Tecnológico Nacional de Mexico/Instituto Tecnológico de Tuxtepec, Calzada Dr. Víctor Bravo Ahuja, 561, Col. Predio el Paraíso, San Juan Bautista Tuxtepec C.P. 68350, Mexico
* Correspondence: taquinones@ciatej.mx

Abstract: Plants are the main sources of bioactive compounds (nutraceuticals) that function under different mechanisms of action for the benefit of human health. Mexico ranks fifth in the world in biodiversity, offering opportunities for healthy food. An important variety of crops are produced in the state of Hidalgo, e.g., based on the 2021 production, alfalfa, oats, maguey, and corn. The present review presents the latest findings of these crops, regarding the benefits they provide to health (bioactivity, nutraceuticals), and presents the compounds and mechanisms identified by which the benefit is provided. The knowledge compiled here is for the benefit of the recovery of the crops, the recognition of their bioactivities, in search of identifying the best routes of action for prevention, treatment and possible cure of chronic degenerative diseases (thereby promoting crop valorization). Exhaustive bibliographic research was carried out by means of engines and scientific databases. Articles published between 2001 and 2022 that included specific keywords (Scopus, EMBASE, EBSCO, PubMed, Science Direct, Web of Science, Google Scholar). Outstanding activities have been identified for the compounds in the crops, such as antiinflammatory, anticholesterolemic, antihypertensive, antidiabetic, anticancer, antimicrobial, antioxidant, and chelating. The compounds that provide these properties are total phenols, phenolic acids, tannins, anthocyanins, carotenoids, iso-flavones, phytosterols, saponins, fructans, glycosides, glucans, avenanthramides, and polysaccharides.

Keywords: crops; bioactivities; functional food; antioxidant compound; bioactive extracts

Citation: Quiñones-Muñoz, T.A.; Villanueva-Rodríguez, S.J.; Torruco-Uco, J.G. Nutraceutical Properties of *Medicago sativa* L., *Agave* spp., *Zea mays* L. and *Avena sativa* L.: A Review of Metabolites and Mechanisms. *Metabolites* **2022**, *12*, 806. https://doi.org/10.3390/metabo12090806

Academic Editors: Ramona Paltinean and Irina Ielciu

Received: 13 August 2022
Accepted: 24 August 2022
Published: 28 August 2022

1. Main Crops of Hidalgo, Mexico, and Their Use in the Gastronomy and Health

Bioactive compounds present in plants (more than 2 million identified), also called phytochemicals, are produced by the defense system that is activated by the presence of biotic and abiotic stress conditions; in addition to improving the general health status of plants, they participate in molecular signaling and in plant-environment interaction [1,2]. The biosynthesis of these secondary metabolites in plants has specific localization organs from which they are transported to the whole plant and the main storage parts (vacuoles) [3]. The recovery of these compounds can be developed from any part of the plant (roots, stems, leaves, somatic embryos, callus, flowers) but it should be considered that they can vary in type and concentration depending on the part used, and therefore, in the bioactivity detected. Bioactivities can be developed from various mechanisms of action, both beneficial to the plant and to humans. Due to the great use that can be made of these compounds, it is necessary to know in depth the factors that influence their production, recovery and maintenance techniques, their mechanisms of action, in addition to an emerging factor in recent years, climate change, which without much explanation, could modify the advances already known about these compounds, their natural sources and possible changes of adaptation.

Mexico ranks fifth in the world in biodiversity, offering opportunities for healthy food. Ninety percent of the population of the state of Hidalgo, Mexico, works in the agricultural sector and, what makes the sector a strategic area for said state, as well as for food sovereignty of Mexico. Hidalgo contributes 2.9% of the national volume of agricultural products [4]. Grain corn has a production value for the year 2021 of $2,756,711.47, green forage oats a value of $104,610.10, of a total annual value for the state of $5,081,511.37 (thousands of Mexican pesos); therefore, the importance of these crops (agricultural year) in production and economy is strengthened. The production values of alfalfa and maguey pulquero, the most important production crops (perennial), are greatest, amounting to $1,448,552.37 and $576,812.51, respectively, of a total annual amount of $2,601,538.05 (thousands of Mexican pesos). The third and fourth place in terms of production value (perennial) have an economic value of $74,165.64 and $187,940.07 (thousands of Mexican pesos), which is a lower production value than alfalfa and pulquero maguey (reported as produced honey water) [5]. The criteria considered for the selection of the crops under review at the bioactive level were focused on the annual production and the economic value they represent, as shown in Table 1 and in the description of the previous lines.

Table 1. Main crops and their production in 2021, in the State of Hidalgo.

No.	Crop	Area (Ha)			Production (Tons)	Yield (Tons Ha^{-1})	Harvest
		Sown	**Harvested**	**Loss**	**Obtained**	**Obtained**	
1	Grain corn (white)	210,343.55	201,453.30	8890.25	618,156.85	3.07	Agricultural year
2	Green forage oats	24,161.78	24,049.78	112.00	324,378.96	13.49	
3	Grain barley	106,956.06	106,113.16	842.90	223,595.43	2.11	
4	Beans	17,420.18	17,381.18	39.00	13,338.79	0.77	
1	Green alfalfa	43,829.00	43,829.00	0.00	4,477,712.05	102.16	Perennial
2	Pulquero maguey (honey water: thousands of liters)	4842.20	1372.20	0.00	110,411.07	80.46	
3	Orange (Valencia)	5747.90	5478.50	0.00	65,627.47	11.98	
4	Cherry coffe	23,069.50	23,014.50	0.00	29,301.60	1.27	

Source: Own elaboration with data from the Agri-Food and Fisheries Information Service [5].

On the other hand, Hidalgo stands out for the diversity of its gastronomy characterized by the presence of dishes made with exotic ingredients, traditionally prepared with sophisticated culinary techniques. Within the list of representative foods, the predominant crops in the region are widely present, maguey and corn. Worms are obtained from the maguey; mixiotes (enchilada meat) and ximbo (cooked meat rolled in stalk) are cooked with a film obtained from the stalk; gualumbos or quiotes are obtained from their flowers; pastes, zacahuil (corn), peanut tamale, guajolotes (telera cakes with beans and enchiladas), moles, tecoquitos, bocoles, and molotes are made from corn. Another regional dish is escamoles, which are the larvae or roe of the scale ant [6].

As can be seen in the preceding paragraphs, in the State of Hidalgo, Mexico, important crops are identified by production and tradition, which have an important impact on the economic development of the State. The present review presents the latest findings of these crops, regarding the benefits they provide to health (bioactivity, nutraceuticals), and presents the compounds and mechanisms identified by which the benefit is provided. No other similar work can be found, neither for the same crops nor with the detailed analysis of properties, mechanisms, concentrations, and types of samples (extracts, plant portions), as discussed in the present work. The knowledge compiled here is, in addition, for the benefit of the recovery of the crops, the recognition of their bioactivities, in search of identifying the best routes of action for prevention, treatment, and possible cure of chronic degenerative diseases (thereby promoting their valorization). In this context, the objective of the present work is to identify the recent findings regarding the main crops of the State of Hidalgo, Mexico, their bioactive properties, and the compounds (nutraceuticals) that develop them, for their better use in the prevention, treatment, or cure of chronic degenerative diseases.

2. Nutrients and Bioactive Compounds in Important Crops

The main crops sown in Hidalgo have a variety of applications in the diet, mainly nutrients and bioactive compounds. Considering and valuing these properties that crops contribute to human health would strengthen the appropriation of regional cultivation, influencing food sovereignty and regional development. By taking advantage of all these crops in an integral way, production would be boosted, generating direct and indirect economic, social, nutritional, and cultural benefits, among others.

It is important to remember that the biological activities determined are not exclusive to a molecule per se (unless the solution has been so prepared) considering its physicochemical and structural characteristics (conjugated double bonds, number and position of methyl and hydroxyl groups), since in extracts, it can be attributed to effects of complexes formed with the individual molecules (synergistic, additive or antagonistic) and to the sensitivity of each molecule or complex to different analytical techniques (free radical inhibition, metal reduction). The bioactive properties discussed below may be a response of individual properties or interactions, depending on the sample analyzed.

2.1. *Alfalfa (Medicago sativa* L.)

According to the Statistical Yearbook of Agricultural Production, SIAP/SADER 2021 [5], the alfalfa production reported for the agricultural year 2020 (perennial cycle) (Table 1) corresponds to the total produced of green alfalfa, not including shrunken alfalfa production. Hidalgo occupies the second national place in alfalfa production (12.95%), only after Chihuahua (7,780,182.40 tons) with a production of 4,477,712 tons, for the same agricultural year (2021). Alfalfa generally is used as fodder, green or dried, in salads and flavored waters. The seed germinates from 2 or 3 °C, with an optimum temperature of 28 to 30 °C, and up to 38 °C, and can survive extreme droughts. It is rich in proteins, minerals, and vitamins [7].

It contains secondary metabolites such as saponins, coumarins, isoflavones, and alkaloids; their content differs with the type of cultivar, tissue, and stage of development. The aerial parts of alfalfa contain mainly glycosides of medicagenic acid substituted at C-3 by glucose or glucuronic acid, zanhic acid, and soyasaponin I tridesmoside [8,9]. It has been reported that the chemical composition of alfalfa sprouts (from the third day) presents water (869.1 g kg^{-1} DM), crude protein (68.2 g kg^{-1}), ethereal extract (5.2 g kg^{-1}), crude fiber (30.9 g kg^{-1}), and ash (20.4 g kg^{-1}). The composition of phytochemicals and bioactive compounds includes phytoestrogens, sterols, tocols, carotenoids, and saturated and unsaturated fatty acids. The main isoflavones found in alfalfa are secoisolariciresinol diglucoside, daidzein, secoisolariciresinol, coumestrol, isolariciresinol, hydroxymatairesinol, and matairesinol. The main sterols found in alfalfa are stigmasterol (1096.8 mg kg^{-1} DM), avenasterol (405.9 mg kg^{-1} DM), β-sitosterol (324.2 mg kg^{-1} DM), and campesterol (49.5 mg kg^{-1} DM). Important antioxidant compounds have also been identified: α-tocopherol (314.1 mg kg^{-1} DM), γ(β)-tocopherol (24.4 mg kg^{-1} DM), α-tocotrienol (4.1 mg kg^{-1} DM), δ-tocopherol (2.7 mg kg^{-1} DM), and γ-tocotrienol (2.1 mg kg^{-1} DM) [10].

Some reported uses for alfalfa seed flour (composition: 14.71% total starch, 37.59% crude protein, 3.74% ash, 26.22% total dietary fiber, 6.71% soluble dietary fiber, 19.51% insoluble dietary fiber) include improvement of the nutritional value of gluten-free biscuits using different levels of substitution for common rice flour (0, 15, 30, 45% *w*/*w*) [11]. In the same context of composition, some studies have reported the factors that may be responsible for variations in the composition and bioactivity of alfalfa (biotic and abiotic) and its biological effect by incorporation in diets. In addition to the inherent factors of plant growth conditions, the effect of light (LED), sound waves, drying, soaking, fermentation, and incorporation of selenium into the crop have been determined in alfalfa, which have diverse effects depending on the stage and potency of incorporation and/or application and the plant's own metabolism. The incorporation of alfalfa in diets has improved the bioactive composition of the diets including polyunsaturated fatty acids (PUFA), isoflavones, tocopherols, anthocyanins, saponins and total polyphenols, with improvements in the antioxidant capacity of the diet, the consumption of some of these diets has improved the oxidative state of the *Longissimus dorsi* muscle of rabbits; also, an anti-cholesterol effect has been observed in chickens (see Table 2).

Table 2. Modification of the composition and/or bioactivity determined in alfalfa (*Medicago sativa* L.) because of different treatments.

Treatments	Conditions	Effects	References
Alfalfa sprout supplementation in a rabbit diet.	Ninety mixed white rabbits were fed for 50 days, divided into 3 groups: • Standard (S) diet, • Standard diet + 20 g d^{-1} of alfalfa sprouts (A), • Standard diet + 20 g d^{-1} of flax sprouts (F).	The alfalfa sprouts presents increase in the total content of fatty acids (PUFA) (linoleic acid by 38.46% and linolenic acid by 70.05%), isoflavones (daidzein) in diets. The linolenic acid content in muscle of the alfalfa group was three times higher than control group. α-tocopherol content and α-tocotrienol, were up.	[12]
Dietary supplementation with alfalfa sprouts.	Dietary supplementation with alfalfa sprouts (40 g d^{-1}) and quantification of bioactive compounds and cholesterol in chicken and chicken eggs.	Decreased cholesterol in chicken plasma from 79.2 to 65.2 mg dL^{-1} and in egg yolk from 11.5 to 10.4 mg g^{-1}.	[13]
Sprouts alfalfa exposed to sound wave	Frequencies (250, 500, 800, 1000, and 1500 Hz) for two 1-h periods until 6 days.	Increase (24–50%) in the expression of genes that promote the production of L-ascorbic acid in sprouts (to 500 and 1000 Hz). The treatments increased the concentration of ascorbic acid and the antioxidant enzyme superoxide dismutase.	[14]
Substitution with alfalfa seed flour.	Adding alfalfa seed flour (0, 15, 30, 45% *w*/*w*) in rice flour biscuits (gluten free).	Increased linearly in: crude protein, total dietary fiber, total polyunsaturated, hardness, total phenolic content (22.9 to 112.9 mg GAE 100 g^{-1} DW for control and 45% substituted flour), and resistant starch. The antioxidant capacity increased proportionally from 14.7 to 194.6 µmol GAE 100 g^{-1} DW (FRAP) and from 739.3 to 3627.7 µmol TE 100 g^{-1} DW (ORAC), for control and 45% of alfalfa flour, respectively.	[11]
Different types of LED lighting in alfalfa sprout composition (FMC: fresh mass of cotyledons).	Four variants using cold white (10,032 K), warm white (3279 K), red green blue (RGB) LEDs with two chips activated: red and blue, which combined gave a violet colour.	Chlorophyll (a) up to 998.9 mg kg^{-1} in FMC with cold LED. β-carotene up to 44.6 mg kg^{-1} in FMC with red-green-blue LED (RGB). Chlorophyll a up to 843.3 mg kg^{-1} FMC, chlorophyll b up to 256.7 mg kg^{-1} FMC, β-carotene up to 21.6 mg kg^{-1} FMC, lutein up to 82.6 mg kg^{-1} FMC, neoxanthin up to 15 mg kg^{-1} FMC and violaxanthin up to 43.7 mg kg^{-1} FMC in shoots with sunlight. Total phenols up to 697 mg GAE kg^{-1} in FMC with blue LED (RGB). Ascorbic acid up to 155 mg kg^{-1} with sunlight.	[15]

Table 2. *Cont.*

Treatments	Conditions	Effects	References
Dry alfalfa sprouts (dry heat, freeze-dried).	Heat-dried samples (HD): stove at 60 °C for 24 h, and the freeze-dried samples (FD) for 24 h under vacuum (50 mTorr).	Greater decrease in isoflavone composition with the application of oven heat drying than with lyophilization. The lyophilisate increased the sterols concentration (41.82% for stigmasterol). The presence of carotenoids (zeaxanthin, β-carotene, retinol, lutein) was only detected after drying processes (not fresh).	[10]
Soak in water of seeds alfalfa.	Seeds were disinfected with hot water: soak at 85 °C, for 10 s; at 85 °C, for 40 s; at 90 °C, for 10 s; and at 100 °C for 10 s.	No effect on: lutein (23.4–26.6 mg kg^{-1}), violaxanthin (16.0–17.2 mg kg^{-1}), neoxanthin (3.5–4.1 mg kg^{-1}), β-carotene (10.1–11.7 mg kg^{-1}), total phenols (486.5–599.4 mg kg^{-1}) and chlorophyll b (64.7–72.8 mg kg^{-1}). The application of 100 °C caused a decrease in the content of ascorbic acid (from 84.5 a 67.5 mg kg^{-1}) and a increased phenolic content (from 537.1 to 599.4 mg kg^{-1}).	[16]
Supplementation, digestion and in vitro fermentation.	Adding alfalfa seed flour (0, 30, 45% *w/w*) in rice flour biscuits (gluten free). Simulated in vitro digestion and fermentation process.	Cookies with 30 and 45% of alfalfa seed flour presented the highest total phenolic content (0.42 and 0.56 mg g^{-1}, respectively) (control 0.15 mg g^{-1}). The in vitro fermentation of 8–48 h increased the concentration of lignans and phenolic acids, whose bioaccessibility at 24 h of in vitro fermentation were 16.2 and 12.2%, respectively.	[17]
Incorporation of selenium to alfalfa crop.	Inorganic selenium was used in two chemical forms: selenite (Na_2SeO_3) or selenate (Na_2SeO_4) (0, 20, or 200 µmol L^{-1}).	Increase in anthocyanins in alfalfa (29%) after of 20 µmol L^{-1} selenite solution (≈8% reduction of DPPH).	[18]
Ultrasound in fresh alfalfa leaves.	Study factors and ranges: Solvent/raw material ratio (mL g^{-1}): 5, 10, 15. Time (h): 1, 2, 3. Temperature (°C): 50, 65, 80. Power (W): 50, 100, 150. Ethanol concentration (%): 60, 75, 90.	Better yield (up to 1.61%) and bioaccessibility (up to 19.7%) of saponins. Conditions: solvent/raw material (9.5 mL g^{-1}), extraction time (2.90 h), extraction temperature (79.1 °C), ultrasound power (111.0 W), ethanol concentration (88.2%).	[8]

On the other hand, it has been reported [19] that alfalfa leaf peptides have a reducer power of 0.69 to 2.00 mg mL^{-1}; they also presented 79.71% (1.60 mg mL^{-1}) and 67.00% (0.90 mg mL^{-1}) scavenging activity of the radicals DPPH (1,1-diphenyl-2-picrylhydrazyl) and superoxide, respectively. In addition, they chelated 65.15% of the ferrous ion at 0.50 mg mL^{-1}. The molecular weight of 67.86% of the peptides was smaller than 1000 Da and was characterized by an amino acid profile with a high nutritional value (glutamic acid, aspartic acid, leucine, arginine, valine, lysine, among others). Alfalfa is a source of isoflavones such as genistein, daidzein, and glycitein [20], and tocopherols (tocols and tocotrienols), compounds with important antioxidant activity. The α-tocopherols are the most abundant of all the tocopherols; their biological activity is double that of the β and γ homologs and 100 times more than the δ homolog [21].

Saponins are a large group of compounds identified in alfalfa, consisting of nonpolar steroidal triterpenoids or aglycones (sapogenins) attached to one or more hydrophilic oligosaccharide moieties via ether or glycosidic ester bond [9]. Saponins function as a chemical protector in the plant's defense system against harmful agents (e.g., pathogens), but bioactive effects such as antioxidants, antimicrobial, anti-inflammatory, antitumor, antidiabetic, anticholesterol, antiviral, immunomodulatory, antibacterial, antiparasitic, and allelopathic activity have also been identified. Due to their properties, saponins are used as natural surfactants in foods, antimicrobial preservatives, and natural emulsifiers [8,9]. The structural differences between saponins have an impact on the bioactivity demonstrated; for example, due to the lack of sugar, sapogenins have shown better chemical properties (lower molecular weight, higher lipophilicity, or lower molecular flexibility) that improve permeability and bioactivity, in comparison with the precursor saponin [9].

In addition, polysaccharides (e.g., hemicellulose and pectin) with important bioactive functions, such as antioxidant, antitumor, immunomodulatory, anti-inflammatory, and growth-promoting properties, have been obtained from fresh alfalfa [22–24]; they have also been recognized as natural alternatives to antibiotics when added to animal diets. A pectic polysaccharide from the alfalfa stem was identified as rhamnogalacturonan I (RG-I; pm 2.38×10^3 kDa) [22]. The polysaccharide (50 μg mL^{-1}) showed a significant anti-inflammatory effect against mRNA expression of the pro-inflammatory genes of the cytokines interleukin (IL)-1β and IL-6, which suggests a potential use in functional foods and supplemented products. Another studied polysaccharide of alfalfa is that formed by galacturonic acid (146.500 μg mg^{-1}), glucose (39.092 μg mg^{-1}), glucuronic acid (29.343 μg mg^{-1}), arabinose (12.282 μg mg^{-1}), galactose (8.649 μg mg^{-1}), mannose (6.791 μg mg^{-1}), xylose (4.811 μg mg^{-1}), and fucose (4.346 μg mg^{-1}). In vitro studies showed that 50 and 100 μg mL^{-1} of the polysaccharide increased the cell viability of macrophages (RAW 264.7) by improving their immune functions, as well as the secretion and gene expression of inflammatory factors (cytokines, NO/iNOS, IL-6, and tumor necrosis factor (TNF)-α) [23]. The same research group reported a characterization for this same polysaccharide, indicating that the molar ratio of the saccharides is 2.6:8.0:4.7:21.3:3.2:1.0:74.2:14.9 for fucose, arabinose, galactose, glucose, xylose, mannose, galacturonic acid, and glucuronic acid, respectively [24]. Furthermore, the polysaccharide markedly increased the proliferation of B cells and the secretion of IgM in a dose- and time-dependent manner but not the proliferation and expression of cytokines (IL-2, -4, and IFN-γ) of T cells. This represents a biological activity that contributes to the immune system [23]. For alfalfa polysaccharides, in a mouse embryonic fibroblast (MEF) model with oxidative stress induced by hydrogen peroxide (150 μM; H_2O_2), the activation of antioxidant capacity (1.0 mM g^{-1} (T-AOC)) was detected as a preventive defense mechanism; and 250 μM/12 h was considered as the optimal concentration to stimulate stress in MEF (because it presents the highest expression of the pro-inflammatory gene related to senescence RIG-I). A concentration of 20 μg mL^{-1} of polysaccharides exhibited the greatest antioxidant effect and the least secretion of inflammatory cytokines [25]. The results demonstrate that alfalfa polysaccharides exert a protective action against oxidative damage induced by hydrogen peroxide.

Another study conducted to determine the effect of alfalfa (*Medicago sativa* L.) polysaccharides considered the growth performance and intestinal health of 200 piglets (35 days old) [26]. Biologically active phytogenic polysaccharides mainly contain carbohydrates comprising β-1,3-D-glycan units. Supplementation with the polysaccharide (0, 300, 500, 800, or 1200 mg polysaccharide kg^{-1} diet for 42 days) increased average daily gain (ADG) and feed ratio (G/F) in a dose-response manner. The experimental group receiving 500 mg kg^{-1} of polysaccharide in the diet showed the highest *Lactobacillus* values in the cecum, colon, and rectum. And the values for *Salmonella* and *Escherichia coli* decreased in all sections of the large intestine. The results showed that supplementation of the diet with alfalfa polysaccharides (500 mg kg^{-1}) improved intestinal morphological development and amylase and protease activity in the small intestine and promoted beneficial microbial populations in the large intestine [26].

It determined that alfalfa fiber (12 and 18% in the diet of piglets) decreased diarrhea and increased the composition and diversity of fecal bacteria (*Bacteroidetes* and *Firmicutes* were the dominant phyla (98% of the total)), and consequently improved the growth performance of weaning piglets [27]. The supplementation of alfalfa fiber (6–12%) in the diet of 48 crossbred piglets significantly increased growth performance and crude protein digestibility [28], particularly that of albumin, globulins, and total protein; however, it decreased levels of glucose (6% supplemented fiber, from 3.87 to 3.75 mmol L^{-1}), cholesterol (12% supplemented fiber, from 2.3 to 2.06 mmol L^{-1}), triglycerides (12% supplemented fiber, from 0.60 to 0.47 mmol L^{-1}), aspartate aminotransferase (6% supplemented fiber, 48 to 46 μ L^{-1}), and alanine aminotransferase (6% supplemented fiber, 42.5 to 39.5 μ L^{-1}).

2.2. *Maguey (Agave* spp.*)*

According to Statistical Yearbook of Agricultural Production, SIAP/SADER 2021 [5], the maguey production reported for the agricultural year 2021 (perennial cycle) (Table 1), corresponds to the total produced of pulquero maguey of aguamiel (58.28% national), not including unclassified pulquero maguey production. Hidalgo occupies the first national place in maguey production, followed by Tlaxcala and Mexico. The same statistical record does not report agave production for the state of Hidalgo, thus recognizing a difference in the use of the terms (maguey and agave) concerning the species and products produced from the plant: the *Agave salmiana* also named pulquero maguey to produce pulque, and *Agave tequilero* known to produce tequila.

Maguey, also called mixiotero magueys (*Agave salmiana*) is a plant with rosette leaves, thick and fleshy, with a short stem, and a lower pineapple that does not protrude from the ground. In Mexico there are about 200 species of maguey, a term applied to species of the genus *Agave* (Asparagaceae). It requires low-humidity soil, intense light, temperatures of 15 to 25 °C, and an approximate altitude of 1700 to 2400 m above sea level [7]. Species of the genus present an important profile of phenolic compounds, such as flavonoids, homoisoflavonoids, and phenolic acids, which have been widely related to important biological, antioxidant, antibacterial, antifungal, antinematode, and immunomodulatory activity [29]. In addition, species of the genus *Agave* are recognized as an important source of monosaccharides as fructose to produce traditional alcoholic beverages, natural fibers, saponins, high-fructose syrups, and fructans; even the different phytocomponents of the thick leaves act as seasoning or flavor sources during the roasting of meat to prepare a barbecue [30,31]. As a source of saponins, the use of agave is emphasized as having antibacterial, antientomological, antifungal, anticholesterolemic, and anticancer effects) [32]. Agave also contains policosanols and sapogenins; agamenone (5,7-dihydroxy-6,5′-dimethoxy-3′,4′-methylenedioxy flavanone), flavonol, or isoflavones have been identified in concentrated honey water [33]. Mature plants contain low concentrations of saponins, and silage reduces their quantity, improving their characteristics for livestock feed [31].

About 30% of the agave plant is made up of leaves, which have few applications. Saponins are mainly present in the leaves and can be used as precursors for sterols of therapeutic importance. Leaves of *A. salmiana* and *A. tequilana* Weber were structurally characterized (light microscopy) and methanolic extracts followed by dichloromethane were recovered, where the presence of saponins was confirmed by hemolytic activity in erythrocytes and a positive reaction with anisaldehyde reagent. *A. salmiana* presented a higher percentage of protein (7.3%) [34,35].

Multiple in vivo tests have been reported to demonstrate the diverse bioactivity of *Agave* spp., not having many outstanding investigations on *A. salmiana*. *A. salmiana* syrup (honey water) is reported to have antioxidant activity of up to 1096.8 μM TE by DPPH, and a total phenolic content of 904.8 μM GAE [31]. The antidiabetic activity of high-fructose agave syrup from *A. salmiana* protected against liver steatosis in rats fed 2 and 5 g of serum kg^{-1} and had a quadratic opposite effect on glycosylated hemoglobin in the blood of diabetic rats (dose 0.5 g kg^{-1}) [36]. Some patents [33,37] reported anticancer activity for the methanolic extract (80%) of concentrated agave syrup (10%) from the species *A. atrovirens*, *A salmiana*, and *A. lehmanni* (15 mg mL^{-1}), with 84.89% inhibition of colon cancer cells (Caco-2) and 67.95% of liver cancer cells (HepG2). In addition, antioxidant activity for the same extract was reported of 61.87 μmol ET g^{-1} sample. These properties are attributed to the composition that includes phytosterols, polyphenols, flavonoids (agamenone), tannins, policosanols, inulin, and saponins (sapogenins).

Another study was carried out on sap concentrated from *A. salmiana*, evaluating the apoptotic activity in HT-29 cells (IC_{50}, of 3.8 mg mL^{-1} for the concentrate) of saponins from the acetonic extract. The most bioactive fractions (up to 80% cell inhibition at 75 μg mL^{-1}) presented an IC_{50} of 108.4, 82.7, and >250 mg mL^{-1}, respectively (partition coefficients (Kd) of 0.23, 0.33 and 0.40); they contained steroidal saponins, mainly magueyoside B (266.4 μg PE mg^{-1}; Kd of 0.33). Flow cytometric analysis has determined that the fraction rich in the glycosides kammogenin and manogenin induces apoptosis and that the presence of gentrogenin and hecogenin is related to a necrotic effect [38]. The phytochemical composition has been proposed as a tool for the classification of different agave syrups (*A. tequilana* (>60% fructose) and *A. salmiana* (sucrose 28–32%)), such as infrared spectroscopy coupled to chemometrics (NIR-MIR-SIMCA-PCA) and high-performance anion exchange chromatography with pulsed amperometric detection (HPAEC-PAD) [39]. The said techniques are reported to identify and classify without significant mistakes. ^{1}H-NMR-spectroscopy-PCA was used to characterize syrup profiles and chemometrics, what allowed the sweeteners' classification by origin and kind of Agave. The agave syrups exhibited appreciable amounts of saponins, cardiac glucosides, and terpenoids (excelled in color intensity in the reaction) followed by glucosides, quinones, flavonoids, and coumarins in moderate amounts. In *A. tequilana* syrup, flavonoids and terpenoids were only detected in a few samples. *A. salmiana* syrup displayed a positive colorimetric reaction for all the evaluated compounds [40].

In addition to the apoptotic activity, it has been reported [41] that antioxidant activity of agave sap (*A. salmiana*) is dependent on storage time and is correlated with the browning developed due to heating and storage time, increasing from 18 to 23 eq Trolox μmol g^{-1} DW in a lot with a high degree of browning (57.7 OD_{490} g^{-1} fw), after 20 weeks of storage. In addition, they reported that the content of saponins (kammogenin glycosides (magueyosides A and B), manogenins, and hecogenin (agavoside C′)) was different per batch, varying from 224.2 to 434.7 PE g^{-1} DW in week 2 and varying up to week 20 of storage (207.7 to 462.4 PE g^{-1} DW). They found no correlation of the browning index and antioxidant capacity (ORAC) with the concentration of free amino acids (serine, phenylalanine, and lysine); the positive correlation found was browning with furosine, an early derivative of the Maillard reaction of lysine, reported as a free radical scavenger [41]. The presence of several compounds in agave syrup (ethanolic extracts), such as saponins, flavonoids, quinones, glycosides, cardiac glycosides, terpenoids, and coumarins, has been identified for several species [40]. The antioxidant activity of agave syrups (*A. tequilana, A. salmiana*) was in the range of 10–53%, while the content of total phenols was from 24 to 300 GAE 100 g^{-1} and that of condensed tannins was from 240 to 1900 mg CE g^{-1}. In addition, a relationship between the color and the antioxidant capacity of the syrups is reported, with dark syrups such as those of *A. salmiana* having the highest antioxidant capacity, about 28.33%, while light syrups show an average capacity of 8.7%. In tests on mice, the consumption of fresh and boiled syrup (*A. salmiana* sap) promoted weight gain (13%) and increased hemoglobin counts to 4.5 and 9%, respectively; the hematocrit count increased from 2.6 to 5.3%; iron, transferrin, ferritin, and phosphorus increase with the consumption of fresh syrup, while iron increase with boiled syrup [42]. The antioxidant capacity of syrup was determined as 7.1 μmol GAE g^{-1} DW, that for commercial coffee being 156.1 μmol GAE g^{-1} DW (commercial coffee). No adverse effects of syrup consumption it is observed.

Due to the importance of maguey, *A. salmiana*, in the production of pulque and functional food industries, processes have been sought to maintain its production and take advantage of its benefits (Table 3). The micropropagation of agave can be an auxiliary process to increase the phytochemicals and bioactivities in the plant. The increase in antioxidant activity in micropropaged plants in vitro has several possible causes: the presence and interaction of cytokines and auxins [43], or the highest concentration of phenols or saponins [44], or bioactive compounds in general, which can be responsive to culture conditions, such as micropropagation [45], which has already been shown to have diverse effects on compositions and properties. Micropropagation needs more research (at different conditions, ages, and physiological states), together with scaling up to primary production levels, so that the effects found in micropropagation can be exploited in the primary production sector and from there to the industrial sector. The study of technologies for the extraction of bioactives from agave bagasse is little to date, the effect of ultrasound assisted with supercritical fluids in *Agave salmiana* has reported antioxidant activity and saponin content [46], which is a promising area for the recovery of bioactives and the valorization of agave processing residues (Table 3).

Table 3. Modification of composition and/or bioactivity determined in *Agave salmiana* by different treatments.

Treatments	Conditions	Effects	Reference
Micropropagation	In vitro of plants from young germinated plantlets by axillary shoots.	Wild plants showed the highest phenolic content (13.06 mg EGA g^{-1}). The antioxidant capacity was higher in vitro (369.84 µmol TE g^{-1} DW) than in normal ex vitro conditions (184.13 µmol TE g^{-1} DW) and with ex vitro irrigation (143.38 µmol TE g^{-1} DW) and than in wild conditions (130.39 µmol TE g^{-1} DW). Glycosylated flavanols were detected in plants with ex vitro irrigation (quercetin) and under normal ex vitro conditions (kaempferol). Saponins were detected: hecogenin (0.418–5.227 mg EHe g^{-1}), tigogenin (18.821–31 mg EHe g^{-1}), mannogenin (0.288–0.861 mg EHe g^{-1}), and chlorogenin (0.339–2.042 mg EHe g^{-1}).	[47]
	Micropropagation was from axillary shoots. Leaf tissue samples were taken from the in vitro plants, ex vitro acclimated plants obtained from open environment conditions, and plants obtained from a natural population.	The total phenolic acids were 35 and 40% higher in plants propagated in vitro (11.8 mg GAE g^{-1} DW) and ex vitro (10.8 mg GAE g^{-1} DW), compared with the wild type (7 mg GAE g^{-1} DW). The saponin content of plants in vitro (77.1 mg PE g^{-1} DW) and ex vitro (63.3 mg PE g^{-1} DW) were higher than those of wild type plants (2.1 mg PE g^{-1} DW). The antioxidant capacity (ORAC) of the plants in vitro (369 µmol TE g^{-1} DW) was higher compared to ex vitro and wild type (184 and 146 µmol TE g^{-1} DW, respectively).	[45]
	Hydrometanolic extraction was applied to the foliar tissues and the content of flavonols and saponins was analyzed.	The plants propagated in vitro presented a higher concentration of flavonols and saponins, quantifying 7 flavanols and 5 saponins. Herbacetin (most abundant flavonol found): wild plants (14.7 mg 100 g^{-1} DW), in vitro (16.3 mg 100 g^{-1} DW), in an open environment (38.4 mg 100 g^{-1}). Tigogenin (most abundant saponin found and only detected in plants propagated): in vitro with 6895.2 mgPE 100 g^{-1} DW and 4997.8 mgPE 100 g^{-1} DW.	[48]
In vitro drought stress effect, generated by polyethylene glycol.	Stress medium: Murashige and Skoog (4.4 g L^{-1}, pH 5.8, 30 g L^{-1} sucrose, and L2 vitamins) with polyethylene-glycol (0, 10, 20, 30%, 27 °C, photoperiod of 12:12 h light:dark, 60 days).	Plants grown with polyethylene glycol (30%) showed the lowest flavonol content, but the highest saponin content (tigogenin glycoside, 163 mg PE g^{-1} DW) and the highest antioxidant capacity (ORAC) (≈1000 mmol TE g^{-1} DW).	[49]
Ultrasonically-assisted supercritical fluid extraction (USFE).	Bagasse of *Agave salmiana* (part not indicated; 10 g). Process factors were pressure (150–450 bar), temperature (40–60 °C), and amount of co-solvent (5–10%).	Increased antioxidant capacity (FRAP) with the use of multiplate (US) transducer geometry of extracts at 20.91 µmol TE g^{-1} and saponin content at 61.59 µg g^{-1}; comparing with the cylinder geometry (with 12.18 TE g^{-1} and 19.05 µg g^{-1}, respectively).	[46]

Antioxidant properties and bioactive compounds of methanolic extracts of *A. salmiana* leaves were evaluated [50] at different stages of development (I–VI, from 1 to 7 years). The total phenolic content from leaves extracts was found to be between 5 (stage VI) and 13 mg GAE g^{-1} (stage II) the maximum; the antioxidant capacity presented a negative trend from stages I to VI (from 146 to 52 µmol TE g^{-1} respectively), the flavonols showing the same behavior (65% reduction from stages I to VI). Five saponins were identified (chlorogenin glycoside 2, chlorogenin glycoside 1, hecogenin glycoside 1, tigogenin glycoside, hecogenin glycoside 2) (also reported [48] in addition to flavonols (maximum concentrations in stage I, kaempferol (0.045 mg g^{-1} DW) and quercetin (0.07 mg g^{-1} DW)). Stage III and IV plants presented the highest content of saponins, mainly chlorogenin glycoside, at 3.19 and 2.90 mg PE g^{-1}, respectively. According to Pearson's correlation, there is a positive relationship between total phenol content and antioxidant activity. Based on these results, it can be said that *A. salmiana* plants from stages I to IV could be a good source of antioxidants and bioactive agents; in addition to that, the concentration of metabolites could be a marker of the developmental stage. In the same way, the content of saponins was evaluated [51], regarding maturity (before (immature) and at the beginning of the reproductive stage (mature)) of the agaves (8 years old) and syrup. Saponins derived from kammogenin, manogenin, gentrogenin, and hecogenin were found. In syrup form immature *A. salmiana*, the saponin content was twice as high (478.4 µg PE g^{-1} DW of syrup) as that of immature *A. americana* (179.0 µg PE g^{-1} DW of mead). For both species, the saponin content decreased when the plants reached sexual maturity (up to 325.7 and 60.5 µg g^{-1} DW of syrup, for *A. salmiana* and *A. americana*, respectively). This finding is important to better select the species and maturity stage of agaves used as a source of bioactive.

In addition to research on stems and syrup, the agave flower has also been studied. The traditional use of the *A. salmiana* flower as an anthelmintic agent, insecticide, and antimicrobial occurs in some regions of Mexico. The antibacterial activity against Gram-negative bacteria of the flower scape was demonstrated [52], these being more susceptible to aqueous and crude extracts than Gram-positive bacteria. The aqueous extract presented better yields (5.935 g 100 g^{-1} DW; young stage of development, upper section), antioxidant activity (78.55% by DPPH; young stage of development, middle part), and total phenolic content (199.12 mg GAE 100 g^{-1} of extract; young stage of development, middle part); the highest antimicrobial activity was found for the aqueous extract of the mid-development and mid-section sample (vs. *E. coli*, *Salmonella typhimurium*, and *Shigella sonnei*) due to the possible presence of saponins, tannins, flavonoids, terpenes, and alkaloids on the flower scape of agave. The best minimum inhibitory concentration (MIC; 6.83 mg mL^{-1}) was found for the ethanolic extract of the adult stage and middle part of the flower, against *Shigella sonnei*.

An important aspect in the extraction of bioactive compounds from underutilized leaves of *Agave* spp. is the structural characteristics of these leaves. A study was reported of structure, water permeability, and resistance to sterilization of the mixiotera leaves or cuticles of agave [53]. The cuticle of *A. salmiana* is a material that resists humid heat when sterilized under pressure conditions (121 °C and 15 lb in^{-2}, 15 min); it has a porous structure that linearly regulates the diffusion of water. The cuticle (mixiote) consists of a lipid matrix based on cutin (40–80%) and is also made up of fatty acids (C16:18) linked by ester-type bonds. In addition, it has been reported that agave leaves are an important source of calcium, particularly *A. salmiana*, one of the species that does not present oxalate crystals; they have been identified in *A. atrovirens*.

Another type of important compound in agaves is fructans, which are the main photosynthetic product in *Agave* spp.; their main function is to store energy and act as an osmoprotectant during periods of drought. Fructans (fructooligosaccharides or inulin) are of nutraceutical interest due to their resistance to human digestion and subsequent fermentation by the colonic microflora, producing short-chain fatty acids [31]. Agave (*A. salmiana*) fructans are involved in the activation and selective differentiation of cells (peripheral blood mononuclear cells (PBMC)) of the immune system (immunomodulator) through interactions with probiotics (*Lactobacillus casei* and *Bifidobacterium lactis*). The agave fructans showed the highest prebiotic activity and increased levels of CD69 expression, proliferative activity and NO production when administered with the probiotic *L. casei* [54]. Short- and medium-chain branched fructans from agave have been tested as prebiotics and have been shown to increase lactobacilli in a dynamic in vitro model of the large intestine. With both substrates, butyrate production was increased, which may have a beneficial impact on health, and the production of potentially harmful putrefaction metabolites was reduced. Based on these findings, agave fructans should be considered as potential prebiotics [55].

Fructans (with β (2→6) linkages; from *Agave tequilana* blue variety) have been shown to attenuate the production of proinflammatory cytokines from stimulated dendritic cells and strongly inhibit Toll-like receptors, thus exerting an anti-inflammatory effect. The structure of fructans should be carefully determined and taken into consideration when intended to be used as food supplement, as the presence of linear or branched structure, the chain-length, as well as the dose of these molecules can exert differential responses. Further studies are needed to establish specifically in which disease state agave fructans could serve as an alternative or supplemental therapeutic option [56]. The exact mechanisms of effect of agave fructans are still under investigation for a better understanding of the benefits to human health and technological developments, for example, in the interactions with the intestinal microbiota that benefit the growth of beneficial bacteria and reduce disorders with glucose metabolism, fats, metabolic syndrome, calcium absorption, mental disorders, oxidative stress and cancer. The properties of agave fructans depend on various structural factors and management conditions, so research on them is a promising and long-term line [57].

2.3. Maize (*Zea mays* L.)

According to the Statistical Yearbook of Agricultural Production, SIAP/SADER 2021 [5], the corn production reported for the agricultural year 2021 (agricultural year) (Table 1), corresponds to white grain corn (617,696.87 tons) plus a portion of yellow (459.98 tons) reaching 618,156.85 tons in total (2.25% national); and green fodder corn with 130,136.92 tons (0.75% national), without presenting the production of shrunken or dry fodder corn. Hidalgo ranks thirteenth in national grain corn production and tenth in national production of green forage corn.

Corn belongs to the kingdom *Plantae*, division *Magnoliophyta*, class *Liliopsida*, subclass *Commelinidae*, order *Poales*, family *Poaceae* (Gramíneae), subfamily *Panicoideae*, genera *Zea* and specie *Z. mays* (https://www.biodiversitylibrary.org/page/358992#page/1/mode/1up (accessed on 12 August 2022)); in Mexico, pre-Hispanic cultures called the ear centli and the grain tlaolli [58]. For its development, grain corn prefers loamy-loamy, loamy-clayey, and loamy-clayey-silty soils with a depth greater than or equal to 1 m, with a pH of 5.5 to 7.5, at an optimum temperature of 18 to 24 °C and requires an average annual rainfall of 700 to 1300 mm [7]. Some of the products made from the transformation of corn are starch, liquid sugar, rectified alcohol, germ, gluten, feed, and bioethanol, in addition to some products called "biorefined" such as sweeteners, polysaccharides, pharmaceuticals, nutraceuticals, fibers, biodegradable films, organic acids, pigments, polyols, and vitamins, among others [59]. In Mexico the products derived from corn that stand out are tortillas, forage, oils, biofuels, starches, glucose, fructose, dextrose, ethanol, atole, and tesquiño [58]. The corn breeds that are grown in Hidalgo include the yellow rice, the cacahuacintle, the chalqueño, the conical and conical norteño, the conical elotes, the mushito, and the negrito, according to the National Commission for the Knowledge and Use of Biodiversity (CONABIO) (https://conabio.shinyapps.io/conabio-pgmaices1/ (accessed on 12 June 2022)). Although, as mentioned before, the statistical data from SADER mention the varieties: white grain corn, yellow grain corn, and green fodder corn.

The composition of blue and white corn flours (g 100 g^{-1}) has been reported [60] as follows: moisture 9.8 and 7.0, protein 9.1 and 8.4, lipids 5.2 and 4.7, ash 1.1 and 1.3, total starch 70.7 and 74.2, total dietary fiber 10.9 and 11.2, and soluble carbohydrates 3.0 and 0.3, respectively. Total carbohydrates represent the majority fraction in purple corn, these being highly available. Starch is the largest component in corn grains (63.48–89.9 g 100 g^{-1} DW), with amylose of 20.70–33.32 g 100 g^{-1} of starch; other components are proteins (6.73–11.37 g 100 g^{-1} DW), lipids (1.80–7.53 g 100 g^{-1} DW), and ash (1.40–2.06 g 100 g^{-1} DW) [61]. The composition of purple corn reported [62] presents humidity (10 g 100 g^{-1}), ash (1.71 g 100 g^{-1}), protein (9.10 g 100 g^{-1}), fat (1.80 g 100 g^{-1}), fiber (11.20 g 100 g^{-1}), starch (57.70 g 100 g^{-1}), amylose (27.10 g 100 g^{-1}). Purple corn has been found to have a lower starch content and a lower glycemic index than other types, such as white corn. The parts of corn differ in composition: the pericarp has high fiber content, the endosperm is rich in starch, and the germ contains proteins, fats, sugars, and ash [58]. The bioactive compound that has been determined as predominant depends on the portion of the grain; for example, anthocyanins have been in the pericarp, aleurone [58,63], and endosperm; and phenolic acids in the pericarp [58].

Mexican corn in its different varieties has shown antioxidant activity, quinone reductase induction, antimutagenic activity (*S. typhimurium* TA98), and antidiabetic activity. Of course, all properties depend on the type of grain, its physiological development, the portion of grain evaluated, the extraction technique used (e.g., solvents), the previous handling of the grain (storage, pretreatment) and the physicochemical properties of the compound to be evaluated, among others. Intense colored grains (purple, blue) stand out in bioactive composition, as well as in their bioactivities. In addition, it has been reported that some processes can modify the composition of compounds and therefore their bioactivities, for example, nixtamalization reduces the antioxidant capacity of ethanolic extracts of white, red, blue, and purple corn, but increases the relative percentage of glycosylated anthocyanins and decreases acylated anthocyanins in raw blue corn grain. On the contrary, air drying of grains did not have a significant effect on desorption isotherms of several varieties (Spanish, white, yellow, yellow, purple). Similarly, cooking in water resulted in an increase in the total phenolic content and antioxidant capacity of blue corn. Loss of anthocyanin content has been observed (83%) in the mass of nixtamalized blue grains and in tortilla processing (64%) [64]. For white grains, nixtamalization also reduced carotenoids by 53 to 56%, but not antioxidant activity and antimutagenicity. Despite the losses in concentration (the blue variety standing out for anthocyanins and the red variety for carotenoids), the pigments of creole corn showed antioxidant and antimutagenic activity after nixtamalization. For details see Table 4.

Table 4. Composition and/or bioactivity determined in maize of different types (*Zea mays* L.) and applied treatments.

Corn	Treatments	Compounds and/or Products	Bioactivity	Reference
Mexican corn (13 pigmented grain): Arrocillo Amarillo (red, blue), Bolita (red, blue), Chihuahua Crystal Blue (blue, red), chalqueño corn (red, blue).	Nixtamalization with alkali (0.8% of the grain weight) for 30 min, followed by resting for 14–16 h, ambient drying, grinding, and sieving (0.5 mm).	Decreases the anthocyanin concentration of corn grains in the pericarp by 73 to 100%, varying according to the type of corn and portion.		[63]
Mexican corn (18 phenotypes).	White corn. Yellow corn.	Total phenolic content: - White corn: 170 mg GA 100 g^{-1} sample. - Yellow corn: 551 mg GA 100 g^{-1} sample. Total anthocyanin content: - White corn: 1.54 mg c3-G 100 g^{-1} sample. - Yellow corn: 70.2 mg c3-G 100 g^{-1} sample.	Antioxidant capacity: - Yellow corn: 89.4% by ABTS. - White corn: 26% by ABTS. - Red colored (phenotype Pinto): 100% by DPPH.	[65]

Table 4. *Cont.*

Corn	Treatments	Compounds and/or Products	Bioactivity	Reference
White, red, blue and purple corn (var Ver 42)	Ethanolic extracts (95%) from nixtamalized grains. Tortilla with nixtamalized grain.	The treatments reduced total phenols, and anthocyanins. Total phenolic content and anthocyanin (respectively): - Purple maize (1760 mg 100 g^{-1}, 325.1 mg 100 g^{-1}), - Red maize (465 mg 100 g^{-1}, 82.3 mg 100 g^{-1}), - Blue maize (343 mg 100 g^{-1}, 63.1 mg 100 g^{-1}), - White maize (170 mg 100 g^{-1}, 1.59 mg 100 g^{-1}).	The processing negatively affected the capacities of the grains. Quinone reductase induction (QR): purple > White > red > blue (anticancer activity). The purple genotype (Ver 42) and its products (dough and tortilla) showed the highest antioxidant capacity (70% by ABTS, 55% by PRAC) and QR (induction twice at 125 g mL^{-1}).	[66]
Creole maize races (*Zea mays* L.) and pigmented varieties (yellow, red and blue).	Nixtamalization (alkaline boiling) and production of dough (grinding, drying) and tortillas.	Carotenoid content (µg of β-carotene eq g^{-1} extract) of raw maize grains and their products (masa, tortilla) respectively: - White (0.39, 0.17, 0.18) - Yellow (0.49, 0.28, 0.56) - Red (1.01, 1.14, 1.01) - Blue (0.18, 0.23, 0.22) For white grains, nixtamalization reduced carotenoids by 53 to 56%. Yellow grain suffered the highest losses from anthocyanins (174.44 to 10.30 mg of c3-GE 100 g^{-1} DW), not detectable in white maize and its products. The anthocyanin content of all grains was 174.44 to 963.00 mg of c3-GE 100 g^{-1} DW.	White corn (≈30%) and products (dough ≈20%, tortilla ≈25%) had higher antiradical (DPPH) activity than BHT (≈10% to 100 µM). Yellow corn (≈22%) and products (dough ≈18%, tortilla ≈29%) had higher antiradical activity (DPPH) than BHT (≈10% to 100 µM). Red (50%) and blue (40%) maize grain showed the highest antiradical activity. The antimutagenic activity (*S. typhimurium* TA98) of the grains: - White grain (35%) - Yellow (4 3%) - Red (53%) - Blue (56%)	[64]
18 samples of blue/purple grain of conical corn (EC), Chalqueño (CHAL), and Bolita (BOL) maize races.	40 grains without the germ, crushed, sieved (0.5 mm), and dried in an oven (40 °C, 18 h). Analysis extract by methanol (acidified to 1% with trifluoroacetic acid) and sonicated for 15 min.	Anthocyanins totals (AT) content (CHAL): varied from 579.4 to 1046.1 mg c3-GE kg^{-1} DW. The total soluble phenols (TSP) (CHAL): varied from 918.9 to 1479.2 mg GAE kg^{-1} DW. AT content (EC): varied from 997.8 to 1332.2 mg c3-GE kg^{-1} DW. TSP (EC): varied from 1328.6 to 1626.7 mg GAE kg^{-1} DW. AT content (BOL): varied from 304.1 to 528.0 mg c3-GE kg^{-1} DW. TSP (BOL): varied from 875.0 to 1276.2 mg GAE kg^{-1} DW.	Antioxidant activity (AA) (Chalqueño): 34 to 60.3% by DPPH. AA (Elote cónico): 46.6 to 60.4% by DPPH. AA (Bolita): 21.0 to 39.5% by DPPH.	[67]

Table 4. *Cont.*

Corn	Treatments	Compounds and/or Products	Bioactivity	Reference
Whole grains of 10 different colored corn (*Zea mays* L.) genotypes (landrace and an inbred line, over the year 2010).	Combined extracts: acetone/methanol/water (7:7:6, $v/v/v$), with alkaline hydrolysis and extracted with ethyl acetate and diethyl ether (1:1, v/v).	White and yellow corn: - Flavonoids (248.64 and 281.20 mg CE kg^{-1} DW), - β-carotene (0.21 and 0.70 mg kg^{-1} DW), - lutein (not detected and 5.91 mg kg^{-1} DW), - total phenolic content (5227.1 and 5393.2 mg GAE kg^{-1} DW), respectively. The light blue genotype had the highest content of total phenols (10,528.8 mg GAE kg^{-1} DW), flavonoids, and ferulic acid.	White and yellow showed antioxidant capacity (ABTS) between 15 and 20 mmol Trolox kg^{-1} DW. The light blue genotype had the highest scavenging activity (ABTS: 35.66 mmol Trolox kg^{-1} DW).	[68]
Blue corn flour	Nixtamalization. Maize grain in cooking (1:2, grain: water), 1.0% (w/w) of calcium hydroxide to 90 °C for 23 min, was soaked for 16 h at ambient temperature, was grounded, and passed through a flash dryer (260 °C for 4 s), the obtained flour was grounded in a mill using a hammer head and a 0.5 mm mesh screen.	Not change the resistant starch content or slow digestion.	The tortilla made with blue corn nixtamalized presented a lower glycemic index (58) and presented antioxidant capacity in the different fractions. They suggest a direct relationship between polyphenol content and antioxidant activity.	[69]
Spanish maize kernels, white (WF, Rebordanes variety), yellow (YF, Sarreaus variety) and purple (PF, Meiro variety).	Air-drying the maize kernels using a pilot-scale tray dryer (45 °C, 2 m s^{-1}, 30% relative humidity, 5 kg m^{-2} of loading density, until an average maize moisture content of 11% DW), crushed, ground, and sieved (200 y 500 µm).	Total starch (TS, % w/w, DW) content of tested maize flours, yellow, white, and purple, ranged from 60.1 (whole flour 500 µm) up to 75.2 (purple 200 µm) and no clear differences between varieties were found.	No significant differences were observed among water desorption isotherms of maize varieties.	[70]
Five blue hybrid maize genotypes and Chalqueño and conic kernels were used as native genotypes cultivated in the highlands of Mexico.	Homogenized with 80% ethanol for 10 min, alkaline digestion (2 M NaOH), acidification (HCl), extraction with ethyl acetate.	The total anthocyanins and anthocyanins in free phenolics of the natives, chalqueño and conic are 646 and 892 mg c3-G kg^{-1} and 48.7 and 60.3%, respectively. The total anthocyanins and anthocyanins in free phenolics of the hybrid genotypes are in the range of 835–1052 mg c3-G kg^{-1} and 62.4–80.6%, respectively.	Antioxidant capacity (free and bound phenols, respectively): - Chalqueño (≈166 and ≈1600 mg TE kg^{-1}). - Cónico (≈166 and ≈1700 mg TE kg^{-1}). The range of antioxidant capacity of free and bound phenols of hybrid genotypes was 166–820 and 862–1533 mg TE kg^{-1}, respectively.	[71]
White corn. Yellow corn.		White corn kernel (anthocyanin free). Yellow corn kernel (702 mg c3-GE kg^{-1})	Antioxidant capacity of: - White corn with 17.4 µM TE g^{-1} sample, - Yellow corn with 90% by ABTS.	[72]
Native Mexican blue corn (*Zea mays* L.).	Nixtamalization (maize kernels were cooked (1:3, maize grains/water) with 5.4 g of $Ca(OH)_2$ L^{-1} water; 31 min, 85 °C, 8.1 h). Wet nixtamal was dried (55 °C/12 h), cooled, and milled to pass through an 80-US mesh (0.180 mm).	Increases the relative percentage of glycosylated anthocyanins and decreases acylated anthocyanins. The most abundant compounds (cyanidin-3-(6″-succinylglucoside) (Cy-Suc-Glu) and cyanidin-3-(6″-disuccinylglucoside) (Cy-diSuc-Glu)).		[73]

Table 4. *Cont.*

Corn	Treatments	Compounds and/or Products	Bioactivity	Reference
Blue and white cornmeal	Cooked samples were prepared in water (1:10 *w*/*v*) by a heating bath with shaking for 30 min.	Extractable polyphenols: - Blue maize flour (165 mg GA g^{-1} DW), - White maize flour (127 mg GA g^{-1} DW). Condensed tannins: - Blue maize flour (198.9 mg GA g^{-1} DW), - White maize flour (38.31 mg GA g^{-1} DW).	Antioxidant capacity and alpha-amylase inhibition (AAI): - Blue maize flour: 6.8 mg ET g^{-1} DW (DPPH), 13.1 mg ET g^{-1} DW (ABTS), 15.5 mg ET g^{-1} DW (FRAP), and 96.8% AAI. - White maize flour: 5.3 mg ET g^{-1} DW (DPPH), 11 mg ET g^{-1} DW (ABTS), 10.6 mg ET g^{-1} DW (FRAP), and 90.9% AAI. Increase in the total phenolic content and antioxidant capacity of cooked blue corn flours, compared to raw ones.	[60]
Purple corn grain flours (control: White corn)	Mixtures of various families (genotypes) of purple corn. Homogenized with ethanol (96%)/HCl (1 N) (85:15 *v*/*v*), 30 min.	White corn presented a total phenol concentration: 319 mg GAE 100 g^{-1}. Total phenols (mixtures genotype) (range): 438 to 1933 mg GAE 100 g^{-1}. Total phenols (original genotype): 1328 mg GAE 100 g^{-1}.		[74]

Corn in its various varieties presents in its composition polyphenols (catechin and epicatechins linked to acylated and non-acylated anthocyanins) [58,73,75–77], flavonoids (naringenin, kaempferol, rutin, morin, and quercetin and hesperidin derivatives), phenolic acids (protocatechic, vanillic, p-coumaric, m-coumaric, o-coumaric, chlorogenic, caffeic, rutin, ferulic, hydroxybenzoic, and sinapic acids, and hydroxycinnamic acid derivatives) [58,78–80], carotenoids, and anthocyanins, compounds whose concentration depends on the coloration [59,72,81]. The average percentages of anthocyanins in blue corn are 90% cyanidin, 8% pelargonidin, and about 2% peonidin [58]. Compounds such as cyanidin-3-glucoside, pelargonidin-3-glucoside, peonidin-3-glucoside, cyanidin-3-(maloyl)-glucoside, pelargonidin-3-(maloyl)-glucoside, and peonidin-3-(maloyl)-glucoside, among others, have been detected in powdered purple corn cultivars [68,77,79,82,83]. Of the total anthocyanins, cyanidin-3-glucoside has been reported as the main anthocyanin in red (51%), blue (49%), multicolored (47%), and purple varieties (31%), while pelargonidin-3-glucoside (43%) predominates in those with pink coloration [61]. It has been detected that the accumulation of anthocyanins in corn leaves can be induced by the application of organic selenium (selenomethionine) to the plant; this could be a symptom of selenium phytotoxicity or indicate a change in the oxidative state of the plant. In food, the presence of anthocyanins is desirable for their nutraceutical and antioxidant properties [18]. The chemical form in which anthocyanins are found has also been reported to exert an effect on the anticancer capacity; thus, the acidified form has the better capacity, at least against the colon (Caco-2), liver (HepG2), breast (MCF7), and prostate (PC3) cancer cells [76].

The seeds and the cob of purple corn have shown antioxidant activity [82], which is affected by the maturation stage (grain) and type of corn, for the reported techniques [78]. It has also been reported that the composition of total polyphenols and anthocyanins (flavonoids) of purple corn powder (cob) cultivars depends on the sowing area, but not on the sowing density or type of fertilizer used [77]. Additionally, it has been seen that the concentration of anthocyanins depends on the degree of maturity of the grains, immature corn being proposed as a better source (67.1–88.2% acylated anthocyanins (with acylated radical) in young plants and 46.2–83.6% at the mature stage). The antioxidant capacity increases with maturation (genotype KKU-WX111031) as determined by the DPPH technique (from 11.7 to 21.6 μmol TE g^{-1} DW), by FRAP iron-reduction techniques (from 69.8 to 159.2 μmol Fe (II) g^{-1} DW), and by Trolox equivalents (from 95.4 to 156.8 μmol TE g^{-1} DW) [83].

In general, it is reported that the concentrations of compounds with antioxidant capacity in dark purple corn kernels are less affected by the application of various soaking and heat treatments; the opposite effect was determined for pearling, which suggests that those compounds are concentrated in the husk of the grain. Purple corn presents higher concentrations and better antioxidant capacity than the reddish variety (Sangre de Cristo) [84]. The total phenolic content and antioxidant capacity of cooked blue corn flours have been reported to be higher than for uncooked flours, so heat treatments appear to cause such behavior. Blue corn flour has a higher resistant starch content and slow digestion; therefore, it exhibits a lower glycemic index (63) than white corn (71) (without cooking). Cooked flour of both varieties has a higher glycemic index, 78 and 82, respectively; in the case of cooked flour of blue corn the interaction with polyphenols, it could be responsible for the decrease of glycemic index, due to the low digestibility of the complexes formed. An analysis of correlations indicates that the increase in polyphenols causes a decrease in the predicted glycemic index in model systems of blue and white corn [60]. Regarding in vivo tests [85], dehydrated and micro pulverized purple corn was administered in 1 g capsules (2 g daily for 30 days) to a group of diabetic patients with mixed dyslipidemia, without previous treatment. They observed better fasting glycemic control (from 211.3 to 112.5 mg dL^{-1}), obtaining a 46.8% reduction from baseline levels. Additionally, purple corn lowers triglycerides and increases HDL cholesterol according to this research.

Purple corn extracts have been proposed as low-cost colorants, as they are concentrates of anthocyanins [86], and as food antioxidant additives, for example in mayonnaise, where the extract (0.4 g kg^{-1}) presented better storage performance than commercial antioxidants (BHT and EDTA) [87].

2.4. Forage Oats (*Avena sativa* L.)

According to the Statistical Yearbook of Agricultural Production, SIAP/SADER 2021 [5], the oat production reported for the agricultural year 2021 (agricultural year) (Table 1) corresponds to green forage oats without classifying (no variety determined), excluding production of shrunken or dried forage oats. Hidalgo ranks ninth in the country (3.24%) in green forage oat production. The grain oats produced in the state of Hidalgo are 2709.49 tons, a quantity much lower than the forage (324,378.96 tons) for the same agricultural year. Oats are used mainly in livestock feed, as a forage plant, in the pasture, and as silage. The oat is an annual herbaceous plant, of the grass family. It is from cold climates and is very sensitive to high temperatures, mainly during flowering and grain formation. It requires a lot of water for its development because it presents great transpiration; it grows better in deep, clay-sandy soils, rich in lime but without excess, and that retain moisture. Oats are highly adapted to acidic soils, which is why they are usually sown in soils rich in organic matter [7].

It is commonly known that oats (mainly *Avena sativa* L.) affect satiety and delay the absorption of nutrients, as well as having a deterrent action against various disorders of the gastrointestinal tract. The effects are mainly attributed to the soluble fiber content [88]. Oatmeal as a functional food has physiological benefits in reducing hyperglycemia, hyperinsulinemia, hypercholesterolemia, hypertension, and cancer; for these actions, the β-glucans of oats are considered beneficial in the prevention, treatment, and control of diabetes and cardiovascular diseases (Table 5).

Table 5. Composition and/or bioactivity determined in oat (*Avena sativa* L.).

Crop	Compounds and/or Products	Conditions of Bioactivity Detected	Bioactivity	Reference
Oat (no variety reported)	Oat bran concentrate containing 43% β-glucan.	Oat β-glucan hydrolysate was prepared by adding Celluclast (840 EGU g^{-1}) to oat bran concentrate suspension (6.25% (*w*/*v*), 50 °C, pH 4.8).	Anti-cholesterol activity: reduced rat serum triglycerides, reduced weight gain, high-density cholesterol (HDL-C) in serum increased up to 42–62% and reduced low-density cholesterol (LDL) by 25–31%.	[89]
Oat (Derby variety)	β-glucan	Two β-glucan extracts were separately added to test solutions at concentrations of 0.1–0.5% (*w*/*w*). β-glucan fractions: 78.5% (E3, E4) content of extracts (*w*/*w*).	Decreased intestinal absorption of fatty acids (18:2 mainly). Inhibition of postprandial rise in glucose and insulin.	[90]
Oat (no variety reported)	β-glucan	Consumption in pigs of 3 and 6% in the diet.	Net glucose absorption reduction from 22 to 51%, relative to the intake percentage.	[91]
Oat (*Avena sativa* L.)	β-glucan	Dosage of 2000 mg kg^{-1} in reduction of hyperglycemia. Intake dose of 70 mg mL^{-1} for 6 weeks for enzyme inhibition.	Reduction of hyperglycemia. Inhibition of intestinal enzymes, sucrase (70.72%), maltase (83.33%) and lactase (89.43%), in diabetic mice. Similar protective effect to the diabetic mice as metformin (1% *w*/*v* metformin solution).	[92]
Oat (no variety reported)	β-glucan	Extract viscosity of 3 mPa, with the presence of starch of 40 g.	Glucose absorption reduction.	[93]
Oat (no variety reported)	β-glucan	Consumption of 3 g d^{-1} of oat or barley β-glucan is sufficient to decrease blood cholesterol.	There was a significant inverse relation in total cholesterol (−0.60 mmol L^{-1}, −0.85 to −0.34), low−density lipoprotein (−0.66 mmol L^{-1}, −0.96 to −0.36), and triglyceride/triacylglycerol (−0.04 mmol L^{-1}, −0.15 to 0.07) after consumption of β-glucan.	[94]
Oat (*Avena sativa* L.)	Extract avenanthramides (EA) is: 6.07% N-(3′,4′-dihydroxycinamoyl)-5-hydroxyanthranilic acid, 4.37% N-(4′-hydroxycinnamoyl)-5-hydroxyanthranilic acid, 4.37% N-(4′-hydroxycinnamoyl)-5-hydroxyanthranilic acid, and 5.36% N-(4′-hydroxy-3′-methoxycinnamoyl)-5-hydroxyanthranilic acid. Phenols: vanillic acid (0.60%), caffeic acid (0.50%), syringic acid (0.54%), p-coumaric acid (0.16%), ferulic acid (0.08%), and sinapic acid (0.03%).	Mice in three experimental groups were (7th week) given EA at 250, 500, and 1000 mg (kg body weight)$^{-1}$ d^{-1} by intragastric gavage (2 weeks). Mice were sacrificed and the liver was collected and stored until analysis.	Antioxidant effect against oxidative stress induced by D-galactose (50 mg kg^{-1} DW d^{-1}) in mice, noted by increased antioxidant enzyme activity (dose-dependent mode) and the regulation of antioxidant gene expression.	[95]
Oat (no variety reported)	β-glucan (low molecular weight)	Concentration of 400 μg mL^{-1}.	Deceased cancer cells viability (human pigmented malignant melanoma (Me45) and the human epidermoid carcinoma A431 cell line), while for the normal cells it was non-toxic.	[96]

Table 5. *Cont.*

Crop	Compounds and/or Products	Conditions of Bioactivity Detected	Bioactivity	Reference
Oat (no variety reported)	β-glucan	β-glucan (200 μg mL^{-1}) with electroporation.	Antitumor activity due to decreased cell viability (human melanoma cell line (Me45)) of 12.5%. Not present toxic effects on normal cells.	[97]
Oat (no variety reported)	β-glucan (high and low molecular weight).		Decreased viability of cancer cells (human lung A549, H69AR) (about 50% decrease at 200 μg mL^{-1}).	[98]
Oat (*Avena sativa* L.)	Avenantramide	100 μL of *Lactobacillus acidophilus* was added to finely powdered oats (solution 1 g/ 50 mL water) for fermented oats. And control was measured (non-fermented).	In vitro studies revealed that fermented and non-fermented oats displayed higher antioxidant activity, having a corresponding IC_{50} value of 201.03 μL and 236.46 μL, respectively. The colon cancer cell (HT29) death percentage, varied in the range of 41.81% and 87.48%, with the highest cytotoxic activity being for non-fermented oats (25 μg mL^{-1}).	[99]

Other bioactive substances that contribute to the medicinal action reported for *Avena sativa* L. are polyphenols (>20), phenolic acids, alkyl resorcinols, and avenanthramides [99–101] that exhibit antioxidant, anti-inflammatory, and antiproliferative activity, which inhibits cancer cell progression [99,102]. Phytosterols in a mixture with β-glucan have shown strong anti-cholesterol properties (low-density lipoprotein (LDL) and total cholesterol). Epidemiological data and clinical trials suggest that a 0.026 mmol L^{-1} increase in LDL cholesterol causes a 1% increase in coronary risk [88] and that for every 1% lowering of serum cholesterol levels, the risk of developing coronary heart disease is reduced by 2–3% and reducing insulin levels reduces the risk of developing insulin insensitivity and metabolic syndrome. Oat bran reduces total serum cholesterol in hypercholesterolemic subjects by as much as 23% with no change in HDL cholesterol [103].

A high fiber intake can improve the conditions of the intestinal environment by diluting carcinogens in the colon and decreasing transfer time. The oat bran β-glucans can regulate glucose metabolism, reducing hyperglycemia, especially in high doses (2000 mg kg^{-1}) like metformin, in diabetic rats and in vitro tests. A dose of β-glucans has fewer adverse effects than traditional therapy for diabetes mellitus and its intake is approved by the Food and Drug Administration (FDA) at 3 g d^{-1}, for nutritional and bioactive effects [92]; dose equal to that concluded by a meta-analysis of 30 research articles on the effect of different levels of exposure to β-glucans reported a dose-response model [94]. The β-glucans in oats are (1,3) or (1,4)-β-D-glucans linear polysaccharides, mainly composed of (1,3) units of cellotriosyl and cellotetraosyl (>90%) [88]. To the mixed-linkage (1,3), (1,4)-β-D-glucans (β-glucans) present in cereals at 30 and 70%, respectively, are attributed a significant number of functionalities and roles that make them unique as part of the plant cell wall and as soluble dietary fiber [103]. The (1,3) bond prevents tight packing of the molecule and causes its partial water solubility. The characteristic molar ratio reported for tri/tetra-oligosaccharides is 2.1–2.4 [88]. Glucans (1,3 and 1,4) from oats are highly soluble in water, have low viscosity and strong biological activity, are part of the human diet, and can contact enterocytes, immune cells, and dendritic cells, that are present in the intestinal immune system and can increase its function [96]. Oat β-glucans are a good agent for reducing total (for each 3 g of dietary fiber per day reduces total cholesterol $\approx$ 2%; for each 2.9 g of β-glucan twice a day reduces the levels of total cholesterol 9.2%) and LDL cholesterol (for each 2.9 g of β-glucan twice a day reduces the LDL cholesterol 10%), improve HDL cholesterol, are a good regulator of blood pressure, improve the lipid profile of the blood, are regulators of postprandial blood glucose (for each gram of β-glucan consumed, the glycemic index decreases by 4 units) and insulin response, and reduce and maintain body weight. Therefore, they help to treat and/or prevent cardiovascular diseases and diabetes; improve immune functions by increasing immunoglobulins, NK cells (natural killer), and killer T cells (lymphocytes) in the blood; they improve resistance to infectious and parasitic diseases; and contribute to a reduction in the risk of cancer and to improve the quality of chemotherapy. The different effects of ingesting β-glucans are due to the type of source, whether they are grains or extracts, frequency, dose, and molecular weight, as well as the age, gender, physiology, and initial levels in individuals [88,103].

The viscosity of β-glucan (in drink mode) could account for 79–96% of the changes in the plasma response of glucose and insulin to 50 g of glucose [103], delaying intestinal glucose absorption as it is trapped in micelles with β-glucan, attenuating the postprandial insulin response (in portal vein), and consequently decreasing the activity of the liver enzyme HMGCoA; these phenomena have beneficial effects in the control and prevention of type 2 diabetes and cholesterol synthesis [88,91]. Another possible mechanism is reducing pancreatic amylase activity and reducing the movement of sugars released into the intestinal wall. It has been identified that a high viscosity of β-glucan extracts (in an in vitro digestive system), impacts on reducing the digestibility of starch, in addition to reducing the absorption of glucose in the blood; this is better presented with 40 g of starch in the formulation, than with 60 g [93]. The intake of purified oat β-glucans (in pigs) reduced the net glucose absorption, which impacts the reduction of the insulin delivery peak (6% β-glucan, 30 min) while maintaining pre-hepatic insulin homeostasis [91]. The cause of the decrease in glucose absorption is reported to be an increase in the water-binding capacity and the viscosity of the gastrointestinal contents, in addition to a possible effect due to gastric emptying. The effect of an increase in the viscosity of the food bolus, from the upper part of the gastrointestinal tract, lengthens gastric emptying, motility, and residence times, and the absorption of nutrients, which is subsequently reflected in a decrease in blood glucose and insulin [88]. Mealtime intake of β-glucans (8.9 g d^{-1}) has been observed to result in carbohydrate- and lipid-like metabolisms, decreasing postprandial glucose and delaying and/or reducing carbohydrate absorption in the intestine [104]. The inhibitory effect of viscous soluble fiber on the postprandial increase in glucose and insulin is decreased when the viscosity of the prepared fiber is reduced by acid treatments [90].

Similar mechanisms of β-glucans are reported to decrease cholesterol absorption [88]: the increase in viscosity reduces the available lipids, slows down diffusion, and modifies the thickness of the non-agitated layer at the intestinal absorption site; entrapment in whole micelles containing bile acid, in the intestine, avoids interaction of the lipid with the luminal membrane transporters in the intestinal epithelium and decreases fat emulsion in the small intestine, actions that increase the excretion of bile acids in feces of 35–65%. The low bile acid, the hepatic conversion of cholesterol to bile acid increases, hepatic cholesterol stores decrease, and, to reach a steady-state, endogenous cholesterol synthesis increases. The restoration of hepatic cholesterol produces a decrease in serum LDL cholesterol. Intestinal absorption of lipids [90] was evaluated in vitro in the presence of β-glucans from oats and barley (0.1–0.5% *w/w*) in an intestinal cell line (NCI-H716) in rats. A decrease in the absorption of fatty acids (18:0, 18:2) was detected due to the effect of β-glucans; this effect varies according to the source of β-glucans, the presence of resistance (aqueous layer, viscosity, agitation), and the portion of intestine evaluated, in addition to the inhibition of regulatory genes for intestinal absorption and lipid synthesis (FAS, ACC, SREBP-1a, SREBP-1c, SREBP-2, i-FABP, FATP4 mRNA), and to the possible participation in the metabolism of enterocytes or transmembrane transport mechanisms. The enzymatic hydrolysis of oat glucans (730,000 g mol^{-1}) improved anticholesterolemic activity, specifically reduced rat serum triglycerides and weight gain, increased HDL cholesterol, and reduced cholesterol. For some tests, there was no significant difference in the effects of hydrolyzed and native β-glucan on LDL [89].

β-glucans may be one of the types of compounds responsible for the immunological effects (innate and adaptive) provided by cereals, fungi, algae, yeasts, and bacteria [96]. The immunomodulatory effect is due to the binding of β-glucans to immune receptors (dectin-1, receptor 3 (CR3; CD11b/CD18), lactosylceramide) that promote a group of immune cells such as monocytes, macrophages, neutrophils, NK cells, and tooth cells. Binding to receptors also promotes the release of cytokines such as IL-12, IL-6, IL-10, and TNF. The antitumor efficacy of (1,3)-β-glucan is related to the type of tumor, the genetic background of the host, the dose, and the route and moment of administration of the glucan, as well as the tumor burden. Proposed mechanisms of anticancer activity include the destruction of tumor cells by macrophages, and modulation of the activity of lymphocytes, neutrophils, and NK cells; in these cases, the action of β-glucans in combination with immune mechanisms is proposed [88]. The antitumor activity of high- and low-molecular-weight oat β-glucans was evaluated in two human lung cancer cell lines (A549 and H69AR) and normal keratinocytes (HaCaT) [98]. High-molecular-weight β-glucans from oats did not show significant toxicity to normal cells but did cause a decrease in cancer cell viability (about a 50% decrease at 200 μg mL^{-1}). The oxidation marker malondialdehyde was increased in both cancer cell lines, indicating a possible induction of oxidative stress by the presence of β-glucan; the high expression of mitochondrial superoxide dismutase and significant changes in the cytoskeleton of cells confirm the hypothesis [88]. In addition, the low-molecular-weight β-glucans from oats significantly decrease the viability of cancer cells with increasing concentration (400 μg mL^{-1}) and incubation time [96]. The mechanism by which β-glucans can kill cancer cells is very complex and not fully understood.

Avenanthramides are phenolic bioactive compounds (in more than 20 chemical forms), exclusive to oats, that exhibit anticancer properties against breast (MDA-MB-231) and human colon cancer cell lines, among others. In vitro anticancer studies of oats (grains) in malignant colon cell lines (HT-29) revealed that fermented (in the presence of *Lactobacillus acidophilus*) and non-fermented oats present high antioxidant and antiproliferative activity in colon cancer cells [99]. Avenanthramides have effects on the prevention and treatment of aging-related human diseases associated with oxidative stress and inflammation, including dermatological, cardiovascular, cerebrovascular, neurodegenerative, and metabolic diseases, and cancer [101]. Avenanthramides are produced by plants as a defense system against pathogens, have antioxidant properties 10 to 30 times greater than vanillin and caffeic acid, promote the activity of superoxide dismutase and glutathione peroxidase in rats fed them (20 mg Avn kg^{-1} body weight), also attenuate the production of free radicals induced by exercise (40 mg Avn kg^{-1} body weight), and have an antihypertensive effect (mediated by the production of nitric oxide) [105]. Avenanthramides are composed of N-cinnamoylanthranilic acids, anthranilic acid, and either cinnamic or avenalumic acid, and vary only in the pattern of substitution in the acidic moieties. Avenanthramides have been shown to significantly increase the expression of heme-oxygenase-1 (HO-1: phase II antioxidant enzyme) in KH-2 cells (adult human renal proximal tubule cells), in both a dose- (250–1000 μM) and time-dependent manner, by mediating reactive oxygen species (ROS) and stimulating the nuclear translocation regulated by Nrf2 (factor 2 related to nuclear factor E2); hydrogenation of the double bond of the carbonyl group in avenanthramides removes this effect. The antioxidant activity has been attributed to the hydroxyl groups of anthranilic and cinnamic acid of avenanthramides; without them, no DPPH radical scavenging has been detected, nor in the FRAP assay. In addition, the ortho-hydroxyl structure, present in cinnamic acid, is presumed to have a greater antioxidant capacity [106].

Form A, B, and C avenanthramides (Avn) are the main ones found in oats and oat bran, the concentration varying according to the processing applied; the Avn content in oat flakes has been reported to be 26–27 mg kg^{-1}, while in bran it is around 13 mg kg^{-1} [102]. An enriched mixture of Avn (4, 20, and 40 ng mL^{-1}; prepared from oat grains) preincubated (24 h) with human aortic endothelial cells significantly decreased the adhesion of monocytic U937 cells to IL-1β in a dose-dependent manner (evidence of antiatherogenic activity). Furthermore, the same mixture significantly suppressed the stimulated IL-1β expression of intracellular adhesion molecule-1 (ICAM-1), vascular cell adhesion molecule-1 (VCAM-1), and E-selectin, and secretion of the pro-inflammatory cytokine IL-6, chemokine IL-8, and monocyte chemoattractant protein (MCP)-1, inflammatory components involved in the formation of fatty portions in arteries. At the same time, the enriched mixture did not show toxicity to human aortic endothelial cells [107]. A possible anticancer mechanism (antiproliferative) of Avn (from oat grains: Avn, Avn-C, methylated derivative of Avn-C (CH3-Avn-C)) is by suppression of cyclooxygenase-2 (COX-2) and prostaglandin (PGE2) activity in colon endothelial cell macrophages [102]; they significantly inhibited the proliferation of human colon cancer cell lines (COX-2-positive HT-29, Caco-2, and LS174T, and COX-2-negative HCT116). The methylated extract of Avn-C had the most powerful anticancer activity (of those evaluated), presumably due to the presence of a single methyl ester group in the structure, which can increase lipid solubility and bioavailability, which means it is easily incorporated into the cell membrane and allows hindrance of the molecular pathways that participate in cell proliferation.

Some research regarding processed oat products has been published. Colloidal oatmeal has a long history of safe use in dermatology in the treatment of atopic dermatitis, psoriasis, drug-induced rash, and other conditions. Some in vitro and in vivo studies show some molecular mechanisms of the anti-inflammatory and antihistamine activity of colloidal oats [108,109]; however, the exact mechanism of action for the anti-inflammatory activity is unclear. The polar extracts of oats (methanol at 80%) significantly inhibited the expression of the cytokine IL-8 (an indicator of itching and prurism) [110]. The 80% acetone extract significantly inhibited levels of the inflammatory promoter NF-κB in the controls treated with TNF-α (25 and 50 $\mu g\ mL^{-1}$) and the production of ROS (in keratinocytes). The groups of compounds recognized in the 80% methanol and 80% acetone extracts evaluated from oats include phenols and alcohol-soluble protein albumins. In the aqueous extract, there are water-soluble proteins such as globulins, and prolamines and carbohydrates, with important anti-inflammatory effects.

"Profermin" is a fermented vegetable product (water, fermented oats, barley malt, lecithin, and *Lactobacillus plantarum* 299v) used for the dietary treatment of ulcerative colitis; it is safe, well-tolerated, acceptable, and capable of reducing the simple clinical colitis activity index (SCCAI) to a statistically and clinically significant level in patients with mild to moderate flare-ups. The decrease in colitis (SCCAI) was 50% greater in the group consuming vegetable ferment (53%) than in the control group (Fresubin: 27%) [111]. A solution prepared with *Avena sativa* L. diluted in vinegar and hydroxyzine was evaluated in the reduction of uremic pruritus, a common complication in patients with chronic kidney disease. The solution significantly decreased the intensity of itching (2.03%), the consequences (4.82%), and the verbal descriptor (2.27%), but had no significant effects on the frequency (1.42%) or surface of the itch (12.40%) [112]. Four colloidal oat extracts were prepared with various solvents and tested in vitro for skin barrier-related gene expression and activity. Colloidal oats promoted the expression of target genes related to the skin barrier and resulted in the recovery of barrier damage in an in vitro model of atopic dermatitis. Clinically, the lotion improved skin dryness, hydration, and barrier [113]. The anti-irritation effect of oats on the skin could be mediated by polyphenols. Evidence indicates that the regular incorporation of oats into the diet can reduce the risk of some diseases associated with inflammation and microbial growth. In addition to these properties, the presence of fiber in oats contributes to reducing the risk of colon cancer.

3. Conclusions and Future Directions

Evidence indicates that representative crops from Hidalgo, Mexico, present important bioactivities which has been shown to reduce total and LDL cholesterol, improve HDL cholesterol, regulate blood pressure, enhance blood lipid profile, regulate glucose in postprandial blood, and insulin response, and reduce and maintain body weight. In addition, there is potential for the treatment of diseases associated with inflammation and microbial proliferation, as well as anticancer treatments. There are sources that help treat and/or prevent cardiovascular diseases and diabetes and improve immune functions by increasing immunoglobulins in the blood. Therefore, it is recommended to promote representative crops of the area, maintain their traditional use, and exploit them in innovative ways to produce compounds and foods beneficial to the health of the population. In addition, it is also worth highlighting that the crops reviewed are distributed throughout the world, where the benefits reviewed here can be used.

Once the bioactive benefits of crops are recognized, further development of plant improvements can be directed towards making plants more resilient to climate change, thereby supporting better control of the biosynthesis process of specific metabolites. Let us also remember that the biosynthesis routes of bioactive compounds (diverse depending on the type of compound) are a response to diverse stimuli specific to the plant, and that external stimuli from the environment also have an influence, thus requiring a constant recognition of the properties of the crops and their fruits, as the changes in the context also develop. The possible decline in plant production and its nutritional and bioactive quality will remain one of the worst consequences of climate change for years to come. With this, food security will be affected and will remain one of the main challenges for humanity. The challenge must be met with full knowledge of plant metabolism and its effects on the environment and humans. Moreover, that knowledge is still in continuous discovery, as demonstrated by the findings presented in this review. In addition to the inherent findings of plants, there are technological developments that can support plant food production, for the improvement of soils, crops, qualities, microbiological context, plant-microorganism relationships, and the effect of all of them on better qualities and higher quantities of phytochemicals. Some of the current tools to support the development of better crops are metabolomics and bioinformatics [2].

Author Contributions: Conceptualization, S.J.V.-R. and T.A.Q.-M.; methodology, S.J.V.-R. and T.A.Q.-M.; software, T.A.Q.-M.; formal analysis, S.J.V.-R. and T.A.Q.-M.; investigation, T.A.Q.-M.; resources, T.A.Q.-M.; data curation, S.J.V.-R. and T.A.Q.-M.; writing—original draft preparation, S.J.V.-R., T.A.Q.-M.; writing—review and editing, S.J.V.-R., J.G.T.-U. and T.A.Q.-M.; visualization, S.J.V.-R. and T.A.Q.-M.; supervision, S.J.V.-R. and T.A.Q.-M. All authors have read and agreed to the published version of the manuscript.

Funding: This research received no external funding.

Conflicts of Interest: The authors declare no conflict of interest.

References

1. Oladipo, A.; Enwemiwe, V.; Ejeromedoghene, O.; Adebayo, A.; Ogunyemi, O.; Fu, F. Production and functionalities of specialized metabolites from different organic sources. *Metabolites* **2022**, *12*, 534. [CrossRef] [PubMed]
2. Razzaq, A.; Wishart, D.S.; Wani, S.H.; Hameed, M.K.; Mubin, M.; Saleem, F. Advances in metabolomics-driven diagnostic breeding and crop improvement. *Metabolites* **2022**, *12*, 511. [CrossRef] [PubMed]
3. Shitan, N. Secondary metabolites in plants: Transport and self-tolerance mechanisms. *Biosci. Biotechnol. Biochem.* **2016**, *80*, 1283–1293. [CrossRef]
4. SIAP/SADER. Hidalgo Infografia Agroalimentaria 2019. Available online: https://agroproductores.com/wp-content/uploads/2020/09/Hidalgo-Infografia-Agroalimentaria-2019.pdf (accessed on 2 July 2022).
5. SIAP/SADER. Anuario Estadístico de la Producción Agrícola. 2022. Available online: https://nube.siap.gob.mx/cierreagricola/ (accessed on 3 July 2022).
6. La Cocina Hidalguense | Mexico Desconocido. (n.d.). 2007. Available online: https://www.mexicodesconocido.com.mx/la-cocina-hidalguense.html (accessed on 15 June 2021).
7. SIAP/SADER. Monografías/SIAP. 2021. Available online: https://www.gob.mx/siap/articulos/monografias-32658 (accessed on 2 July 2022).
8. Hadidi, M.; Ibarz, A.; Pagan, J. Optimisation and kinetic study of the ultrasonic-assisted extraction of total saponins from alfalfa (*Medicago sativa*) and its bioaccessibility using the response surface methodology. *Food Chem.* **2020**, *309*, 125786. [CrossRef] [PubMed]
9. Navarro del Hierro, J.; Herrera, T.; García-Risco, M.R.; Fornari, T.; Reglero, G.; Martin, D. Ultrasound-assisted extraction and bioaccessibility of saponins from edible seeds: Quinoa, lentil, fenugreek, soybean and lupin. *Food Res. Int.* **2018**, *109*, 440–447. [CrossRef]

10. Mattioli, S.; Dal Bosco, A.; Castellini, C.; Falcinelli, B.; Sileoni, V.; Marconi, O.; Mancinelli, A.C.; Cotozzolo, E.; Benincasa, P. Effect of heat- and freeze-drying treatments on phytochemical content and fatty acid profile of alfalfa and flax sprouts. *J. Sci. Food Agric.* **2019**, *99*, 4029–4035. [CrossRef] [PubMed]
11. Giuberti, G.; Rocchetti, G.; Sigolo, S.; Fortunati, P.; Lucini, L.; Gallo, A. Exploitation of alfalfa seed (*Medicago sativa* L.) flour into gluten-free rice cookies: Nutritional, antioxidant and quality characteristics. *Food Chem.* **2018**, *239*, 679–687. [CrossRef]
12. Dal Bosco, A.; Castellini, C.; Martino, M.; Mattioli, S.; Marconi, O.; Sileoni, V.; Ruggeri, S.; Tei, F.; Benincasa, P. The effect of dietary alfalfa and flax sprouts on rabbit meat antioxidant content, lipid oxidation and fatty acid composition. *Meat Sci.* **2015**, *106*, 31–37. [CrossRef]
13. Mattioli, S.; Dal Bosco, A.; Martino, M.; Ruggeri, S.; Marconi, O.; Sileoni, V.; Falcinelli, B.; Castellini, C.; Benincasa, P. Alfalfa and flax sprouts supplementation enriches the content of bioactive compounds and lowers the cholesterol in hen egg. *J. Funct. Foods* **2016**, *22*, 454–462. [CrossRef]
14. Kim, J.Y.; Lee, S.I.; Kim, J.A.; Park, S.C.; Jeong, M.J. Sound waves increases the ascorbic acid content of alfalfa sprouts by affecting the expression of ascorbic acid biosynthesis-related genes. *Plant Biotechnol. Rep.* **2017**, *11*, 355–364. [CrossRef]
15. Fiutak, G.; Michalczyk, M.; Filipczak-Fiutak, M.; Fiedor, L.; Surówka, K. The impact of LED lighting on the yield, morphological structure and some bioactive components in alfalfa (*Medicago sativa* L.) sprouts. *Food Chem.* **2019**, *285*, 53–58. [CrossRef] [PubMed]
16. Michalczyk, M.; Fiutak, G.; Tarko, T. Effect of hot water treatment of seeds on quality indicators of alfalfa sprouts. *LWT-Food Sci. Technol.* **2019**, *113*, 108270. [CrossRef]
17. Rocchetti, G.; Senizza, A.; Gallo, A.; Lucini, L.; Giuberti, G.; Patrone, V. In vitro large intestine fermentation of gluten-free rice cookies containing alfalfa seed (*Medicago sativa* L.) flour: A combined metagenomic/metabolomic approach. *Food Res. Int.* **2019**, *120*, 312–321. [CrossRef] [PubMed]
18. Woch, W.; Hawrylak-Nowak, B. Selected antioxidant properties of alfalfa, radish, and white mustard sprouts biofortified with selenium. *Acta Agrobot.* **2019**, *72*, 1768. [CrossRef]
19. Xie, Z.; Huang, J.; Xu, X.; Jin, Z. Antioxidant activity of peptides isolated from alfalfa leaf protein hydrolysate. *Food Chem.* **2008**, *111*, 370–376. [CrossRef]
20. Thomas, R.; Butler, E.; Macchi, F.; Williams, M. Phytochemicals in cancer prevention and therapy? *BJMP.* **2015**, *8*, a815.
21. Nanditha, B.; Prabhasankar, P. Antioxidants in bakery products: A review. *Crit. Rev. Food Sci. Nutr.* **2009**, *49*, 1–27. [CrossRef]
22. Chen, L.; Liu, J.; Zhang, Y.; Dai, B.; An, Y.; Yu, L. Structural, thermal, and anti-inflammatory properties of a novel pectic polysaccharide from Alfalfa (*Medicago sativa* L.) stem. *J. Agric. Food Chem.* **2015**, *63*, 3219–3228. [CrossRef]
23. Xie, Y.; Wang, L.; Sun, H.; Wang, Y.; Yang, Z.; Zhang, G.; Jiang, S.; Yang, W. Polysaccharide from alfalfa activates RAW 264.7 macrophages through MAPK and NF-κB signaling pathways. *Int. J. Biol. Macromol.* **2019**, *126*, 960–968. [CrossRef]
24. Zhang, C.; Li, Z.; Zhang, C.Y.; Li, M.; Lee, Y.; Zhang, G.G. Extract methods, molecular characteristics, and bioactivities of polysaccharide from Alfalfa (*Medicago sativa* L.). *Nutrients* **2019**, *11*, 1181. [CrossRef]
25. Wang, L.; Xie, Y.; Yang, W.; Yang, Z.; Jiang, S.; Zhang, C.; Zhang, G. Alfalfa polysaccharide prevents H_2O_2-induced oxidative damage in MEFs by activating MAPK/Nrf2 signaling pathways and suppressing NF-κB signaling pathways. *Sci. Rep.* **2019**, *9*, 1782. [CrossRef] [PubMed]
26. Zhang, C.Y.; Gan, L.P.; Du, M.Y.; Shang, Q.H.; Xie, Y.H.; Zhang, G.G. Effects of dietary supplementation of alfalfa polysaccharides on growth performance, small intestinal enzyme activities, morphology, and large intestinal selected microbiota of piglets. *Livest. Sci.* **2019**, *223*, 47–52. [CrossRef]
27. Adams, S.; Xiangjie, K.; Hailong, J.; Guixin, Q.; Sossah, F.L.; Dongsheng, C. Prebiotic effects of alfalfa (*Medicago sativa*) fiber on cecal bacterial composition, short-chain fatty acids, and diarrhea incidence in weaning piglets. *RSC Adv.* **2019**, *9*, 13586–13599. [CrossRef] [PubMed]
28. Adams, S.; Kong, X.; Che, D.; Qin, G.; Jiang, H. Effects of dietary supplementation of alfalfa (*Medicago Sativa*) fibre on the blood biochemistry, nitrogen metabolism, and intestinal morphometry in weaning piglets. *Appl. Ecol. Environ. Res.* **2019**, *17*, 2275–2295. [CrossRef]
29. Almaráz-Abarca, N.; Delgado-Alvarado, E.A.; Ávila-Reyes, J.A.; Uribe-Soto, J.N.; González-Valdez, L.S. The phenols of the genus Agave (*Agavaceae*). *J. Biomater. Nanobiotechnol.* **2013**, *4*, 9–16. [CrossRef]
30. Soto-Simental, S.; Caro, I.; Quinto, E.J.; Mateo, J. Effect of cooking lamb using maguey leaves (*Agave salmiana*) on meat volatile composition. *Int. Food Res. J.* **2016**, *23*, 1212–1216.
31. Santos-Zea, L.; Leal-Diaz, A.; Cortes-Ceballos, E.; Gutierrez- Uribe, J. Agave (*Agave* spp.) and its traditional products as a source of bioactive compounds. *Curr. Bioact. Compd.* **2012**, *8*, 218–231. [CrossRef]
32. Narvaez-Zapata, J.A.; Sánchez-Teyer, L.F. Agaves as a raw material: Recent technologies and applications. *Recent Pat. Biotechnol.* **2009**, *3*, 185–191. [CrossRef]
33. Gutiérrez-Uribe, J.A.; Serna-Saldivar, S. Agave Syrup Extract Having Anticancer Activity. U.S. Patent 20090124685, B2, 14 May 2009. (Grant Date 26 June 2013).
34. Ortiz-Torres, D.L.; Galván-Valencia, M.; Delgadillo-Ruíz, L.; Cabral-Arellano, F.J.; Bañuelos-Valenzuela, R.; León-Esparza-Ibarra, E. Extraction of saponins from leaves of Agaves. In *Sustainable and Integral Exploitation of Agave*; Gutiérrez-Mora, A., Ed.; CIATEJ: Guadalajara, Mexico, 2014; pp. 100–103. Available online: http://www.ciatej.net.mx/agave/1.7agave.pdf (accessed on 10 January 2022).

35. Ortiz-Torres, D.; Galván-Valencia, M.; Delgadillo-Ruíz, L.; Huerta-García, J.; Cabral-Arellano, F.; Bañuelos-Valenzuela, R.; Esparza-Ibarra, E. Extracción y obtención de fracciones de saponinas a partir de hojas de los Agaves salmiana y tequilana weber de Zacatecas. In Proceedings of the XIII Simposium-Taller Nacional y VI Internacional "Producción y Aprovechamiento Del Nopal y Maguey", Monterrey, Nuevo León, Mexico, 10 October 2014; pp. 169–182.
36. García-Pedraza, L.G.; Juárez-Flores, B.I.; Aguirre-Rivera, J.R.; Pinos-Rodríguez, J.M.; Martínez, J.F.; Santoyo, M.E. Effects of *Agave salmiana* Otto ex Salm-Dick high-fructose syrup on non-diabetic and streptozotocin-diabetic rats. *J. Med. Plants Res.* **2009**, *3*, 932–940.
37. Gutiérrez-Uribe, J.A.; Santos-Zea, L.; Serna-Saldivar, S.R.O. Agave Syrup Extract Having Anticancer Activity. U.S. Patent 20130209588, B2, 2013. (Grant Date 7 March 2017).
38. Santos-Zea, L.; Fajardo-Ramírez, O.R.; Romo-López, I.; Gutiérrez-Uribe, J.A. Fast centrifugal partition chromatography fractionation of concentrated Agave (*Agave salmiana*) sap to obtain saponins with apoptotic effect on colon cancer cells. *Plant Foods Hum. Nutr.* **2016**, *71*, 57–63. [CrossRef]
39. Mellado-Mojica, E.; López, M.G. Identification, classification, and discrimination of agave syrups from natural sweeteners by infrared spectroscopy and HPAEC-PAD. *Food Chem.* **2015**, *167*, 349–357. [CrossRef] [PubMed]
40. Velázquez-Ríos, I.O.; González-García, G.; Mellado-Mojica, E.; Veloz García, R.A.; Dzul Cauich, J.G.; López, M.G.; García-Vieyra, M.I. Phytochemical profiles and classification of *Agave* syrups using 1H-NMR and chemometrics. *Food Sci. Nutr.* **2018**, *7*, 3–13. [CrossRef]
41. Santos-Zea, L.; Leal-Díaz, A.M.; Jacobo-Velázquez, D.A.; Rodríguez-Rodríguez, J.; García-Lara, S.; Gutiérrez-Uribe, J.A. Characterization of concentrated agave saps and storage effects on browning, antioxidant capacity and amino acid content. *J. Food Compos. Anal.* **2016**, *45*, 113–120. [CrossRef]
42. Tovar-Robles, C.L.; Perales-Segovia, C.; Cedillo, A.N.; Valera-Montero, L.L.; Gómez-Leyva, J.F.; Guevara-Lara, F.; Hernández-Duque, J.L.M.; Silos-Espino, H. Effect of aguamiel (agave sap) on hematic biometry in rabbits and its antioxidant activity determination. *Ital. J. Anim. Sci.* **2011**, *10*, e21. [CrossRef]
43. Dakah, A.; Zaid, S.; Suleiman, M.; Abbas, S.; Wink, M. In vitro propagation of the medicinal plant *Ziziphora tenuior* L. and evaluation of its antioxidant activity. *Saudi J. Biol. Sci.* **2014**, *21*, 317–323. [CrossRef]
44. Campos, H.; Trejo, C.; Peña-Valdivia, C.B.; García-Nava, R.; Conde-Martínez, F.V.; Cruz-Ortega, M.D.R. Photosynthetic acclimation to drought stress in *Agave salmiana* Otto ex Salm-Dyck seedlings is largely dependent on thermal dissipation and enhanced electron flux to photosystem I. *Photosynth. Res.* **2014**, *122*, 23–39. [CrossRef]
45. Puente-Garza, C.A.; Gutiérrez-Mora, A.; García-Lara, S. Micropropagation of *Agave salmiana*: Means to production of antioxidant and bioactive principles. *Front. Plant Sci.* **2015**, *6*, 1026. [CrossRef]
46. Santos-Zea, L.; Gutiérrez-Uribe, J.A.; Benedito, J. Effect of ultrasound intensification on the supercritical fluid extraction of phytochemicals from *Agave salmiana* bagasse. *J. Supercrit. Fluids* **2019**, *144*, 98–107. [CrossRef]
47. Puente-Garza, C.A.; Gutiérrez-Mora, A.; García-Lara, S. Effects on saponin, flavonol and antioxidant activity *in vitro* plants of *Agave salmiana*. In *Sustainable and Integral Exploitation of Agave*; CIATEJ: Guadalajara, Mexico, 2014; pp. 27–31. Available online: https://www.ciatej.mx/files/divulgacion/divulgacion_5b084050b05ed.pdf (accessed on 10 January 2022).
48. Puente-Garza, C.A.; García-Lara, S.; Gutiérrez-Uribe, J.A. Enhancement of saponins and flavonols by micropropagation of *Agave salmiana*. *Ind. Crops Prod.* **2017**, *105*, 225–230. [CrossRef]
49. Puente-Garza, C.A.; Meza-Miranda, C.; Ochoa-Martínez, D.; García-Lara, S. Effect of in vitro drought stress on phenolic acids, flavonols, saponins, and antioxidant activity in *Agave salmiana*. *Plant Physiol. Biochem.* **2017**, *115*, 400–407. [CrossRef] [PubMed]
50. Puente-Garza, C.A.; Espinosa-Leal, C.A.; García-Lara, S. Steroidal saponin and flavonol content and antioxidant activity during sporophyte development of maguey (*Agave salmiana*). *Plant Foods Hum. Nutr.* **2018**, *73*, 287–294. [CrossRef] [PubMed]
51. Leal-Díaz, A.M.; Santos-Zea, L.; Martínez-Escobedo, H.C.; Guajardo-Flores, D.; Gutiérrez-Uribe, J.A.; Serna-Saldivar, S.O. Effect of *Agave americana* and *Agave salmiana* ripeness on saponin content from aguamiel (agave sap). *J. Agric. Food Chem.* **2015**, *63*, 3924–3930. [CrossRef] [PubMed]
52. Medina-Galván, M.I.; Bernardino-Nicanor, A.; Castro-Rosas, J.; Negrete-Rodríguez, M.L.X.; Conde-Barajas, E.; González-Cruz, L. Antimicrobial and antioxidant activity of flower scape extracts of *Agave salmiana*: Effect of the extraction solvent and development stage. *Res. J. Biotechnol.* **2018**, *13*, 12.
53. Vargas-Rodríguez, L.; García-Vieyra, M.I.; León-Bata, B.I. Lozano-Sotomayor, P. Physical properties and microscopic structure of the *Agave Salmiana* cuticle (mixiote). *Rev. Chapingo Ser. Zonas Áridas* **2018**, *17*, 1–9. [CrossRef]
54. Moreno-Vilet, L.; Garcia-Hernandez, M.H.; Delgado-Portales, R.E.; Corral-Fernandez, N.E.; Cortez-Espinosa, N.; Ruiz-Cabrera, M.A.; Portales-Perez, D.P. In vitro assessment of agave fructans (*Agave salmiana*) as prebiotics and immune system activators. *Int. J. Biol. Macromol.* **2014**, *63*, 181–187. [CrossRef]
55. Koenen, M.E.; Cruz Rubio, J.M.; Mueller, M.; Venema, K. The effect of agave fructan products on the activity and composition of the microbiota determined in a dynamic *in vitro* model of the human proximal large intestine. *J. Funct. Foods* **2016**, *22*, 201–210. [CrossRef]
56. Fernández-Lainez, C.; Akkerman, R.; Oerlemans, M.M.P.; Logtenberg, M.J.; Schols, H.A.; Silva-Lagos, L.A.; López-Velázquez, G.; de Vos, P. $\beta(2\rightarrow6)$-Type fructans attenuate proinflammatory responses in a structure dependent fashion via Toll-like receptors. *Carbohydr. Polym.* **2022**, *277*, 118893. [CrossRef]

57. Espinosa-Andrews, H.; Urías-Silvas, J.E.; Morales-Hernández, N. The role of agave fructans in health and food applications: A review. *Trends Food Sci. Technol.* **2021**, *114*, 585–598. [CrossRef]
58. Castañeda-Sánchez, A. Propiedades nutricionales y antioxidantes del maíz azul (*Zea mays* L.). *TSIA* **2011**, *5*, 75–83.
59. Gyori, Z. Chapter 11, Corn: Grain-quality characteristics and management of quality requirements. In *Cereal Grains: Assessing and Managing Quality*, 2nd ed.; Woodhead Publishing Series in Food Science, Technology and Nutrition; Wrigley, C., Batey, I., Miskelly, D., Eds.; Woodhead Publishing: Amsterdam, The Netherlands, 2017; pp. 257–290. ISBN 9780081007198. [CrossRef]
60. Camelo-Méndez, G.A.; Agama-Acevedo, E.; Tovar, J.; Bello-Pérez, L.A. Functional study of raw and cooked blue maize flour: Starch digestibility, total phenolic content and antioxidant activity. *J. Cereal Sci.* **2017**, *76*, 179–185. [CrossRef]
61. Sáyago-Ayerdi, S.; Álvarez-Parrilla, E. Alimentos Vegetales Autóctonos Iberoamericanos Subutilizados. Copyright © Red ALSUB-CYTED, Fabro Editores. 2018. Available online: https://alimentos-autoctonos.fabro.com.mx/legal.html (accessed on 16 February 2022).
62. Nascimento, A.C.; Mota, C.; Coelho, I.; Gueifão, S.; Santos, M.; Matos, A.S.; Gimenez, A.; Lobo, M.; Samman, N.; Castanheira, I. Characterisation of nutrient profile of quinoa (*Chenopodium quinoa*), amaranth (*Amaranthus caudatus*), and purple corn (*Zea mays* L.) consumed in the North of Argentina: Proximates, minerals and trace elements. *Food Chem.* **2014**, *148*, 420–426. [CrossRef] [PubMed]
63. Salinas-Moreno, Y.; Martínez-Bustos, F.; Soto-Hernández, M.; Ortega-Paczka, R.; Arellano-Vázquez, J.L. Efecto de la nixtamalización sobre las antocianinas del grano de maíces pigmentados. *Agrociencia* **2003**, *37*, 617–628.
64. Mendoza-Díaz, S.; Ortiz-Valerio, M.C.; Castaño-Tostado, E.; Figueroa-Cárdenas, J.D.; Reynoso-Camacho, R.; Ramos-Gómez, M.; Campos-Vega, R.; Loarca-Piña, G. Antioxidant capacity and antimutagenic activity of anthocyanin and carotenoid extracts from nixtamalized pigmented creole maize races (*Zea mays* L.). *Plant Foods Hum. Nutr.* **2012**, *67*, 442–449. [CrossRef]
65. Lopez-Martinez, L.X.; Oliart-Ros, R.M.; Valerio-Alfaro, G.; Lee, C.H.; Parkin, K.L.; Garcia, H.S. Antioxidant activity, phenolic compounds and anthocyanins content of eighteen strains of Mexican maize. *LWT-Food Sci. Technol.* **2009**, *42*, 1187–1192. [CrossRef]
66. Lopez-Martinez, L.X.; Parkin, K.L.; Garcia, H.S. Phase II-Inducing, polyphenols content and antioxidant capacity of corn (*Zea mays* L.) from phenotypes of white, blue, red and purple colors processed into masa and tortillas. *Plant Foods Hum. Nutr.* **2011**, *66*, 41–47. [CrossRef] [PubMed]
67. Salinas-Moreno, Y.; Pérez-Alonso, J.J.; Vázquez-Carrillo, G.; Aragón-Cuevas, F.; Velázquez-Cardelas, G.A. Anthocyanins and antioxidant act in maize grains of chalqueño elotes cónicos and bolita races. *Agrociencia* **2012**, *46*, 693–706.
68. Žilić, S.; Serpen, A.; Akillioğlu, G.; Gökmen, V.; Vančetović, J. Phenolic compounds, carotenoids, anthocyanins, and antioxidant capacity of colored maize (*Zea mays* L.) kernels. *J. Agric. Food Chem.* **2012**, *60*, 1224–1231. [CrossRef]
69. Bello-Pérez, L.A.; Flores-Silva, P.C.; Camelo-Méndez, G.A.; Paredes-López, O.; de Figueroa-Cárdenas, J.D. Effect of the nixtamalization process on the dietary fiber content, starch digestibility, and antioxidant capacity of blue maize tortilla. *Cereal Chem.* **2015**, *92*, 265–270. [CrossRef]
70. Moreira, R.; Chenlo, F.; Arufe, S.; Rubinos, S.N. Physicochemical characterization of white, yellow and purple maize flours and rheological characterization of their doughs. *J. Food Sci. Technol.* **2015**, *52*, 7954–7963. [CrossRef]
71. Urias-Lugo, D.A.; Heredia, J.B.; Serna-Saldivar, S.O.; Muy-Rangel, M.D.; Valdez-Torres, J.B. Total phenolics, total anthocyanins and antioxidant capacity of native and elite blue maize hybrids (*Zea mays* L.). *CYTA-J. Food* **2015**, *13*, 336–339. [CrossRef]
72. Bello-Pérez, L.A.; Camelo-Mendez, G.A.; Agama-Acevedo, E.; Utrilla-Coello, R.G. Aspectos nutracéuticos de los maíces pigmentados: Digestibilidad de los carbohidratos y antocianinas. *Agrociencia* **2016**, *50*, 1041–1063.
73. Mora-Rochín, S.; Gaxiola-Cuevas, N.; Gutiérrez-Uribe, J.A.; Milán-Carrillo, J.; Milán-Noris, E.M.; Reyes-Moreno, C.; Serna-Saldivar, S.O.; Cuevas-Rodríguez, E.O. Effect of traditional nixtamalization on anthocyanin content and profile in Mexican blue maize (*Zea mays* L.) landraces. *LWT-Food Sci. Technol.* **2016**, *68*, 563–569. [CrossRef]
74. Mansilla, P.S.; Nazar, M.C.; Pérez, G.T. Flour functional properties of purple maize (*Zea mays* L.) from Argentina. Influence of environmental growing conditions. *Int. J. Biol. Macromol.* **2020**, *146*, 311–319. [CrossRef] [PubMed]
75. González-Manzano, S.; Pérez-Alonso, J.J.; Salinas-Moreno, Y.; Mateus, N.; Silva, A.M.S.; de Freitas, V.; Santos-Buelga, C. Flavanol-anthocyanin pigments in corn: NMR characterisation and presence in different purple corn varieties. *J. Food Compost. Anal.* **2008**, *21*, 521–526. [CrossRef]
76. Urias-Lugo, D.A.; Heredia, J.B.; Muy-Rangel, M.D.; Valdez-Torres, J.B.; Serna-Saldívar, S.O.; Gutiérrez-Uribe, J.A. Anthocyanins and phenolic acids of hybrid and native blue maize (*Zea mays* L.) extracts and their antiproliferative activity in mammary (MCF7), liver (HepG2), colon (Caco2 and HT29) and prostate (PC3) cancer cells. *Plant Foods Hum. Nutr.* **2015**, *70*, 193–199. [CrossRef]
77. Jing, P.; Noriega, V.; Schwartz, S.J.; Giusti, M.M. Effects of growing conditions on purple corncob (*Zea mays* L.) anthocyanins. *J. Agric. Food Chem.* **2007**, *55*, 8625–8629. [CrossRef]
78. Hu, Q.P.; Xu, J.G. Profiles of carotenoids, anthocyanins, phenolics, and antioxidant activity of selected color waxy corn grains during maturation. *J. Agric. Food Chem.* **2011**, *59*, 2026–2033. [CrossRef]
79. Pedreschi, R.; Cisneros-Zevallos, L. Phenolic profiles of andean purple corn (*Zea mays* L.). *Food Chem.* **2007**, *100*, 956–963. [CrossRef]
80. Ramos-Escudero, F.; Muñoz, A.M.; Alvarado-Ortíz, C.; Alvarado, Á.; Yáñez, J.A. Purple corn (*Zea mays* L.) phenolic compounds profile and its assessment as an agent against oxidative stress in isolated mouse organs. *J. Med. Food* **2012**, *15*, 206–215. [CrossRef]
81. Ruiz Canizales, J.; Heredia, J.B.; Domínguez Avila, J.A.; Madera Santana, T.J.; Villegas Ochoa, M.A.; Robles Sánchez, R.M.; González Aguilar, G.A. Microencapsulation of blue maize (*Zea mays* L.) polyphenols in two matrices: Their stability during storage and in vitro digestion release. *J. Food Meas. Charact.* **2019**, *13*, 892–900. [CrossRef]

82. Yang, Z.; Zhai, W. Identification and antioxidant activity of anthocyanins extracted from the seed and cob of purple corn (*Zea mays* L.). *Innov. Food Sci. Emerg. Technol.* **2010**, *11*, 169–176. [CrossRef]
83. Harakotr, B.; Suriharn, B.; Tangwongchai, R.; Scott, M.P.; Lertrat, K. Anthocyanins and antioxidant activity in coloured waxy corn at different maturation stages. *J. Funct. Foods* **2014**, *9*, 109–118. [CrossRef]
84. Villacres Poveda, C.E.; Tanquina Páramo, I.M.; Yáñez Guzmán, C.F.; Quelal Tapia, M.B.; Alvarez Murillo, M.J.; Ramos Moya, M.R. Impacto del procesamiento sobre los compuestos con propiedades antioxidantes de dos variedades de maíz (*Zea mays* L.). *ACI Adv. Cienc. Ing.* **2019**, *11*, 104–115. [CrossRef]
85. Ronceros, G.; Ramos, W.; Arroyo, J.; Galarza, C.; Gutiérrez, E.L.; Ortega-Loayza, A.G.; La Rosa, C.; Cucho, C.; Palma, L. Estudio comparativo del maíz morado (*Zea mays* L.) y simvastatina en la reducción de lípidos séricos de pacientes diabéticos normotensos con dislipidemia. *An. Med.* **2012**, *73*, 113. [CrossRef]
86. De Mejia, E.G.; Dia, V.P.; West, L.; West, M.; Singh, V.; Wang, Z.; Allen, C. Temperature dependency of shelf and thermal stabilities of anthocyanins from corn distillers' dried grains with solubles in different ethanol extracts and a commercially available beverage. *J. Agric. Food Chem.* **2015**, *63*, 10032–10041. [CrossRef]
87. Li, C.Y.; Kim, H.W.; Li, H.; Lee, D.C.; Rhee, H.I. Antioxidative effect of purple corn extracts during storage of mayonnaise. *Food Chem.* **2014**, *152*, 592–596. [CrossRef]
88. Daou, C.; Zhang, H. Oat beta-glucan: Its role in health promotion and prevention of diseases. *Compr. Rev. Food Sci. Food Saf.* **2012**, *11*, 355–365. [CrossRef]
89. Bae, I.Y.; Kim, S.M.; Lee, S.; Lee, H.G. Effect of enzymatic hydrolysis on cholesterol-lowering activity of oat β-glucan. *New Biotechnol.* **2010**, *27*, 85–88. [CrossRef]
90. Drozdowski, L.A.; Reimer, R.A.; Temelli, F.; Bell, R.C.; Vasanthan, T.; Thomson, A.B.R. β-Glucan extracts inhibit the in vitro intestinal uptake of long-chain fatty acids and cholesterol and down-regulate genes involved in lipogenesis and lipid transport in rats. *J. Nutr. Biochem.* **2010**, *21*, 695–701. [CrossRef]
91. Hooda, S.; Matte, J.J.; Vasanthan, T.; Zijlstra, R.T. Dietary purified oat β-glucan reduces peak glucose absorption and portal insulin release in portal-vein catheterized grower pigs. *Livest. Sci.* **2010**, *134*, 15–17. [CrossRef]
92. Dong, J.; Cai, F.; Shen, R.; Liu, Y. Hypoglycaemic effects and inhibitory effect on intestinal disaccharidases of oat beta-glucan in streptozotocin-induced diabetic mice. *Food Chem.* **2011**, *129*, 1066–1071. [CrossRef] [PubMed]
93. Regand, A.; Chowdhury, Z.; Tosh, S.M.; Wolever, T.M.S.; Wood, P. The molecular weight, solubility and viscosity of oat beta-glucan affect human glycemic response by modifying starch digestibility. *Food Chem.* **2011**, *129*, 297–304. [CrossRef] [PubMed]
94. Tiwari, U.; Cummins, E. Meta-analysis of the effect of β-glucan intake on blood cholesterol and glucose levels. *Nutrition* **2011**, *27*, 1008–1016. [CrossRef] [PubMed]
95. Ren, Y.; Yang, X.; Niu, X.; Liu, S.; Ren, G. Chemical characterization of the avenanthramide-rich extract from oat and its effect on D-galactose-induced oxidative stress in mice. *J. Agric. Food Chem.* **2011**, *59*, 206–211. [CrossRef]
96. Choromanska, A.; Kulbacka, J.; Rembialkowska, N.; Pilat, J.; Oledzki, R.; Harasym, J.; Saczko, J. Anticancer properties of low molecular weight oat beta-glucan—An in vitro study. *Int. J. Biol. Macromol.* **2015**, *80*, 23–28. [CrossRef]
97. Choromańska, A.; Kulbacka, J.; Harasym, J.; Dubińska-Magiera, M.; Saczko, J. Anticancer activity of oat β-glucan in combination with electroporation on human cancer cells. *Acta Pol. Pharm.* **2017**, *74*, 616–623.
98. Choromanska, A.; Kulbacka, J.; Harasym, J.; Oledzki, R.; Szewczyk, A.; Saczko, J. High- and low-molecular weight oat beta-glucan reveals antitumor activity in human epithelial lung cancer. *Pathol. Oncol. Res.* **2018**, *24*, 583–592. [CrossRef]
99. Sunderam, V.; Mohammed, S.S.S.; Madhavan, Y.; Dhinakaran, M.; Sampath, S.; Patteswaran, N.; Thangavelu, L.; Lawrence, A.V. Free radical scavenging activity and cytotoxicity study of fermented oats (*Avena sativa*). *Int. J. Res. Pharm. Sci.* **2020**, *11*, 1259–1262. [CrossRef]
100. Vetvicka, V.; Vetvickova, J.B. 1,3-Glucan in cancer treatment. *Am. J. Immunol.* **2012**, *8*, 38–43. [CrossRef]
101. Perrelli, A.; Goitre, L.; Salzano, A.M.; Moglia, A.; Scaloni, A.; Retta, S.F. Biological activities, health benefits, and therapeutic properties of avenanthramides: From skin protection to prevention and treatment of cerebrovascular diseases. *Oxidative Med. Cell. Longev.* **2018**, *2018*, 6015351. [CrossRef]
102. Guo, W.; Nie, L.; Wu, D.; Wise, M.L.; Collins, F.W.; Meydani, S.N.; Meydani, M. Avenanthramides inhibit proliferation of human colon cancer cell lines in vitro. *Nutr. Cancer* **2010**, *62*, 1007–1016. [CrossRef]
103. Wood, P.J. Cereal β-glucans in diet and health. *J. Cereal Sci.* **2007**, *46*, 230–238. [CrossRef]
104. Battilana, P.; Ornstein, K.; Minehira, K.; Schwarz, J.M.; Acheson, K.; Schneiter, P.; Burri, J.; Jéquier, E.; Tappy, L. Mechanisms of action of β-glucan in postprandial glucose metabolism in healthy men. *Eur. J. Clin. Nutr.* **2001**, *55*, 327–333. [CrossRef]
105. Meydani, M. Potential health benefits of avenanthramides of oats. *Nutr. Rev.* **2009**, *67*, 731–735. [CrossRef] [PubMed]
106. Fu, J.; Zhu, Y.; Yerke, A.; Wise, M.L.; Johnson, J.; Chu, Y.; Sang, S. Oat avenanthramides induce heme oxygenase-1 expression via Nrf2-mediated signaling in HK-2 cells. *Mol. Nutr. Food Res.* **2015**, *59*, 2471–2479. [CrossRef] [PubMed]
107. Liu, L.; Zubik, L.; Collins, F.W.; Marko, M.; Meydani, M. The antiatherogenic potential of oat phenolic compounds. *Atherosclerosis* **2004**, *175*, 39–49. [CrossRef]
108. Cerio, R.; Dohil, M.; Downie, J.; Magina, S.; Mahé, E.; Stratigos, A.J. Mechanism of action and clinical benefits of colloidal oatmeal for dermatologic practice. *J. Drugs Dermatol.* **2010**, *9*, 1116–1120.
109. Wu, J. Anti-inflammatory ingredients. *J. Drugs Dermatol.* **2008**, *7*, s13-6.

110. Reynertson, K.A.; Garay, M.; Nebus, J.; Chon, S.; Kaur, S.; Mahmood, K.; Kizoulis, M.; Southall, M.D. Anti-inflammatory activities of colloidal oatmeal (*Avena sativa*) contribute to the effectiveness of oats in treatment of itch associated with dry, irritated skin. *J. Drugs Dermatol.* **2015**, *14*, 43–48.
111. Krag, A.; Munkholm, P.; Israelsen, H.; Von Ryberg, B.; Andersen, K.K.; Bendtsen, F. Profermin is efficacious in patients with active ulcerative colitis—A randomized controlled trial. *Inflamm. Bowel Dis.* **2013**, *19*, 2584–2592. [CrossRef] [PubMed]
112. Nakhaee, S.; Nasiri, A.; Waghei, Y.; Morshedi, J. Comparison of *Avena Sativa*, vinegar, and hydroxyzine for uremic pruritus of hemodialysis patients a crossover randomized clinical trial. *Iran. J. Kidney Dis.* **2015**, *9*, 316–322. [PubMed]
113. Ilnytska, O.; Kaur, S.; Chon, S.; Reynertson, K.A.; Nebus, J.; Garay, M.; Mahmood, K.; Southall, M.D. Colloidal oatmeal (*Avena Sativa*) improves skin barrier through multi-therapy activity. *J. Drugs Dermatol.* **2016**, *15*, 684–690. [PubMed]

metabolites

Article

Luteolin-7-O-rutinoside Protects RIN-5F Cells from High-Glucose-Induced Toxicity, Improves Glucose Homeostasis in L6 Myotubes, and Prevents Onset of Type 2 Diabetes

Pandurangan Subash-Babu, Sahar Abdulaziz AlSedairy, Manal Abdulaziz Binobead and Ali A. Alshatwi *

Adipogenesis and Immunobiology Research Lab, Department of Food Sciences and Nutrition, College of Food and Agricultural Sciences, King Saud University, P.O. Box 2460, Riyadh 11451, Saudi Arabia
* Correspondence: alshatwi@ksu.edu.sa

Abstract: Luteolin-7-*O*-rutinoside (lut-7-*O*-rutin), a flavonoid commonly present in *Mentha longifolia* L. and *Olea europaea* L. leaves has been used as a flavoring agent with some biological activity. The present study is the first attempt to analyze the protective effect of lut-7-*O*-rutin on high-glucose-induced toxicity to RIN-5F cells in vitro. We found that lut-7-*O*-rutin improved insulin secretion in both normal and high-glucose conditions in a dose-dependent manner, without toxicity observed. In addition, 20 µmol of lut-7-*O*-rutin improves insulin sensitization and glucose uptake significantly ($p \leq 0.01$) in L6 myotubes cultured in a high-glucose medium. Lut-7-*O*-rutin has shown a significant ($p \leq 0.05$) effect on glucose uptake in L6 myotubes compared to the reference drug, rosiglitazone (20 µmol). Gene expression analysis confirmed significantly lowered CYP1A, TNF-α, and NF-κb expressions in RIN-5F cells, and increased mitochondrial thermogenesis-related LPL, Ucp-1 and PPARγC1A mRNA expressions in L6 myotubes after 24 h of lut-7-*O*-rutin treatment. The levels of signaling proteins associated with intracellular glucose uptakes, such as cAMP, ChREBP-1, and AMPK, were significantly increased in L6 myotubes. In addition, the levels of the conversion rate of glucose to lactate and fatty acids were raised in insulin-stimulated conditions; the rate of glycerol conversion was found to be higher at the basal level in L6 myotubes. In conclusion, lut-7-*O*-rutin protects RIN-5F cells from high-glucose-induced toxicity, stimulates insulin secretion, and promotes glucose absorption and homeostasis via molecular mechanisms.

Keywords: luteolin-7-*O*-rutinoside; RIN-5F cells; L6 myotubes; glucose homeostasis; mitochondria

check for updates

Citation: Subash-Babu, P.; Abdulaziz AlSedairy, S.; Abdulaziz Binobead, M.; Alshatwi, A.A. Luteolin-7-O-rutinoside Protects RIN-5F Cells from High-Glucose-Induced Toxicity, Improves Glucose Homeostasis in L6 Myotubes, and Prevents Onset of Type 2 Diabetes. *Metabolites* **2023**, *13*, 269. https://doi.org/10.3390/metabo13020269

Academic Editors: Ramona Paltinean and Irina Ielciu

Received: 22 December 2022
Revised: 1 February 2023
Accepted: 4 February 2023
Published: 14 February 2023

1. Introduction

Noninsulin-dependent diabetes mellitus (NIDDM), or type 2 diabetes, is a complex metabolic disorder resulting from either insulin insufficiency or insulin dysfunction, leading to hyperglycemia [1]. Hyperglycemia is a significant factor in inducing β-cell apoptosis. However, the precise mechanisms underlying β-cell dysfunction in type 2 diabetes are interrelated with inflammation and β-cell glucotoxicity [2]. In hyperglycemic conditions, pancreatic β cells are exposed to increased metabolic flux and associated cellular stress, leading to impairment of β-cell function and survival, a process called glucotoxicity [3], including prolonged hyperglycemia, which increases superoxide, oxidative stress, endoplasmic reticulum (ER) stress and IL-1β cytokine in islets, which collectively activate c-Jun N-terminal kinase (JNK) [4]. The ER stress may trigger islet JNK activation, which plays a significant role in glucose-induced β-cell dysfunction [1].

NIDDM is characterized by insulin resistance, in which the primary insulin-target organs, such as skeletal muscle, liver, and adipocytes are poorly responsive to insulin action, which may combine with reduced insulin secretion caused by a progressive loss of β-cell function [5]. Insulin-dependent glucose uptake and glycogen storage have been found to be decreased in skeletal muscle, a key tissue site for insulin resistance [6]. A diminished insulin level modulates the expression of inflammatory adipokines and myokines in muscle cells, which anticipates pancreatic β-cell apoptosis [7]. Combatting hyperglycemia-induced inflammation and oxidative injury may prevent pancreatic β-cell toxicity, metabolic stress, and diabetic complications [8]. Cultured human muscle cells, or myotubes, are considered to be a valuable model for studying the morphological, metabolic, and biochemical properties of adult skeletal muscle [9].

Modern antidiabetic drugs, such as rosiglitazone, a thiazolidinedione that is highly recommended by physicians to control hyperglycemia, contributes to the amelioration of whole-body insulin resistance. Meanwhile, rosiglitazone intake causes the development of cardiovascular risk, osteoporosis, and there is evidence that it increase weight gain and fluid retention [10]. So, the need for a potential drug to increase insulin sensitivity without developing side effect remains. Medicinal-plant-derived flavonoids are antioxidants and play a significant role in cellular pro- and antioxidant regulations without producing side effects [11]. Dietary flavonoids are considered biologically essential molecules. Luteolin (3′,4′,5,7-tetrahydroxyflavone), a widely distributed flavonoid found in many herbs, was recently shown to reduce oxidative stress and inflammatory responses [12]. Luteolin down-regulates the expression of the inflammatory cytokine NF-κb, which is further associated with an Nrf-2-mediated antioxidant response in C57BL/6 mice [13].

Chronic hyperglycemia increases intracellular reactive oxygen species (ROS), which subsequently induce cellular stress and inflammation in pancreatic β cells, and this process in aerobic and anaerobic cells has been well-explored [14]. Flavonoids, such as luteolin isolated from the mulberry leaf, have been identified for their pancreatic protection and glucose homeostasis potential [15,16]. In addition, medicinal-plant-derived luteolin, flavanones, and quercetin have been highlighted for their effect on insulin secretion, glucose uptake, and glucose homeostasis [9,11,17].

Luteolin-7-*O*-rutinoside (Figure 1), a flavonoid found in many traditional dietary plants, such as *Mentha longifolia* L., *Artemisia Montana* (Nakai) Pamp., *Olea europaea* L., and *Argyreia nervosa* (Burm. f.) Bojer [18–20], has been used globally as an appetizing agent in regular diets. However, the effects of lut-7-*O*-rutin on the regulation of pancreatic β-cell oxidative stress, cellular protection, and glucose homeostasis mechanisms remain unexplored. The present study is the first attempt to analyze the bioefficacy of lut-7-*O*-rutin on the prevention of glucotoxicity and the regulation of insulin secretion in response to low- and high-glucose conditions in RIN-5F cells. The insulin sensitivity of lut-7-*O*-rutin was analyzed via the stimulation of glucose uptake in basal and insulin-stimulated conditions in L6 myotubes. The bioefficacy was compared with the presently used reference drug, rosiglitazone. In addition, the effect of lut-7-*O*-rutin on glucose-stimulated cellular mitochondrial oxidation, lactate conversion, and de novo fatty acid synthesis in L6 myotubes was determined. A detailed in vitro study was designed to analyze the effect of lut-7-*O*-rutin on insulin secretion, insulin action, and insulin response on lipid metabolism in intramuscular-protein and gene-level functions, which mimic in vivo models.

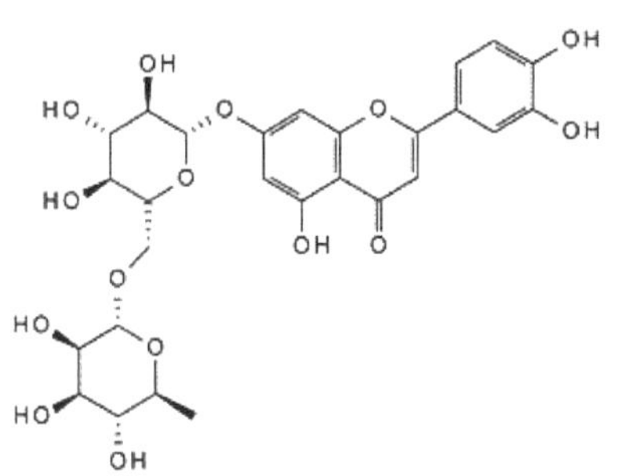

Figure 1. Chemical structure of luteolin-7-*O*-rutinoside (Cas No: 20633-84-5).

2. Materials and Methods

2.1. Cell Lines and Molecular Biology Chemicals

RIN-5F pancreatic β cells and L6 myotubes were purchased from the American Type Culture Collection (ATCC, Rockville, MD, USA). DMEM (Dulbecco's Modified Eagle Medium) was used as a cell-culture growth medium. DMEM and cell-culture reagents were purchased from Invitrogen, Carlsbad, Germany. Deionized water was obtained using a Direct-QUV 3 multipore water purification system (Millipore, Burlington, MA, USA). Luteolin-7-*O*-rutinoside (CAS No: 20633-84-5) was purchased from Sigma-Aldrich, St. Louis, MO, USA. All other chemicals related to the molecular biology experiments were purchased from Sigma-Aldrich (St. Louis, MO, USA). All spectrophotometric measurements were performed with a UV2010 spectrophotometer (Hitachi, Düsseldorf, Germany).

2.2. In Vitro Cell-Culture Studies

2.2.1. In Vitro Cell Viability Assay Using RIN-5F Cells

The cytotoxic effect of lut-7-*O*-rutin against RIN-5F pancreatic β cells and L6 myotubes was analyzed using the MTT (3-(4, 5-Dimethylthiazol-2-yl)-2, 5-Diphenyltetrazolium Bromide) method as described by Mosmann [21]. Briefly, RIN-5F cells were cultured with low-glucose (5.5 mM), normal-glucose (11.1 mM) and high-glucose (25 mM)-containing growth medium (DMEM) and then treated with increasing concentrations of lut-7-*O*-rutin (such as 0, 20, 40, 60, 80, and 100 μM) and incubated for 24 h and 48 h. After incubation, the cell-growth-inhibition curve was analyzed using the formation of formazan crystals from MTT. The formazan crystals were dissolved with 100% DMSO, and the optical density was measured at 570 nm using a 96-well-microplate reader (Bio-Rad, Model 680, Hercules, CA, USA).

2.2.2. Measurement of Intracellular ROS

The production of intracellular reactive oxygen species (ROS) was measured using 2′, 7′-dichlorofluorescein diacetate (DCFH-DA) [22]. DCFH-DA passively enters the cell and reacts with ROS to form dichlorofluorescein (DCF), a highly fluorescent compound. Briefly, 10 mM DCFH-DA stock solution (in methanol) was diluted 500-fold in Hanks' Balanced Salt Solution (HBSS), without serum or other additives, to yield a 20 μM working solution. RIN-5F cells were cultured in low-glucose (5.5 mM), normal-glucose (11.1 mM), and high-glucose (25 mM) concentrations of the growth medium in 24-well plates and were treated with 5, 10, and 20 μM of lut-7-*O*-rutin for 24 h. Lut-7-*O*-rutin-treated RIN-5F cells were washed twice with HBSS and then incubated in 2 mL of 20 μM DCFH-DA at 37 °C for 30 min. Fluorescence was determined at an excitation of 485 nm and an emission of 520 nm using a microplate reader.

2.2.3. Nuclear Damage Analysis Using Propidium Iodide Fluorescence Staining

Morphological changes in the nucleus or nuclear damage in the RIN-5F cells maintained in the normal- and high-glucose media were observed after 24 h using propidium iodide (PI) staining (Sigma-Aldrich, St. Louis, MO, USA) [23]. Briefly, RIN-5F cells (50,000) were seeded and cultured in normal- and high-glucose conditions in a 24-well plate, and then treated with 5, 10, and 20 μM of lut-7-*O*-rutin for 24 h. Further, treated cells were fixed on the same plate using 4% paraformaldehyde and stained with 1 mg/mL of propidium iodide at 37 °C for 15 min in the dark. Randomly, 300 stained cells were analyzed using an inverted fluorescence microscope (at 20× magnification), and the pathological changes in the cells were calculated manually.

2.2.4. Assay of Insulin Secretion Activity

RIN-5F cells derived from rat pancreatic β cells were used to determine the insulin secretion level in low- (5.5 mmol) and high-glucose (25 mmol) conditions. A quantity of 2.0×10^5 cells per well was seeded in a 24-well plate in RPMI-1640 medium. After incubation for 72 h, the medium was replaced with fresh medium, and the cells were incubated for another 24 h. The medium was removed from the wells, and the cells were washed with fresh medium (supplemented with 1% FBS) containing a low level of glucose (5.5 mM) or a high level of glucose (25 mM) [24]. Lut-7-O-rutin in 5, 10, and 20 μM concentrations was added to the respective wells. After incubation for 3 h, the condition media from all the wells was collected after the cells were separated by centrifugation. The concentration of insulin in the condition media was determined by ELISA. The insulin secretion level of lut-7-O-rutin was evaluated and compared with the control (without lut-7-O-rutin). Each experiment was performed in triplicate, and the results are presented as the means ± SD.

2.2.5. Determination of Glucose Uptake by Cultured L6 Myotubes

A glucose uptake assay in L6 myotubes was carried out via the modified method described by Doi et al. [25]. Briefly, L6 myoblasts (2×10^4 cells/well) were cultured into 24-well plates and grown for 11 days to form myotubes in 0.5 mL of 10% FBS/DMEM. The medium was replaced once every 3 days. Later, the 11-day-old myotubes were kept for 2 h in filter-sterilized Krebs–Henseleit buffer (pH 7.4, 0.141 g/L $MgSO_4$, 0.16 g/L KH_2PO_4, 0.35 g/L KCl, 6.9 g/L NaCl, 0.373 g/L $CaCl_2$-$2H_2O$, and 2.1 g/L $NaHCO_3$) containing 0.1% bovine serum albumin, 10 mM HEPES, and 2 mM sodium pyruvate (KHH buffer). The myotubes were further cultured in KHH buffer containing normal (11.1 mM) and high (25 mM) levels of glucose, with or without lut-7-O-rutin (5, 10, 20 μM), for another 3 h. The same experimental set-up was used to treat L6 myotubes with the reference drug rosiglitazone (20 μM) to determine the comparative glucose uptake level. Glucose concentrations in the KHH buffer were determined with a glucose assay kit using a microplate reader (Thermo Fisher Scientific Inc., Waltham, MA, USA) at 508 nm. The amounts of glucose consumption were calculated from the differences in glucose concentrations before and after the culture.

2.2.6. Determination of Intracellular Lactate, Glycerol, and Fatty Acids in L6 Myotubes

L6 myotubes (~4×10^5 cells/well) were incubated in Krebs–Ringer phosphate buffer (pH 7.4) containing BSA (1%) and glucose (2 mM) for 2 h at 37 °C. The myotubes were cultured in KHH buffer containing normal (11.1 mM) and high (25 mM) levels of glucose, with or without lut-7-O-rutin (10 and 20 μM), for another 12 h. At the end of incubation, the medium was collected to measure the levels of lactate (ab282923), glycerol (ab65337) and free fatty acid (ab65341) concentrations using an enzymatic calorimetric kit purchased from Abcam (Cambridge, UK).

2.2.7. Assay of de Novo Fatty Acid Synthesis Enzymes in L6 Myotubes

The activities of glucose-6-phosphate dehydrogenase (G6PDH) (EC 1.1.1.49), ATP citrate lyase (ACL) (EC 4.1.3.8), and fatty acid synthase (FAS) (EC 2.3.1.85) were analyzed in L6 myotubes after treatment with lut-7-O-rutin (10 and 20 μM) for 12 h. L6 myotubes (~4×10^5 cells/well) were homogenized in extraction buffer containing sucrose (0.25 mM), EDTA (1 mM), DTT (1 mM), a protease inhibitor (leupeptin) (20 μg/mL), and bovine pancreatic trypsin inhibitor (aprotinin) (5 μg/mL) (1:1, pH 7.4), and then centrifuged at 20,000× *g* at 4 °C for 5 min. The fat-free supernatant fraction was used for the quantification of enzyme activities according to the protocol provided with the ELISA kit. The ELISA kits for glucose-6-phosphate dehydrogenase (G6PDH) (CAT. #MBS035211), ATP citrate lyase (ACL) (CAT. #MBS938549), and fatty acid synthase (FAS) (CAT. #MBS2883650) were obtained from Life science, Biotech company (San Diego, CA, USA).

2.2.8. Gene Expression

The cDNA was directly prepared from cultured cells using a Fastlane® Cell cDNA kit (QIAGEN, Hilden, Germany) after 24 h, respectively, from a 20 μM dose of lut-7-O-rutin-treated RIN-5F cells and L6 myotubes. Then, the transcription of the proinflammatory genes (CYP1A, TNF-α, and NF-κB) in RIN-5F cells and lipid-metabolism-related genes (Lpl,

Ucp-1, and PPARγC1A) were quantified using a QIAGEN real-time SYBR Green/ROX assay kit according to the kit protocol in a real-time PCR instrument (Applied Biosystems, 7500 Fast, Waltham, MA, USA). β-actin was used as a reference gene. We used the $2^{-\Delta\Delta Ct}$ calculations to determine a specific gene and relative mRNA expression level, such as where ΔΔCt = (Ct, target gene of an experimental group—Ct, β-actin of experimental group)—(Ct, target gene of control group—Ct, β-actin of control group) [26].

2.2.9. Quantification of Signaling Proteins Using ELISA Method

ELISA was performed to quantify the cellular-biogenesis-related signaling cascade proteins, such as cAMP, ChREBP-1, and AMPK, in 10 and 20 μM doses of lut-7-O-rutin-treated L6 myotubes after 24 h. Cellular proteins were extracted to determine the cAMP, ChREBP-1, and AMPK protein levels using the quantitative ELISA method using a multiwell-plate reader. The cAMP (Cat. #ab65355), ChREBP-1 (Cat. #ab162408), and AMPK (Cat. #ab181422) ELISA kits were purchased from Abcam (Cambridge, UK). The values were expressed as pg/mL cells for all the proteins.

2.3. Statistical Analysis

All of the grouped data were statistically evaluated using the SPSS/26.0 software package. The values were analyzed using a one-way analysis of variance (ANOVA) followed by Tukey's test [27]. All results were presented as six biological replicates (mean ± SD), and the differences were presented as statistically significant at $p \leq 0.01$ and $p \leq 0.05$.

3. Results

3.1. Determination of Cell-Viability Percentages of RIN-5F Cells Cultured in Low-, Normal-, and High-Glucose Conditions

RIN-5F cells cultured in high-glucose (25 mM) medium for 48 h produced a significant decline in the percentage of live cells or decreased cell-growth, as compared to the normal medium (Figure 2a). The increasing concentration of lut-7-O-rutin (such as 5, 10, 20, 40, 80, and 100 μM) was used to determine the viability or proliferation potential on RIN-5F cells cultured in low-, normal-, and high-glucose conditions for 48 h. We found that the RIN-5F cells treated with lut-7-O-rutin prevented growth inhibition, as compared to the untreated control (Figure 2b). In general, the tested concentration of lut-7-O-rutin did not produce toxicity to RIN-5F cells cultured in the normal medium, as compared with the control.

3.2. Quantification of Reactive Oxygen Species (ROS) Levels in RIN-5F Cells

ROS generation was quantified in RIN-5F cells growing in low-, normal-, and high-glucose media (as compared with the untreated control), with and without treating with lut-7-O-rutin. RIN-5F cells cultured in a high-glucose (25 mM) medium showed significantly increased ROS generation when compared to low- and normal-glucose conditions (Figure 2b). Cell-growth inhibition in high-glucose media may be due to the observed high ROS production (Figure 2a). Treatment with lut-7-O-rutin (20 μM) reduced the cellular stress and ROS generation in RIN-5F cells when compared to 10 μM or 5 μM doses of lut-7-o-rutin (Figure 2c).

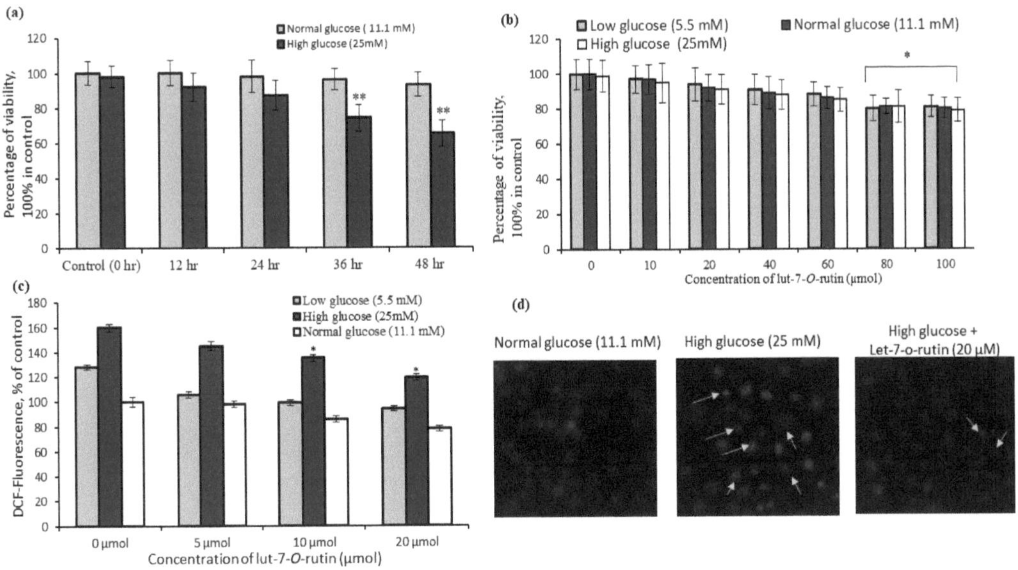

Figure 2. Comparative cell viability of RIN-5F cells cultured in normal- and high-glucose medium (**a**); the effect of increasing concentrations of lut-7-*O*-rutinoside on cytotoxicity (**b**); High-glucose-induced intracellular ROS generation (**c**); propidium iodide staining assay to determine the protective effect of lut-7-*O*-rutinoside on high-glucose-induced nuclear damage in RIN-5F cells (**d**). Each value is presented as mean ± SD for 6 replicates. In (**a**), **: $p \leq 0.001$, lut-7-*O*-rutinoside-treated RIN-5F cells as compared between high-glucose and normal-glucose conditions. In (**b**), *: $p \leq 0.05$, lut-7-*O*-rutinoside-treated RIN-5F cells as compared with untreated cells. In (**c**), *: $p \leq 0.05$, lut-7-*O*-rutinoside-treated RIN-5F cells, as compared with low- or high-glucose concentrations.

3.3. *Identification of Nuclear Damage in RIN-5F Cells Using Fluorescence Microscopy*

Figure 2d shows the propidium iodide staining of the normal morphology of the control cells, but the cells grown in high-glucose media were shown as abnormal, characteristically irregular, and horseshoe-shaped nuclei, which confirmed the nuclear damage and the initiation of glucotoxicity and cell death. Lut-7-*O*-rutin (20 µM)-treated-RIN-5F cells cultured in high-glucose media were observed with a normal morphology with intact-nucleus-containing cells (Figure 2d).

3.4. *Insulin Secretory Effect in RIN-5F Cells*

To identify the influence of lut-7-*O*-rutin on insulin secretion, we examined insulin secretion by RIN-5F cells in a rat islet tumor cell line. Dose-dependently, lut-7-*O*-rutin stimulated insulin secretory potential, and significant ($p < 0.05$) effects were seen at 10 and 20 µM, in both normal- and high-glucose medium (Figure 3). With treatment with lut-7-*O*-rutin in varying glucose concentrations, there was no significant change in the proliferation of RIN-5F cells or ROS production, indicating that the stimulatory effect of lut-7-*O*-rutin on insulin secretion fully depended on the severity of the glucose level.

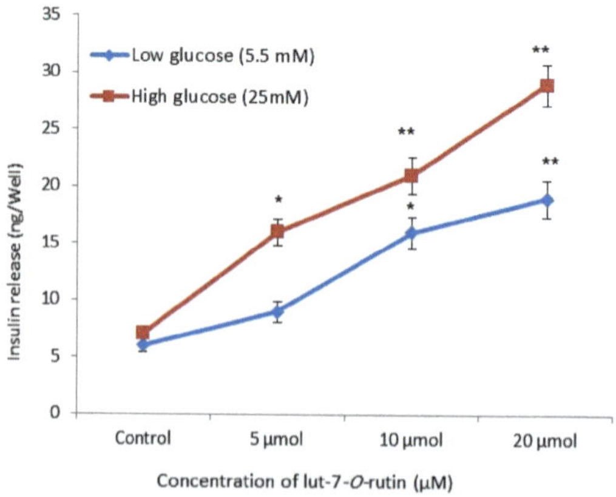

Figure 3. Effect of lut-7-*O*-rutinoside on glucose-stimulated insulin secretion in RIN-5F cells. Each value is presented as the mean ± SD for 6 replicates. *: $p \leq 0.05$ and **: $p \leq 0.001$, lut-7-*O*-rutinoside-treated groups compared with the control and the low-glucose medium.

3.5. Stimulation of Glucose Uptake in L6 Myotubes

Glucose uptake was determined in L6 myotubes cultured under normal- and high-glucose conditions. Th normoglycemic and hyperglycemic conditions of diabetes were achieved in L6 myotubes cultured in normal-glucose (11.1 mM) and high-glucose (25 mM) media. Lut-7-*O*-rutin's effect on glucose uptake under basal and insulin-stimulated conditions were determined in normal-glucose levels; the glucose uptake was found to be significantly higher in 20 µM concentration ($p < 0.05$) (Figure 4a). However, in the insulin-stimulated condition, the glucose uptake was high in both the 10 and 20 µM concentrations of lut-7-*O*-rutin (Figure 4b). Lut-7-*O*-rutin significantly improved the glucose uptake in the high-glucose medium compared to the normal-glucose medium in the L6 myotubes; the glucose uptake level was higher than that with rosiglitazone (Figure 4c).

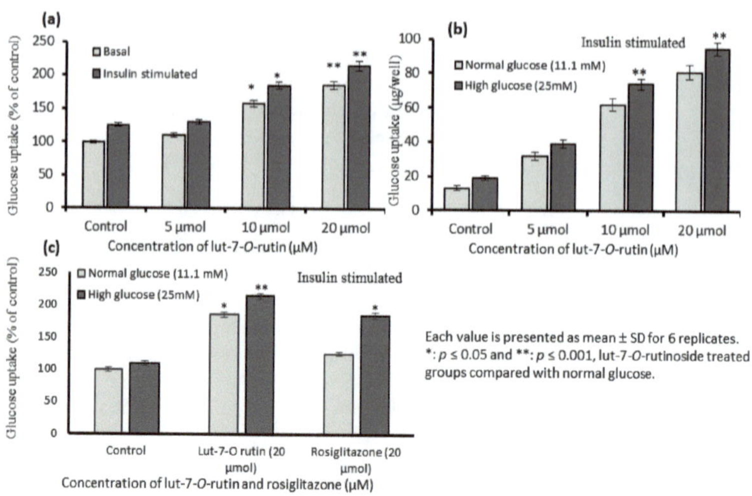

Figure 4. Effect of lut-7-*O*-rutinoside on dose-dependent glucose uptake in L6 myotubes during basal (**a**), insulin-stimulated conditions (**b**), and compared with the untreated control and rosiglitazone (**c**). Each value is presented as the mean ± SD for 6 replicates. *: $p \leq 0.05$ and **: $p \leq 0.001$, lut-7-*O*-rutinoside-treated groups as compared with a normal-glucose medium.

3.6. Levels of Lactate, Glycerol, and Fatty Acids in L6 Myotubes

We investigated the intracellular effect of lut-7-*O*-rutin on the metabolic conversion of glucose in L6 myotubes. Figure 5a depicts the way that lut-7-*O*-rutin significantly converts glucose into lactate in normal and insulin-stimulated conditions, as compared with the control (65% and 78%, respectively). As shown in Figure 5b, treatment with lut-7-*O*-rutin significantly ($p \leq 0.05$) increased the generation of glycerol from glucose in both basal and insulin-stimulated conditions (41% and 32%, respectively). Lut-7-*O*-rutin treatment decreased insulin-stimulated de novo lipogenesis from glucose or other sources as estimated by acetate and glucose incorporated into fatty acids of TAG. As shown in Figure 5c, treatment with lut-7-*O*-rutin reduced fatty acid synthesis from glucose in insulin-stimulated conditions. Our observation evidenced that the higher rate of glucose incorporated in the glycerol fraction of TAG under basal, but not insulin-stimulated, conditions, as compared to the control cells (Figure 5b). In addition, there was no significant difference between the 10 and 20 μM concentrations.

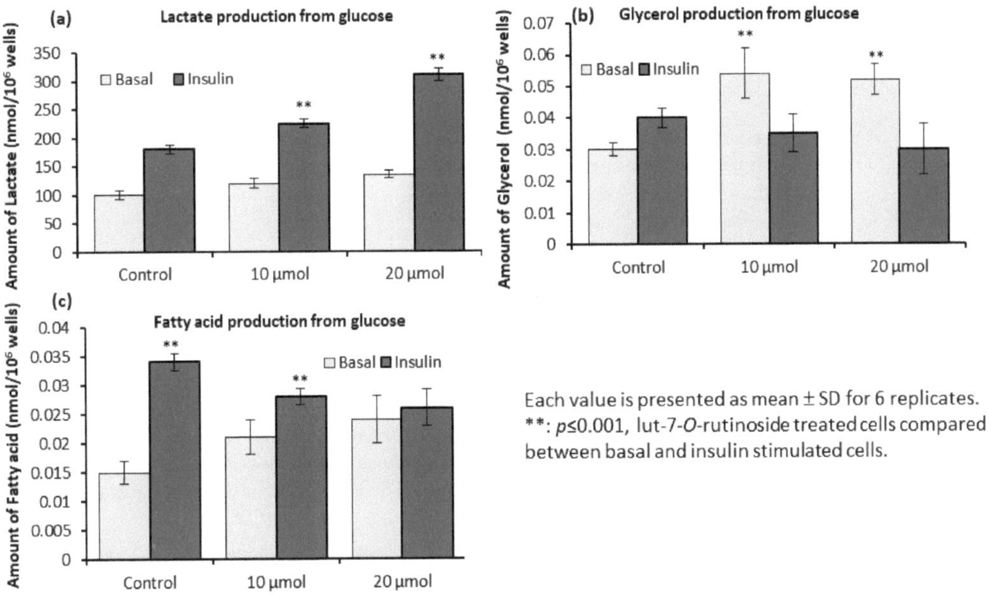

Figure 5. Effect of lut-7-*O*-rutinoside on intracellular production of lactate (**a**), glycerol (**b**), and fatty acids (**c**) from glucose in basal and insulin-stimulated conditions in L6 myotubes. Each value is presented as the mean ± SD for 6 replicates. **: $p \leq 0.001$, lut-7-*O*-rutinoside-treated cells compared to basal and insulin-stimulated cells.

Figure 6 shows the changes in the lipogenesis-related parameters after lut-7-*O*-rutin treatment. The maximal activity of ATP citrate lyase (ACL) was significantly reduced (Figure 6a). There was no change in the activities of fatty acid synthase (Figure 6b) or G6PDH (Figure 6c), which catalyzes the generation of NADPH required for lipogenesis.

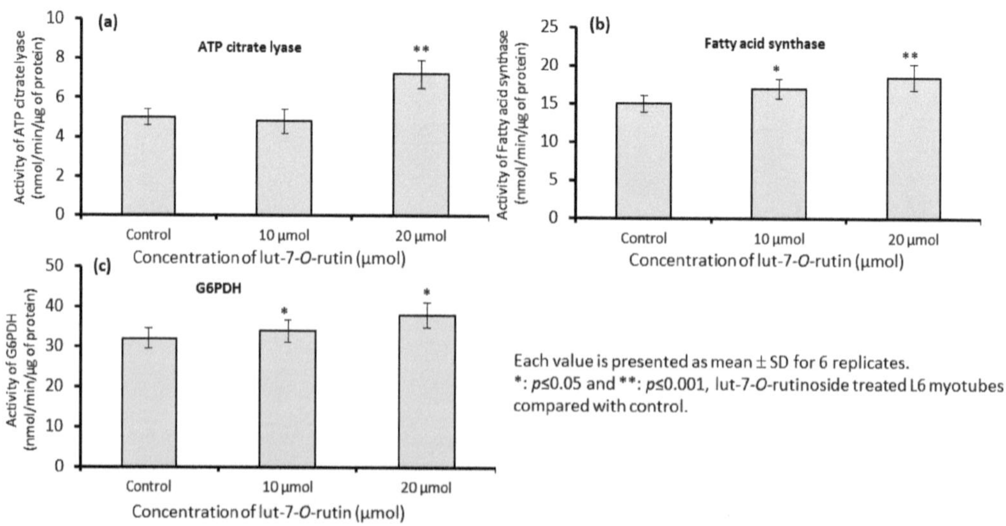

Figure 6. Effect of lut-7-*O*-rutinoside on mitochondrial maximal activity level of enzymes ATP citrate lyase (**a**), fatty acid synthase (**b**), and G6PDH (**c**) in L6 myotubes. Each value is presented as the mean ± SD for 6 replicates. *: $p \leq 0.05$ and **: $p \leq 0.001$, lut-7-*O*-rutinoside-treated L6 myotubes, as compared with the control.

3.7. Gene Expression Levels

Figure 7 depicts the proinflammatory mRNA expression levels of lut-7-*O*-rutin-treated RIN-5F cells. We found that 20 µM of lut-7-*O*-rutin significantly reduced CYP1A, TNF-α, and NF-κB expression when compared to the untreated control or 10 µM of lut-7-*O*-rutin in high-glucose conditions within 24 h. In addition, lipid metabolism of biological thermogenesis-related genes (Lpl, Ucp-1, and PPARGC1A) expression levels in L6 myotubes are presented in Figure 8a. Most interestingly, treatment with 20 µM of lut-7-*O*-rutin significantly increased the mitochondrial-thermogenesis-associated Lpl, Ucp-1, and PPARGC1A expressions by two-fold. The results confirm the cellular uptake of glucose and further mitochondrial metabolic active progress of L6 myotubes.

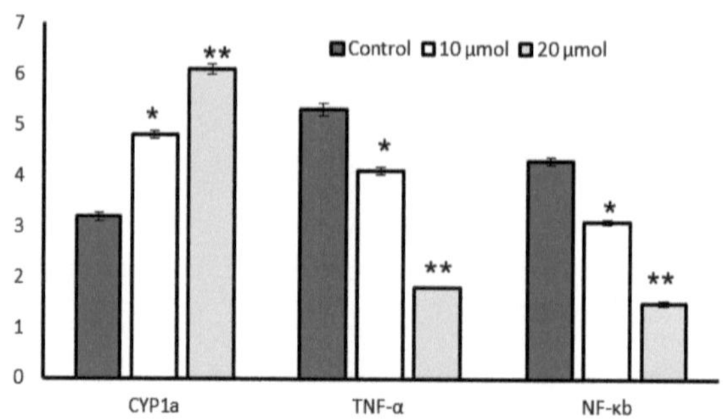

Figure 7. Effect of lut-7-*O*-rutinoside on CYP1a, TNF-α, and NF-κb expression levels in RIN-5F cells. Each value is presented as the mean ± SD for 6 replicates. *: $p \leq 0.05$ and **: $p \leq 0.001$, lut-7-*O*-rutinoside-treated RIN-5F cells, as compared with the control.

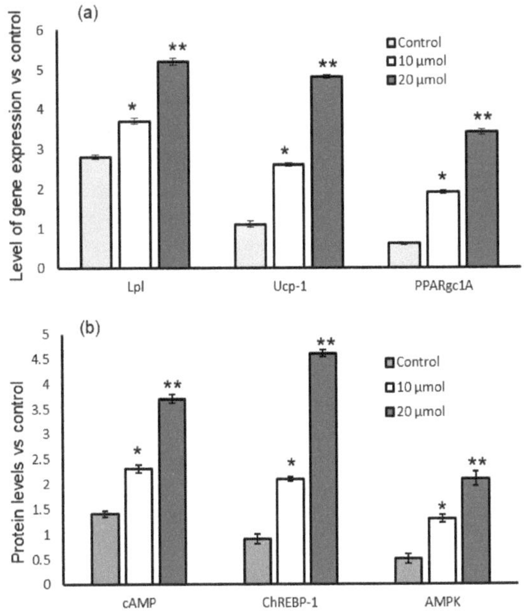

Figure 8. Effect of lut-7-*O*-rutinoside on mRNA expression levels of Lpl, Ucp-1, and PPARGC-1A (**a**) and protein levels of cAMP, chREBP, and AMPK (**b**) in L6 myotubes. Each value is presented as the mean ± SD for 6 replicates. *: $p \leq 0.05$ and **: $p \leq 0.001$, lut-7-*O*-rutinoside-treated L6 myotubes compared with control.

3.8. Quantification of Protein Levels

Cellular biogenesis-related signaling cascade proteins, such as cAMP, ChREBP-1, and AMPK, were quantified in 10 and 20 µM doses of lut-7-*O*-rutin-treated insulin-stimulated L6 myotubes. We found a two-fold increase in cAMP and ChREBP-1 in 20 µM of lut-7-*O*-rutin-treated L6 myotubes after 24 h (Figure 8b). In addition, AMPK protein levels were also found to have a one-fold increase with 20 µM of lut-7-*O*-rutin-treated L6 myotubes. This significant effect was not observed with 10 µM of lut-7-*O*-rutin, as compared to the untreated control.

4. Discussion

Chronic hyperglycemia and the subsequent augmentation of reactive oxygen species (ROS) deteriorate β-cell functions, which leads to hyperglycemia and increased insulin resistance [28,29]. Insulin resistance is an important predictor of the future development of type 2 diabetes [30]. During the progression of type 2 diabetes, glucotoxicity is an important factor contributing to advancing pancreatic β-cell failure and the development of diabetes [31].

Pancreatic β-cell mitochondria play a significant role in insulin secretion [32]. Mitochondrial failure in β-cells has emerged as an important step in the pathogenesis of type 2 diabetes [33,34]. The generation of ROS in response to high concentrations of glucose also causes mitochondrial dysfunction and triggers β-cells apoptosis [35]. In the present study, high-glucose-induced glucotoxicity was confirmed by nuclear damage using PI staining, but treatment with 20 µM of lut-7-*O*-rutin did not cause nuclear damage or glucotoxicity. Additionally, lut-7-*O*-rutin treatment of RIN-5F cells produced significantly elevated insulin levels in glucose-stimulated conditions, with a relatively low production of ROS, as compared to the control cells. The stimulation of insulin secretion by lut-7-*O*-rutin depends on the severity of the hyperglycemic condition, which is not related to cellular stress or altered mitochondrial dynamics. Our previous study confirmed that nymphayol could improve early-phase glucose-stimulated insulin secretion and restore cellular insulin sensitivity in vitro [24]. Similarly, Du et al. [36] reported that lut-7-*O*-rutin stimulates insulin release and improves glucose homeostasis in db/db mice.

Li et al. [37] also reported that natural products might stimulate the islet cells to secrete insulin and increase insulin sensitivity, thereby enhancing glucose uptake. The lut-7-*O*-rutin treatment restored the β-cell and functional mitochondrial content to near normalcy. In the present findings, lut-7-*O*-rutin significantly increased the insulin levels in RIN-5F cells and increased insulin sensitivity in the hyperinsulinemic–euglycemic clamp, which is often combined with the hyperglycemic clamp to determine the adequacy of compensatory cellular hypersensitivity. Diminished skeletal-muscle glucose uptake and attenuated skeletal-muscle insulin sensitivity are important precursors for the pathogenesis of type 2 diabetes mellitus [32]. The skeletal muscle of insulin-resistant subjects has diminished mitochondrial density and reduced mitochondrial oxidative phosphorylation [38]. During glucose-stimulated insulin secretion, glucose metabolism generates ATP in mitochondria and increases the ATP/ADP ratio in β cells [38]. Our study found that basal and insulin-stimulated conditions achieved significant glucose uptake in L6 myotubes. In addition, we observed the increased production of lactate and glycerol by lut-7-*O*-rutin in L6 myotubes in basal and insulin-stimulated conditions; at the same time, there was no alteration in the fatty acid conversion rate.

The pancreatic protective effect of lut-7-*O*-rutin achieved through glucose-stimulated insulin secretion and cellular glucose uptake might be due to the reduction of oxidative stress and increased insulin sensitivity. This effect was confirmed by the molecular-level findings, such as decreased TNF-α and NF-κb, and increased cytochrome P4501A (CYP1A) mRNA expressions. CYP1A involves uncoupling reactions in ETC cycles, which release reactive oxygen species, such as superoxide radicals, hydrogen peroxide, and hydroxy radicals. An imbalance of pro- and antioxidants results in oxidative stress, which leads to inflammation and toxicity [39]. Lingappan et al. [40] have found that $CYP1A1^{-/-}$ mice produced a high level of lipid peroxides and the oxidative damage caused by these molecules, which increase DNA adducts and mutations. However, CYP1A expression and the presence of CYP1A enzymes reduced hypoxic conditions and detoxified the DNA-reactive species, resulting in the reduction of oxidative DNA adducts.

Further, intracellular mitochondrial biogenesis was confirmed by the enhanced expression of lipid metabolism or biological thermogenesis-related genes (Lpl, Ucp-1, and PPARGC1A) expression levels in L6 myotubes. The Lpl–Ucp1–PPARGC1A signaling axis was identified as the candidate pathway involved in maintaining the balance of thermogenesis and energy metabolism. Higher expressions of the Lpl–Ucp1–PPARGC1 signaling axis were confirmed by higher levels of the cAMP, ChREBP-1, and AMPK signaling proteins involved in the mitochondrial biogenesis pathway [41].

5. Conclusions

The mechanism of the action of lut-7-*O*-rutin may be due to enhanced stabilization of mitochondrial potential via quenching ROS development after high-glucose-induced cellular stress. It might protect RIN-5F cells from glucotoxicity and restore insulin secretion and function. Further, glucose utilization was improved in L6 myotubes via increased skeletal-muscle glucose disposal, which consequently increases mitochondrial thermogenesis or energy biogenesis, which was confirmed by the high levels of functional proteins, such as cAMP, ChREBP-1, and AMPK. Lut-7-*O*-rutin may be used as a therapeutic agent to control glucotoxicity and insulin resistance in hyperglycemic conditions or to control the progression of type 2 diabetes.

Author Contributions: Conceptualization, P.S.-B. and A.A.A.; methodology, P.S.-B. and A.A.A., formal analysis, P.S.-B., M.A.B. and S.A.A.; investigation, P.S.-B., M.A.B. and S.A.A.; resources, P.S.-B. and A.A.A.; validation, P.S.-B. and A.A.A.; writing—original draft preparation and writing—review and editing, P.S.-B. and A.A.A.; supervision, P.S.-B. and A.A.A.; project administration, P.S.-B., M.A.B., S.A.A. and A.A.A.; funding acquisition, A.A.A. All authors have read and agreed to the published version of the manuscript.

Funding: This research was funded by the 'Researchers Supporting Project' number (RSP2023R178), King Saud University, Riyadh, Saudi Arabia.

Institutional Review Board Statement: This study did not require ethical approval.

Informed Consent Statement: Not applicable.

Data Availability Statement: The data presented in this study are available in the main article.

Acknowledgments: The authors would like to acknowledge the financial support provided by the Researchers Supporting Project (RSP2023R178), King Saud University, Riyadh, Saudi Arabia.

Conflicts of Interest: The authors declare that they have no conflict of interest.

Abbreviations

ALT, alanine aminotransferase; AST, aspartate aminotransferase; ACP, acid phosphatase; ALP, alkaline phosphatase; NIDDM, noninsulin dependent diabetes mellitus; Lut-7-*O*-rutin, Luteolin-7-*O*-rutinoside; G6PDH, glucose-6-phosphate dehydrogenase; RPMI medium, Roswell Park Memorial Institute medium; DCFH-DA, 2,7-dichlorofluorescin diacetate; ROS, reactive oxygen species; cDNA, complementary deoxyribonucleic acid; CYP1A, cytochrome P450 family 1 subfamily A member 1; TNF-α, tissue necrosis factor-alpha; NF-κb, nuclear factor kappa B subunit 1; Il-1β, interleukin 1-beta; LPL, lipoprotein lipase; Ucp-1, uncoupling protein-1; PPARγC1A, peroxisome proliferator-activated receptor gamma coactivator 1-alpha; cAMP, cyclic adenosine monophosphate; ChREBP-1, carbohydrate-response element-binding protein-1, and AMPK, adenosine monophosphate (A) activated protein kinase.

References

1. Tang, C.; Yeung, L.S.N.; Koulajian, K.; Zhang, L.; Tai, K.; Volchuk, A.; Giacca, A. Glucose-Induced *β*-Cell Dysfunction in Vivo: Evidence for a Causal Role of C-jun *N*-terminal Kinase Pathway. *Endocrinology* **2018**, *11*, 3643–3654. [CrossRef] [PubMed]
2. Prentki, M.; Nolan, C.J. Islet beta cell failure in type 2 diabetes. *J. Clin. Investig.* **2006**, *3*, 1802–1812. [CrossRef] [PubMed]
3. Böni-Schnetzler, M.; Meier, D.T. Islet inflammation in type 2 diabetes. *Semin. Immunopathol.* **2019**, *41*, 501–513. [CrossRef]
4. Maedler, K.; Schulthess, F.T.; Bielman, C.; Berney, T.; Bonny, C.; Prentki, M.; Donath, M.Y.; Roduit, R. Glucose and leptin induce apoptosis in human beta-cells and impair glucose-stimulated insulin secretion through activation of c-Jun N-terminal kinases. *FASEB J.* **2008**, *22*, 1905–1913. [CrossRef]
5. Hurrle, S.; Hsu, W.H. The etiology of oxidative stress in insulin resistance. *Biomed. J.* **2017**, *40*, 257–262. [CrossRef]
6. Röder, P.V.; Wu, B.; Liu, Y.; Han, W. Pancreatic regulation of glucose homeostasis. *Exp. Mol. Med.* **2016**, *48*, e219. [CrossRef] [PubMed]
7. Sattar, A.A.; Sattar, R. Insulin-regulated expression of adiponectin receptors in muscle and fat cells. *Cell Biol. Int.* **2012**, *36*, 1293–1297. [CrossRef]
8. Kittla, M.; Beyreisa, M.; Tumurkhuua, M.; Fürst, J.; Helm, K.; Pitschmann, A.; Gaisberger, M.; Glasl, S.; Ritter, M.; Jakab, M. Quercetin stimulates insulin secretion and reduces the viability of rat INS-1 beta cells. *Cell Physiol. Biochem.* **2016**, *39*, 278–293. [CrossRef] [PubMed]
9. Gaster, M.; Kristensen, S.R.; Beck-Nielsen, H.; Schrøderm, H.D. A cellular model system of differentiated human myotubes. *APMIS* **2001**, *109*, 735–744. [CrossRef]
10. Vallon, V.; Hummler, E.; Rieg, T.; Pochynyuk, O.; Bugaj, V.; Schroth, J.; Dechenes, G.; Rossier, B.; Cunard, R.; Stockand, J. Thiazolidinedione-induced fluid retention is independent of collecting duct alphaENaC activity. *J. Am. Soc. Nephrol.* **2009**, *20*, 721–729. [CrossRef]
11. Park, H.S.; Lee, K.; Kim, S.H.; Hong, M.J.; Jeong, N.J.; Kim, M.S. Luteolin improves hypercholesterolemia and glucose intolerance through LXRα-dependent pathway in diet-induced obese mice. *J. Food Biochem.* **2020**, *29*, e13358. [CrossRef]
12. Hytti, M.; Piippo, N.; Korhonen, E.; Honkakoski, P.; Kaarniranta, K.; Kauppinen, A. Fisetin and luteolin protect human retinal pigment epithelial cells from oxidative stress-induced cell death and regulate inflammation. *Sci. Rep.* **2015**, *5*, 17645. [CrossRef] [PubMed]

13. Li, L.; Luo, W.; Qian, Y.; Zhu, W.; Qian, J.; Li, J.; Jin, Y.; Xu, X.; Liang, G. Luteolin protects against diabetic cardiomyopathy by inhibiting NF-κB-mediated inflammation and activating the Nrf2-mediated antioxidant responses. *Phytomedicine* **2019**, *59*, 152774. [CrossRef]
14. Kayama, Y.; Raaz, U.; Jagger, A.; Adam, M.; Schellinger, I.S.; Sakamoto, M.; Suzuki, H.; Toyama, K.; Spin, J.M.; Tsao, P.S. Diabetic cardiovascular disease induced by oxidative stress. *Int. J. Mol. Sci.* **2015**, *16*, 25234–25263. [CrossRef] [PubMed]
15. Zhang, M.; Lv, X.Y.; Li, J.; Xu, Z.G.; Chen, L. The characterization of high-fat diet and multiple low-dose streptozotocin induced type 2 diabetes rat model. *Exp. Diabetes Res.* **2008**, *2008*, 704045. [CrossRef] [PubMed]
16. Jia, Z.H.; Liu, Z.H.; Zheng, J.M.; Zeng, C.H.; Li, L.S. Combined therapy of Lut-7-*O*-rutin and benazepril on the treatment of diabetic nephropathy in db/db mice. *Exp. Clin. Endocrinol. Diabetes* **2007**, *115*, 571–576. [CrossRef]
17. Lieder, B.; Hoi, J.K.; Holik, A.K.; Geissler, K.; Hans, J.; Friedl, B.; Liszt, K.; Krammer, G.E.; Ley, J.P.; Somoza, V. The flavanone homoeriodictyol increases SGLT-1-mediated glucose uptake but decreases serotonin release in differentiated Caco-2 cells. *PLoS ONE* **2017**, *13*, e0171580. [CrossRef]
18. Mikaili, P.; Mojaverrostami, S.; Moloudizargari, M.; Aghajanshakeri, S. Pharmacological and therapeutic effects of *Mentha longifolia* L. and its main constituent, menthol. *Anc. Sci. Life* **2013**, *33*, 131–138.
19. Wei, Y.; Zhang, T.; Ito, Y. Preparative separation of Lut-7-*O*-rutin from Chinese traditional herb by repeated high-speed counter-current chromatography. *J. Chromatogr. A* **2003**, *1017*, 125–130. [CrossRef]
20. Kim, N.M.; Kim, J.; Chung, H.Y.; Choi, J.S. Isolation of Luteolin 7-O-rutinoside and esculetin with potential antioxidant activity from the aerial parts of *Artemisia montana*. *Arch. Pharm. Res.* **2000**, *23*, 237–239. [CrossRef]
21. Mosmann, T. Rapid colorimetric assay for cellular growth and survival: Application to proliferation and cytotoxicity assays. *J. Immunol. Methods* **1983**, *65*, 55–63. [CrossRef] [PubMed]
22. Joseph, J.; Ametepe, E.S.; Haribabu, N.; Agbayani, G.; Krishnan, L.; Blais, A.; Sad, S. Inhibition of ROS and upregulation of inflammatory cytokines by FoxO3a promotes survival against Salmonella typhimurium. *Nat. Commun.* **2016**, *7*, 12748. [CrossRef] [PubMed]
23. Leite, M.; Quinta-Costa, M.; Leite, P.S.; Guimaraes, J.E. Critical evaluation of techniques to detect and measure cell death–study in a model of UV radiation of the leukaemic cell line HL60. *Anal. Cell. Pathol.* **1999**, *19*, 139–151. [CrossRef]
24. Subash-Babu, P.; Ignacimuthu, S.; Alshatwi, A.A. Nymphayol increases glucose-stimulated insulin secretion by RIN-5F cells and GLUT4-mediated insulin sensitization in type 2 diabetic rat liver. *Chem. Biol. Interact.* **2015**, *25*, 72–81. [CrossRef]
25. Doi, M.; Yamaoka, I.; Fukunaga, T.; Nakayama, M. Isoleucine, a potent plasma glucose-lowering amino acid, stimulates glucose uptake in C_2C_{12} myotubes. *Biochem. Biophys. Res. Commun.* **2003**, *312*, 1111–1117. [CrossRef] [PubMed]
26. Yuan, J.S.; Reed, A.; Chen, F.; Stewart, C.N. Statistical analysis of real–time PCR data. *BMC Bioinform.* **2006**, *7*, 85. [CrossRef]
27. Kim, H.Y. Analysis of variance (ANOVA) comparing means of more than two groups. *Restor. Dent. Endod.* **2014**, *39*, 74–77. [CrossRef]
28. Kaneto, H.; Katakami, N.; Matsuhisa, M.; Matsuoka, T.A. Role of reactive oxygen species in the progression of type 2 diabetes and atherosclerosis. *Mediat. Inflamm.* **2010**, *2010*, 453892. [CrossRef]
29. Robertson, R.P.; Harmon, J.; Tran, P.O.; Poitout, V. Beta-cell glucose toxicity, lipotoxicity, and chronic oxidative stress in type 2 diabetes. *Diabetes* **2004**, *53*, 119–124. [CrossRef]
30. Legaard, G.E.; Feineis, C.S.; Johansen, M.Y.; Hansen, K.B.; Vaag, A.A.; Larsen, E.L.; Poulsen, H.E.; Almdal, T.P.; Karstoft, K.; Pedersen, B.K.; et al. Effects of an exercise-based lifestyle intervention on systemic markers of oxidative stress and advanced glycation end products in persons with type 2 diabetes: Secondary analysis of a randomised clinical trial. *Free Radic. Biol. Med.* **2022**, *25*, 240–244.
31. D'Angelo, C.V.; West, H.L.; Whitticar, N.B.; Corbin, K.L.; Donovan, L.M.; Stiadle, B.I.; Nunemaker, C.S. Similarities in calcium oscillations between neonatal mouse islets and mature islets exposed to chronic hyperglycemia. *Endocrinology* **2022**, *163*, bqac066. [CrossRef] [PubMed]
32. Liu, J.; Zen, K.; Chen, Z.; Zhang, Y.; Zhang, M.; Zhu, X.; Fan, Y.; Shi, S.; Liu, Z. Lut-7-*O*-rutin protects pancreatic beta-cells from dynamin related protein-1–mediated mitochondrial fission and cell apoptosis under hyperglycemia. *Diabetes* **2013**, *62*, 3927–3935. [CrossRef] [PubMed]
33. Maechler, P.; Wollheim, C.B. Mitochondrial function in normal and diabetic beta-cells. *Nature* **2001**, *414*, 807–812. [CrossRef] [PubMed]
34. Lowell, B.B.; Shulman, G.I. Mitochondrial dysfunction and type 2 diabetes. *Science* **2005**, *307*, 384–387. [CrossRef]
35. Hodgin, J.B.; Nair, V.; Zhang, H.; Randolph, A.; Harris, R.C.; Nelson, R.G.; Weil, E.J.; Cavalcoli, J.D.; Patel, J.M.; Brosius, F.C.; et al. Identification of cross-species shared transcriptional networks of diabetic nephropathy in human and mouse glomeruli. *Diabetes* **2013**, *62*, 299–308. [CrossRef]
36. Du, H.; Shao, J.; Gu, P.; Lu, B.; Ye, X.; Liu, Z. Improvement of glucose tolerance by Lut-7-*O*-rutin with restored early-phase insulin secretion in db/db mice. *J. Endocrinol. Investig.* **2012**, *35*, 607–612. [CrossRef]
37. Li, W.L.; Zheng, H.C.; Bukuru, J.; De Kimpe, N. Natural medicines used in the traditional Chinese medical system for therapy of diabetes mellitus. *J. Ethnopharmacol.* **2004**, *92*, 1–21. [CrossRef]
38. Morino, K.; Petersen, K.F.; Dufour, S.; Befroy, D.; Frattini, J.; Shatzkes, N.; Neschen, S.; White, M.F.; Bilz, S.; Sono, S.; et al. Reduced mitochondrial density and increased IRS-1 serine phosphorylation in muscle of insulin-resistant offspring of type 2 diabetic parents. *J. Clin. Investig.* **2005**, *115*, 3587–3593. [CrossRef]

39. Stading, R.; Chu, C.; Couroucli, X.; Lingappan, K.; Moorthy, B. Molecular role of cytochrome P4501A enzymes inoxidative stress. *Curr. Opin. Toxicol.* **2020**, *20–21*, 77–84. [CrossRef]
40. Lingappan, K.; Maity, S.; Jiang, W.; Wang, L.; Couroucli, X.; Veith, A.; Zhou, G.; Coarfa, C.; Moorthy, B. Role of cytochrome p450 (cyp)1a in hyperoxic lung injury: Analysis of the transcriptome and proteome. *Sci. Rep.* **2017**, *7*, 642. [CrossRef]
41. Mao, X.; Liu, Y.; Li, W.; Wang, K.; Li, C.; Wang, Q.; Chen, W.; Ma, Z.; Wang, X.; Ding, Z.; et al. A promising drug combination of mangiferin and glycyrrhizic acid ameliorates disease severity of rheumatoid arthritis by reversing the disturbance of thermogenesis and energy metabolism. *Phytomedicine* **2022**, *27*, 154216. [CrossRef] [PubMed]

metabolites

MDPI

Article

Widely Targeted Metabolomics Reveals Metabolite Diversity in Jalapeño and Serrano Chile Peppers (*Capsicum annuum* L.)

Dennis N. Lozada [1,2,*], Sahithi Reddy Pulicherla [3] and Francisco Omar Holguin [1]

1 Department of Plant and Environmental Sciences, New Mexico State University, Las Cruces, NM 88003, USA
2 Chile Pepper Institute, Department of Plant and Environmental Sciences, New Mexico State University, Las Cruces, NM 88003, USA
3 Molecular Biology Interdisciplinary Life Sciences Program, Department of Entomology, Plant Pathology, and Weed Science, New Mexico State University, Las Cruces, NM 88003, USA
* Correspondence: dlozada@nmsu.edu; Tel.: +1-575-646-5171

Abstract: Chile peppers (*Capsicum annuum* L.) are good sources of vitamins and minerals that can be included in the diet to mitigate nutritional deficiencies. Metabolomics examines the metabolites involved in biological pathways to understand the genes related to complex phenotypes such as the nutritional quality traits. The current study surveys the different metabolites present in jalapeño ('NuMex Pumpkin Spice') and serrano ('NuMex LotaLutein') type chile peppers grown in New Mexico using a widely targeted metabolomics approach, with the 'NuMex LotaLutein' as control. A total of 1088 different metabolites were detected, where 345 metabolites were differentially expressed; 203 (59%) were downregulated and 142 (41%) were upregulated (i.e., relative metabolite content is higher in 'NuMex Pumpkin Spice'). The upregulated metabolites comprised mostly of phenolic acids (42), flavonoids (22), and organic acids (13). Analyses of principal component (PC) and orthogonal partial least squares demonstrated clustering based on cultivars, where at least 60% of variation was attributed to the first two PCs. Pathway annotation identified 89 metabolites which are involved in metabolic pathways and the biosynthesis of secondary metabolites. Altogether, metabolomics provided insights into the different metabolites present which can be targeted for breeding and selection towards the improvement of nutritional quality traits in *Capsicum*.

Keywords: flavonoids; human health; 'NuMex LotaLutein'; 'NuMex Pumpkin Spice'; nutrition; phenolic acids; ultra-performance liquid chromatography tandem mass spectrometry

check for updates

Citation: Lozada, D.N.; Pulicherla, S.R.; Holguin, F.O. Widely Targeted Metabolomics Reveals Metabolite Diversity in Jalapeño and Serrano Chile Peppers (*Capsicum annuum* L.). *Metabolites* **2023**, *13*, 288. https://doi.org/10.3390/metabo13020288

Academic Editors: Ramona Paltinean and Irina Ielciu

Received: 27 January 2023
Revised: 12 February 2023
Accepted: 13 February 2023
Published: 16 February 2023

1. Introduction

The recent global health crisis necessitates the improvement of diets by consuming healthier and more nutritious foods. Chile peppers (*Capsicum* spp.) are good sources of vitamins and minerals that can mitigate nutritional deficiencies. These include phenols; ascorbic acid; carotenoids; chlorophylls; vitamins A, C, and E; capsaicinoids; and flavonoids [1,2]. Capsaicinoids, which give chile peppers their unique 'heat' sensation, possess anti-cancer properties [3] and have been shown to have the potential of alleviating cough, rheumatism, sore throat, and toothache [4]. Phenolic compounds show great potential as strong antioxidants that can aid the human body against free radicals resulting from reactive oxygen species (ROS) to help prevent cancer, cardiovascular disorders, and neurodegenerative diseases [5–7]. These metabolites have also demonstrated anti-aging and depigmentation properties, anti-inflammatory potential, and antimicrobial activity [8]. A survey on the variation of vitamin contents in a *Capsicum* germplasm identified eight cultivars that have higher vitamin A concentration than sweet potato (*Ipomea batatas*), and 16 pepper types that have vitamin C content higher than kiwi (*Actinidia deliciosa*), indicating a great potential for improving nutrition through genetic improvement [9].

A deeper understanding of the different metabolites present in chile peppers will provide avenues to understand the complex array of genes resulting in quantitative phenotypes such as the nutritional quality traits. Metabolomics studies link the phenotype and genotype, acting as a bridge between the phenome and the genome [10]. Previous estimates revealed at least 200,000 metabolites in the plant kingdom, significantly higher than those found in animals; a potential consequence of a wide variety of metabolic pathways developed by plants to thrive in diverse environmental conditions [11]. In chile peppers, metabolomics approaches have profiled metabolites involved in growth, development, and adaptation to different environments [12,13]. Numerous changes in the biochemical and physiological processes involving enzyme activity, gene expression, and metabolite synthesis result in developmental changes in the fruit in response to the environment [14]. For example, during fruit development in *C. chinense* Jacq, a metabolomics approach revealed distinct patterns of metabolite distribution at 16 days post-anthesis across the orange and red ripening periods [15]. In another study, fruit discoloration was attributed to the reduction in carotenoid content in discolored red peppers compared to normal peppers, where lipid and flavonoid synthesis were significantly associated with discoloration [16]. Increased lipid metabolism and decrease in capsaicinoid, flavones, flavanol compounds, and terpenoids were further observed in domesticated chile peppers compared to their wild ancestors [17], indicating that genetic improvement for increased nutritional quality through breeding and selection have been implemented for the cultivated chile peppers.

Widely targeted metabolomics is the next generation of metabolomics, combining the benefits of the targeted and untargeted approaches, and has several advantages, including being high-throughput, highly sensitive, and having a broad coverage [18], in contrast to targeted approaches where some metabolites may be left unidentified [19]. The detection and identification of broadly targeted metabolites are made possible by Q-TRAP mass spectrometry based on multiple reaction monitoring mode, which allows for the simultaneous quantification of hundreds of known metabolites over a thousand of unknown metabolites [20,21].

The objectives of this study were to: (1) survey the metabolite diversity in jalapeño and serrano peppers using a nontargeted metabolomics approach; (2) identify differentially expressed metabolites; and (3) perform the functional annotation of genes involved in different metabolic pathways. A widely targeted metabolomics approach through an ultra-performance liquid chromatography tandem mass spectrometry (UPLC-MS/MS) was employed to determine differentially expressed metabolites in jalapeño and serrano pepper types cultivated in a New Mexico growing environment. The results from this study will be relevant for future metabolite-based genomics-assisted mapping approaches for the genetic improvement of nutritional quality traits in *Capsicum*.

2. Materials and Methods

2.1. Plant Material and Collection of Fruit Samples

Two *C. annuum* L. cultivars previously released by the New Mexico State University (NMSU) Chile Pepper Breeding and Genetics Program, viz., 'NuMex Pumpkin Spice' (a jalapeño type) [22] and 'NuMex LotaLutein' (a serrano type), [23] were used in this study (Figure 1). The 'NuMex Pumpkin Spice' originated from the hybridization between the 'Permagreen' bell pepper and 'Early Jalapeno' and was developed using a pedigree breeding approach with three generations of backcrossing to a jalapeño fruit type, followed by multiple cycles of single plant selections [22]. 'NuMex LotaLutein' is a biofortified serrano pepper with improved lutein content and was derived from a segregating F_2 population resulting from selfing a commercial F_1 hybrid serrano cultivar [23]. 'NuMex Pumpkin Spice' has a characteristic deep orange (i.e., 'pumpkin'-like) color, whereas 'NuMex LotaLutein' has a distinct yellow color.

The cultivars were transplanted at the Amy Goldman Flower Chile Pepper Institute Teaching Garden, Fabián García Science Center, Las Cruces, NM, USA. Three mature fruits from three individual plants (representing three biological replicates) from each cultivar were sampled ~130 days after transplanting (~190 days after initial sowing in the greenhouse). Fruit samples were subsequently freeze-dried for ~48 h using a Labconco

Freeze Dry System (Labconco Corporation, Kansas City, MO, USA) and were sent to MetWare® Bio, MA, USA (https://www.metwarebio.com/ (accessed on 4 November 2022)) for processing and metabolomics analyses.

(a) (b)

Figure 1. 'Numex Pumpkin Spice' (**a**) and 'NuMex LotaLutein' (**b**) fruits showing their characteristic orange and yellow colors, respectively. 'NuMex Pumpkin Spice' is a colored jalapeño, whereas 'NuMex LotaLutein' is a biofortified serrano type with improved lutein content. Adapted with permission from Refs. [22,23]. ©2015, 2020, American Society for Horticultural Science.

2.2. UPLC-MS/MS Based Widely Targeted Metabolomics Approach

Biological samples were re-freeze-dried using a vacuum lyophilizer (Scientz-100 F) at MetWare® Bio. A mixer mill (MM 400, Retsch, Verder Scientific, Inc., Newtown, PA, USA) with zirconia beads was used to grind the samples for 1.5 min at 30 Hz. The freeze-dried powder samples (~50 mg) were dissolved with 70% methanol solution (1.2 mL; 0.84 mL methanol; 0.36 mL distilled water), mixed using a vortex for 30 sec every 30 min, and stored at 4 °C overnight. The extracts were later centrifuged at 12,000 rpm for 10 min and filtered using a 0.22 μm filter before liquid chromatography and mass spectrometry.

Ultra-performance liquid chromatography (UPLC) (ExionLC™ AD, https://sciex.com.cn/ (accessed on 4 November 2022)) and tandem mass spectrometry (MS/MS) (Applied Biosystems QTRAP 6500, https://sciex.com.cn/ (accessed on 4 November 2022)) were performed with the following conditions (a): liquid phase chromatographic column: Agilent SB-C18 1.8 μm, 2.1 mm × 100 mm; (b) mobile phase: A phase—ultrapure water (0.1% formic acid added), B phase—acetonitrile (0.1% formic acid added); (c) elution gradient: 0.00 min, the proportion of B phase was 5%, within 9.00 min, the proportion of B phase increased linearly to 95% and remained at 95% for 1 min, 10.00–11.10 min, the proportion of B phase decreased to 5% and balanced at 5% up to 14 min; (d) flow rate: 0.35 mL/min; (e) column temperature: 40 °C; and (f) injection volume: 2 μL.

Mass spectrum scans were acquired using both the Linear Ion Trap (LIT) and Triple Quadrupole (QqQ) modes of a hybrid QqQ LIT Mass Spectrometer (Q TRAP ®) (AB6500 Q TRAP® UPLC/MS/MS system) with an ESI Turbo ion spray interface, and both cation and anion modes were controlled by Analyst 1.6.3 software (AB Sciex, MA, USA). Operating conditions consisted of the following: turbine spray (ion source); source temperature 550 °C; ion spray voltage (IS) 5500 V (cation mode)/−4500 V (anion mode); ion source gas I (GSI), gas II (GSII), and curtain gas (CUR) set to 50, 60, and 25 psi, respectively; collision-induced ionization parameter set to high. The multiple reaction monitoring (MRM) mode was used for the QqQ scan, where a specific set of MRM ion pairs was monitored based on the eluted metabolites at each period.

2.3. Qualitative and Quantitative Analyses of Metabolites

An in-house database (MetWare® Database) was used to identify metabolites based on their secondary spectrum information. The following were excluded during the qualitative analyses: signals from isotopes, and repeated signals containing cations such as K^+, Na^+, and $NH_4{}^+$. Semi-quantitative analyses of the samples were performed by MRM using QqQ MS. Peak integration on the mass spectrum peaks of all the metabolites identified was performed after obtaining spectrum analyses data from the different samples. Integral correction was implemented on the peaks of similar metabolites in different samples according to Fraga et al. [24].

2.4. Sample Quality Control and Correlation Analysis

A quality control (QC) sample was prepared from a mixture of sample extracts to examine the repeatability of the sample under the same treatment conditions. Metabolite extraction and detection repeatability were evaluated by analyzing the overlapping display of total ion chromatogram (TIC) of different QC samples. The high stability of the instrument was related to the high repeatability and reliability of the data. The correlation between samples was established using the 'cor' function in R software (www.r-project.org (accessed on 4 November 2022)).

2.5. Analysis of Principal Components and Discriminant Analysis by Orthogonal Partial Least Squares

Metabolite data were compressed into N principal components (PCs) to describe the features of the dataset, where the first principal component (PC1) represented the highest degree of variation, followed by the second principal component (PC2), and so on. Principal components analysis (PCA) was performed using the 'prcomp' function in the R program, where parameter scale = 'True' demonstrated the unit variance (UV) scaling for normalizing the data. Biplots were constructed using the first two PCs (PC1 and PC2). Orthogonal partial least squares discriminant analysis (OPLS-DA) for the metabolome data was further implemented to show the differences between each group. During the modeling of OPLS-DA, $\boldsymbol{X}$ matrix information is divided into information that is either related or unrelated (i.e., noise) to $\boldsymbol{Y}$, where the variables associated with $\boldsymbol{Y}$ are the PC, and the information unrelated to $\boldsymbol{Y}$ is the orthogonal PC [25]. The 'OPLSR.Anal' function in the package 'MetaboAnalystR' in R was used for OPLS-DA. S plots for the OPLS-DA were constructed by plotting the correlation coefficient values between the PCs and the metabolites against the covariance between the PCs and metabolites. The OPLS-DA model validation was performed using permutation tests, represented as R^2X, R^2Y, and Q^2, where R^2X and R^2Y denote the explanatory rate of the model to the $\boldsymbol{X}$ and $\boldsymbol{Y}$ matrices, respectively; and Q^2 signifies the prediction ability of the model. Values closer to 1 indicate a more stable and reliable model for validation.

2.6. Identification and Functional Annotation of Differentially Expressed Metabolites

Variable importance in projection (VIP) from modeling using OPLS-DA was used to determine metabolites that were differentially expressed. Given that the number of biological replicates, n, is ≥ 3, the different metabolites were screened by combining the fold change and VIP of the discriminant analysis model, where metabolites with VIP ≥ 1 and with fold change ≥ 2 or ≤ 0.5 at $p < 0.05$ were considered to be differentially expressed. Volcanic plots of the differentially expressed metabolites were constructed by plotting VIP values against the $\log_2$ (fold change; FC). Functional annotation of the differential metabolites was performed using the KEGG (Kyoto Encyclopedia of Genes and Genomes) database [26] based on the KO (KEGG Orthology) of molecular functions. The PCA, OPLS-DA model validations, and functional annotation of the different metabolites were all performed at MetWare® Bio. Screening results were based on performing comparisons of metabolite contents where the 'NuMex LotaLutein' was used as the control. Positive and negative values for log_2FC indicate upregulation and downregulation, respectively. Upregulation signifies that the relative content for the metabolite is higher in 'NuMex Pumpkin Spice' than in 'NuMex LotaLutein' and vice versa.

3. Results

3.1. Fruit Sample Collection and Metabolite Identification

Mature fruit samples of 'NuMex LotaLutein' and 'NuMex Pumpkin Spice' were collected for metabolomic profiling. After the initial freeze drying, the weight of the 'NuMex LotaLutein' samples ranged between 1.72 and 2.77 g, whereas for 'NuMex Pumpkin Spice' weights ranged between 1.28 and 2.35 g. Using a UPLC-MS/MS system and the MetWare® in-house database, a total of 1088 metabolites were identified. Flavonoids and phenolic acids comprised 16.82% and 15.99% of the identified metabolites, respectively. Along with them, amino acids and derivatives (12.96%), alkaloids (11.76%), and lipids (10.94%) were also present in considerable quantities (Figure 2).

Figure 2. Proportion of different classes of compounds for the metabolites detected from jalapeño and serrano chile peppers using a widely targeted metabolomics approach.

3.2. Analysis of Principal Components and Orthogonal Partial Least Squares

PCA has identified two distinct groups based on the cultivars, where the biological replicates from each sample were grouped together (Figure 3a). Group 1, which consisted of the 'NuMex LotaLutein' samples, clustered on the left side, whereas Group 2, comprising 'NuMex Pumpkin Spice' samples, clustered on the right side of the PCA diagram. The first principal component (PC1) contributed to 47.19% of variation, whereas PC2 was associated with 15.95% of variation. Model prediction by OPLS-DA modeling revealed consistent clustering for the samples according to cultivars (Figure 3b). A significant model for partial LS discriminant analyses was supported by the high values of the explanatory rates, R^2X (0.67), R^2Y (1.0), and of the prediction ability of the model, Q^2 (0.94). Excluding the QC samples, the overall contributions of the T-score and orthogonal T-score for the OPLS-DA were 48.3 and 19.1%, respectively. The results from correlation analyses further supported the differentiation observed from PCA. Mean Pearson correlation coefficients of 0.92 were observed within cultivars, whereas an average of 0.66 was observed across treatments (Figure 4). Coefficient of variation (CV) values across all the QC samples also demonstrated the stability of the experimental data, with the proportion of samples having a CV less than 0.3 and 0.5 being higher than 75% and 85%, respectively. The CV across all metabolites ranged between 1.34% and 115.54%, with a mean value of 36.50%.

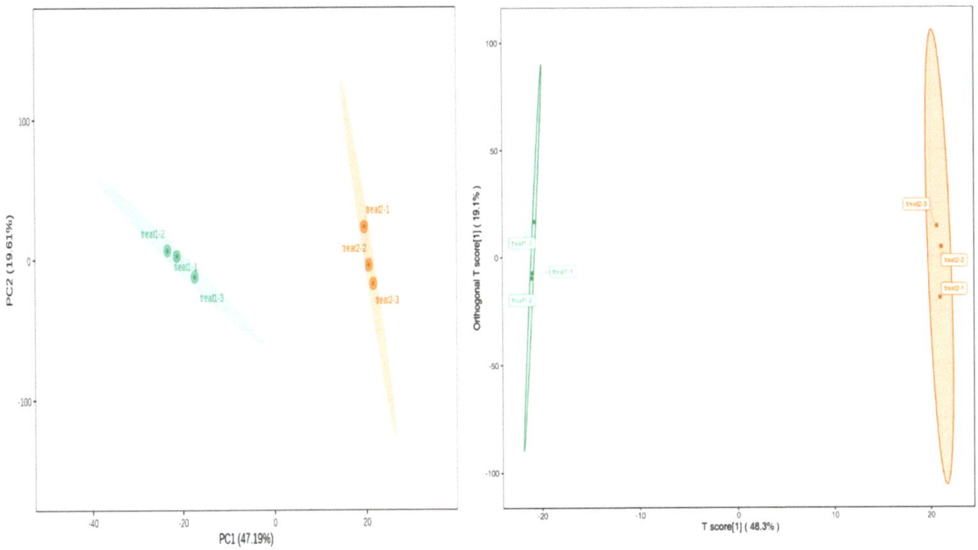

(a) (b)

Figure 3. (**a**) Principal component (PC) biplot for the PC1 and PC2 reveal the overall variation between the cultivars in terms of metabolomic profiles. (**b**) Orthogonal partial least squares discriminant plot for the chile pepper cultivars. Treatment 1: 'NuMex LotaLutein'; treatment 2: 'NuMex Pumpkin Spice'. Numbers 1, 2, and 3 correspond to the first, second, and third biological replications, respectively.

3.3. Identification and Functional Annotation of Differentially Expressed Metabolites

A total of 345 metabolites (32%) were differentially expressed, where 203 (59%) of these were downregulated and 142 (41%) were upregulated (Table S1). The upregulated metabolites were classified into 10 different classes of compounds, which mostly comprised phenolic acids (42; 30%), flavonoids (22; 15%), organic acids (13; 9%), and amino acids and derivatives (11; 8%). The downregulated metabolites mainly consisted of flavonoids (43; 21%) phenolic acids (41; 20%), and alkaloids (32; 16%), and were grouped into nine different classes. Values for log_2FC of the upregulated metabolites ranged between 1 and 18, with a mean value of 2.59. An average VIP value of 1.31 was observed for all the upregulated metabolites. Capsaicin and related compounds were not differentially expressed.

Among the top 10 upregulated metabolites, eight were classified as phenolic acids, and two were flavonoids (Table 1). The average VIP and log_2FC values for the top 10 upregulated metabolites were 1.42 and 9.52, respectively. 1-Phenylethanol, Dicaffeoylshikimic acid, 2-Methoxycinnamic acid, 1,3-O-Dicaffeoylquinic Acid (Cynarin), and 4-O-(4′-O-alpha-D-Glucopyranosyl)caffeoylquinic acid, all phenolic acids, were the top five upregulated metabolites, with log_2FC values of 18.04, 16.4, 13.33, 8.06, and 7.52, respectively. For the top ten downregulated metabolites, three were phenolic acids, six were flavonoids, and one was an alkaloid. Mean values for the VIP and log_2FC for the top 10 downregulated metabolites were 1.42 and -91.68, respectively. Phenolic acids and flavonoids comprised the top five downregulated metabolites; namely, 3,6-Di-O-caffeoyl glucose (log_2FC value = −16.73), Rosmarinic acid methyl ester (−15.36), 2,3-Dimethoxybenzaldehyde(−14.93) (all three are phenolic acids), and Quercetin-3-O-(2″-O-rhamnosyl)rutinoside-7-O-glucoside (−11.96) and Quercetin-3-O-(6″-O-acetyl)glucosyl-(1→3)-Galactoside (−6.01) (both flavonoids).

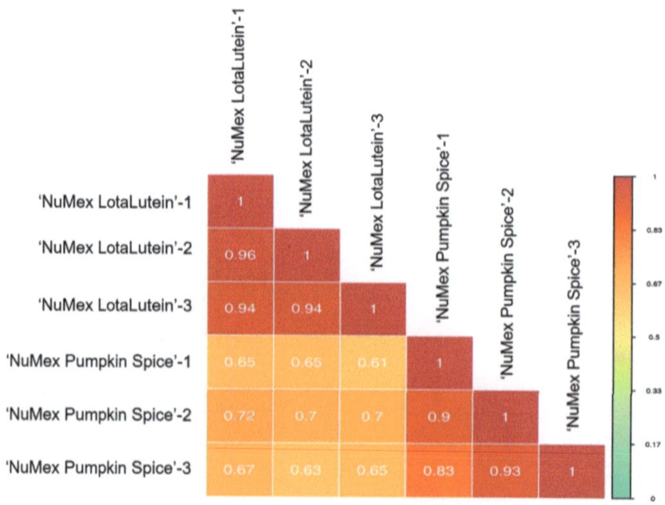

Figure 4. Pearson correlation coefficients among the biological replicates for the cultivars. Numbers 1, 2, and 3 correspond to the first, second, and third biological replications, respectively. Correlation coefficients were significant at $p < 0.0001$.

Table 1. Top 10 most highly upregulated and downregulated metabolites.

Metabolite	Class	VIP [1]	Fold Change (FC)	log_2FC [2]
1-Phenylethanol	Phenolic acids	1.44	269,754.80	18.04
Dicaffeoylshikimic acid	Phenolic acids	1.44	86,594.42	16.40
2-Methoxycinnamic acid	Phenolic acids	1.44	10,264.53	13.33
1,3-O-Dicaffeoylquinic Acid (Cynarin)	Phenolic acids	1.42	266.71	8.06
4-O-(4′-O-alpha-D-Glucopyranosyl)caffeoylquinic acid	Phenolic acids	1.43	182.93	7.52
Chlorogenic acid methyl ester	Phenolic acids	1.44	103.94	6.70
Cyanidin-3-O-(6″-O-caffeoyl-2″-O-xylosyl)glucoside	Flavonoids	1.43	82.45	6.37
Tri-O-galloyl Methyl gallate	Phenolic acids	1.41	82.20	6.36
Luteolin-7-O-Sophoroside-5-O-arabinoside	Flavonoids	1.43	74.07	6.21
5-O-Feruloyl quinic acid glucoside	Phenolic acids	1.37	73.70	6.20
3,6-Di-O-caffeoyl glucose	Phenolic acids	1.44	9.18×10^{-6}	−16.73
Rosmarinic acid methyl ester	Phenolic acids	1.44	2.37×10^{-5}	−15.36
2,3-Dimethoxybenzaldehyde	Phenolic acids	1.44	3.2×10^{-5}	−14.93
Quercetin-3-O-(2″-O-rhamnosyl)rutinoside-7-O-glucoside	Flavonoids	1.44	2.5×10^{-4}	−11.96
Quercetin-3-O-(6″-O-acetyl)glucosyl-(1→3)-Galactoside	Flavonoids	1.41	0.016	−6.01
p-Coumaroylagmatine	Alkaloids	1.43	0.020	−5.64
Quercetin-7-O-(6″-malonyl)glucosyl-5-O-glucoside	Flavonoids	1.42	0.023	−5.43
Quercetin-3-O-rutinoside-7-O-rhamnoside	Flavonoids	1.40	0.0251	−5.31
Quercetin-3-O-xylosyl(1→2)glucosyl(1→2)glucoside	Flavonoids	1.40	0.0253	−5.30
Quercetin-3-O-(6″-O-malonyl)glucosyl-5-O-glucoside	Flavonoids	1.38	0.031	−5.00

[1] VIP—Variable importance in projection value. [2] Positive and negative values for log_2FC indicate upregulation and downregulation, respectively. Screening results were based on a comparison where 'NuMex LotaLutein' was used as the control. Upregulation signifies that the relative content of the metabolite is higher in 'NuMex Pumpkin Spice' than in 'NuMex LotaLutein' and vice versa.

The KEGG database was used in order to understand the metabolite content as a whole network in the form of metabolic pathways. A total of 89 differentially expressed metabolites had annotation from the KEGG database, where 36 metabolites were upregulated and 53 were downregulated (Table S2). The top 10 upregulated metabolites comprised mostly phenolic acids, whereas the top 10 downregulated metabolites with annotation consisted mostly of flavonoids (Table 2). Classification based on annotation from the KEGG database revealed that of the differentially expressed metabolites, 70 (78.65%) were annotated to be involved in metabolic pathways. A total of 42 (47.19%) and 17 (19.1%) metabolites were associated with the biosynthesis of secondary metabolites and of cofactors, respectively (Figure 5). The upregulated metabolites with KEGG annotation were classified into seven different groups that comprised organic acids, nucleotides and derivatives, and flavonoids, among others. Similarly, downregulated metabolites comprised eight different groups, including amino acids and derivatives, organic acids, alkaloids, and phenolic acids. Flavonoids, phenolic acids, organic acids, and nucleotides and derivatives comprised the top 10 differentially expressed upregulated metabolites with KEGG database map annotation. The significant enrichment of metabolites involved in terpenoid backbone biosynthesis was demonstrated (rich factor (RF) value of 1). Differential metabolites involved in photosynthesis also showed significant enrichment, with an RF value equal to 0.65 (Figure 6). Functions related to the biosynthesis of secondary metabolites including lutein (β,ε-carotene-3,3′-diol), the predominant compound in the 'NuMex LotaLutein' serrano pepper, have been annotated in the KEGG Orthology database for a number of differentially expressed metabolites identified.

Table 2. Top 10 differentially expressed upregulated metabolites with KEGG database annotation.

Metabolite	Class	Variable Importance in Projection (VIP) Value	Fold Change (FC)	log_2FC	KEGG Orthology (KO) Index	Annotation
Chlorogenic acid (3-O-Caffeoylquinic acid)	Phenolic acids	1.43	20.46	4.35	KO00940; KO00941; KO00945; KO01110	Phenylpropanoid biosynthesis; flavonoid biosynthesis; stilbenoid, diarylheptanoid and gingerol biosynthesis; metabolic pathways
3,7-Di-O-methylquercetin	Flavonoids	1.34	11.21	3.49	KO00944; KO01110	Flavone and flavonol biosynthesis; metabolic pathways
Phenol	Phenolic acids	1.42	11.01	3.46	KO00350; KO01100	Tyrosine metabolism; metabolic pathways
ATP; Adenosine 5′-Triphosphate	Nucleotides and derivatives	1.42	8.39	3.07	KO00190; KO00195; KO00230; KO00908 KO01100; KO01110; KO01232; KO01240	Oxidative phosphorylation; photosynthesis; purine metabolism; zeatin biosynthesis; metabolic pathways; nucleotide metabolism; biosynthesis of cofactors
2-Furoic acid	Organic acids	1.38	5.77	2.53	KO01100	Metabolic pathways
Kaempferol-3-O-rutinoside(Nicotiflorin)	Flavonoids	1.18	5.61	2.49	KO00944; KO01110	Flavone and flavonol biosynthesis; metabolic pathways
2-Oxoheptanedionic acid	Organic acids	1.37	5.55	2.47	KO01100; KO01210; KO01240	Metabolic pathways; 2-oxocarboxylic acid metabolism; biosynthesis of cofactors
Kaempferol-3-O-rhamnoside (Afzelin)(Kaempferin)	Flavonoids	1.31	4.92	2.30	KO00944	Flavone and flavonol biosynthesis
Luteolin-7-O-glucuronide	Flavonoids	1.26	4.76	2.25	KO00944	Flavone and flavonol biosynthesis
3-O-Methylquercetin	Flavonoids	1.40	4.56	2.19	KO00944; KO01110	Flavone and flavonol biosynthesis; metabolic pathways

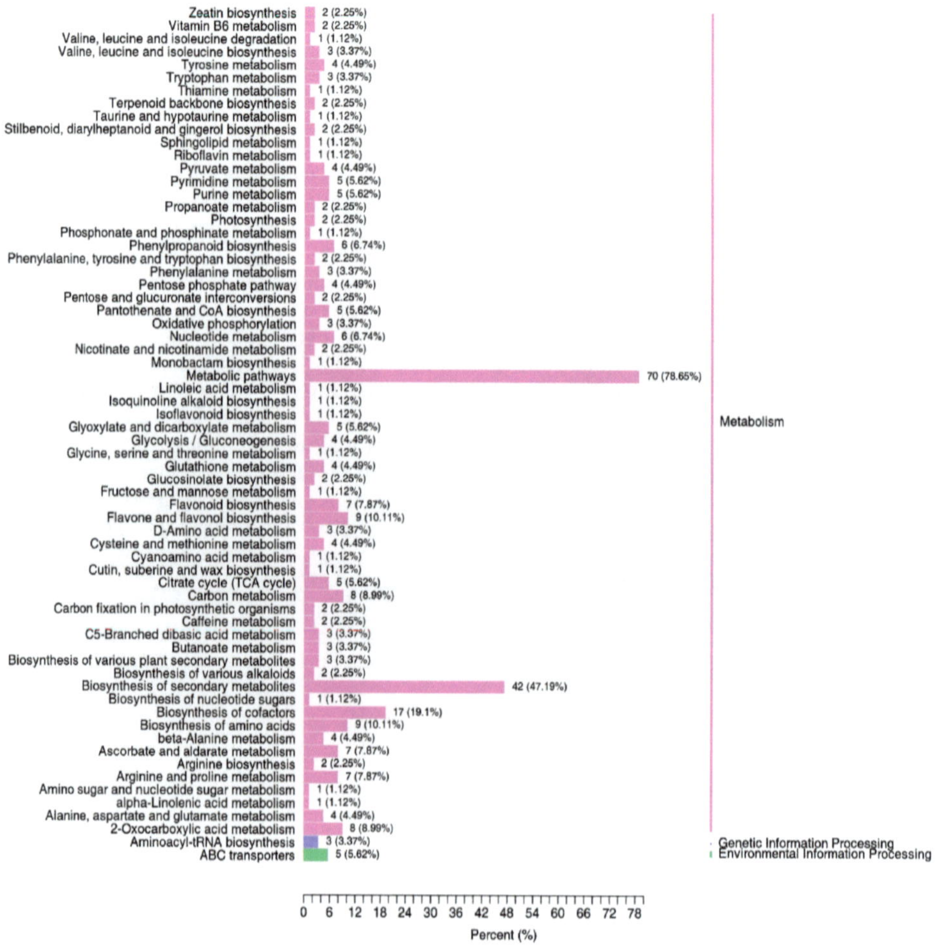

Figure 5. Summary of the different functions annotated for the differentially expressed metabolites (DEMs) using the KEGG database. Annotations revealed that the majority of the DEMs are involved in metabolic pathways and in the biosynthesis of secondary metabolites and cofactors. Note: A DEM can have multiple annotated functions.

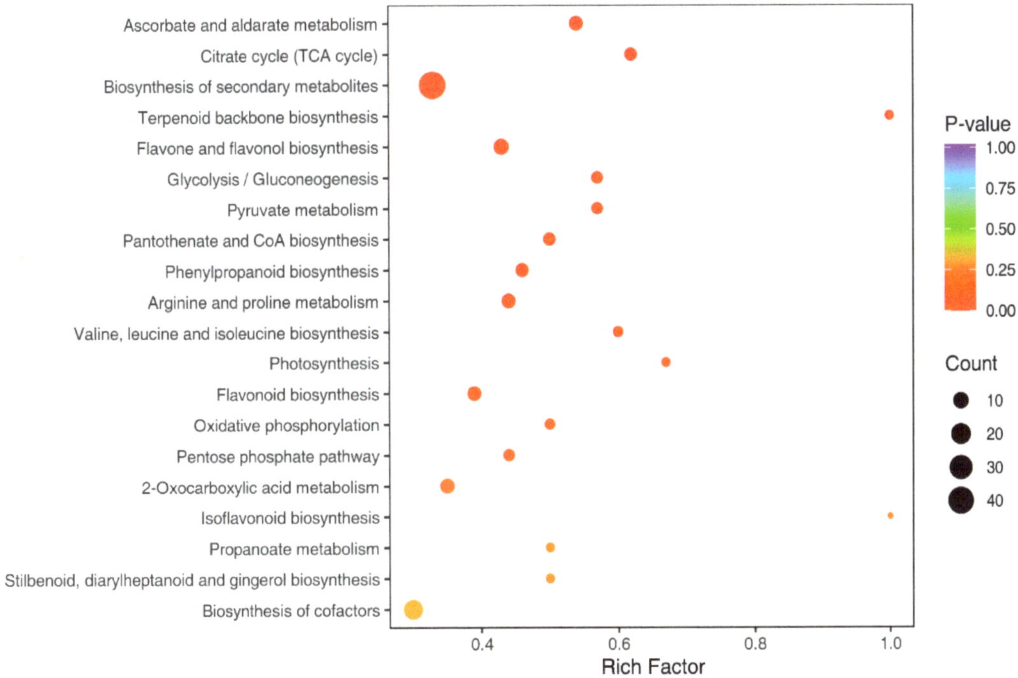

Figure 6. Rich factor (RF) values for the differentially expressed metabolites with annotated functions on the KEGG database. RF is the ratio of the number of differentially expressed metabolites in a pathway to the total number of metabolites annotated in the same pathway (RF for a pathway = no. of differentially expressed metabolites/total number of metabolites). Terpenoid backbone biosynthesis had significant enrichments, with an RF value of 1.

4. Discussion

Chile pepper fruits possess many compounds of high nutritive values which can be integrated in the human diet to mitigate nutritional deficiencies. Crop nutrition is important not only for offering quality and economic value to growers, but also for providing nutritional security to consumers [27]. Since their initial application to examine plant functional genomics in the early 2000s [28–30], metabolomics approaches have been used to understand the different biochemical and physiological pathways for the selection of optimal alleles for the improvement of crops [31]. In the current study, a representative jalapeño and a serrano cultivar grown in New Mexico, USA, were subjected to widely targeted metabolomics to survey the different metabolites present in their fruit samples. The identified metabolites were able to distinguish the representative cultivars. The majority of the differentially expressed metabolites comprised the phenolic acids and the flavonoids. Metabolomics could be integrated with genomics-assisted approaches to facilitate the breeding and selection of chile pepper cultivars with improved nutritive value for consumers.

4.1. Metabolite Profiles Successfully Discriminate Jalapeño and Serrano Type Chile Peppers

Though crop metabolite profiling is an important area of nutritional and health research, this information can also be used for the classification of peppers. Principal components and partial least discriminant analyses are relevant for the classification of metabolomic datasets based on different factors such as variety, location, and their interactions [32]. In the current study, the jalapeño and serrano type chile peppers were differentiated based on their metabolite profiles. Both PCA and OPLS-DA showed consistent clustering based on cultivars, with total variation explained by the first two PCs reaching a total of 63%, indicating that the cultivars were significantly different from each other. The values for Q^2, R^2X, and R^2Y derived from performing OPLS-DA further support the good predictive ability and reproducibility of the model in classifying the cultivars based on metabolomic information. Correlation analyses within sample treatments were significant ($p < 0.0001$), demonstrating clustering based on the cultivars. Genomic information using 66,650 single nucleotide polymorphism markers, nonetheless, previously classified 'NuMex LotaLutein' and 'NuMex Pumpkin Spice' in the same cluster [33], indicating high genetic relatedness; yet, implementing a metabolomic approach can resolve differences between the cultivars.

Metabolite profiles have been previously used for species differentiation and cultivar identification in chile peppers. Capsaicinoid and flavonoid contents, for example, were recently used as biochemical markers to distinguish Jize from Korean chile peppers [34]. In another study, an untargeted metabolomics approach identified biomarkers, including flavonoids, steroids, terpenoids, anthocyanin, and carboxylic acids, and characterized three *Capsicum* species (*C. baccatum*, *C. chinense*, and *C. frutescens*) from a collection of native South American chile peppers [35]. Eleven Mexican *C. annuum* cultivars were differentiated based on their metabolomic fingerprints, which consisted of sugars, amino acids, organic acids, polyphenolic acids, and alcohols [36]. Serrano peppers from Mexico were grouped in two different clusters based on their metabolite profiles, which included organic acids, sugars, amino acids, and polyphenols [37]. In contrast, using a single metabolite (i.e., capsaicin) alone did not successfully discriminate 'Superhot' chile peppers cultivated across multi-year trials in New Mexico [38]. Taken together, our results and those of previous studies indicate that a 'macroscale' metabolomic profiling would be more advantageous than using only a single metabolite in cultivar identification and species differentiation in *Capsicum*.

4.2. *Differentially Expressed Metabolites Have Implications on Human Health and Nutrition*

Widely targeted metabolomics for jalapeño and serrano peppers revealed differentially expressed metabolites comprising mostly of upregulated phenolic acids and flavonoids. Phenolic acids are among the most commonly found compounds in *Capsicum* fruits and have been observed to possess vital functions in human health, nutrition, and the prevention of diseases because of their antioxidant, antiviral, and antibacterial properties [5,39]. Phenolics have been previously identified in different species of peppers, including *C. annuum* cultivars from India [40]; *C. annuum*, *C. chinense*, and *C. baccatum* cultivated in New Mexico [41]; *C. baccatum* [42]; and *C. frutescens* and *C. chinense* cultivars from Peru, South America [35]. The human body cannot produce phenolic compounds; hence, they must be acquired from food sources [5]. Major phenolic acid derivatives, caffeic acid (CA), sinapic acid (SA), and ferulic acid (FA) [5], were observed to be differentially expressed in the present work; CA and FA were upregulated, whereas SA was downregulated. Flavonoids (e.g., quercetin and luteolin) were also detected to be differentially expressed, consistent with previous studies [43–46]. A total of 23 quercetin metabolites were found to be differentially expressed, where five metabolites (namely, quercetin-3-O-sambubioside-5-O-glucoside, 3-O-methylquercetin, 3,7-di-O-methylquercetin, quercetin-3-O-(2″-O-malonyl)sophoroside-7-O-arabinoside, and 6-methoxyquercetin-3-O-rhamnoside) were upregulated and the remaining 18 metabolites were downregulated. A total of 10 differentially expressed metabolites belong to the luteolin class, comprising mostly upregulated compounds (luteolin-4′-O-glucoside, luteolin-7-O-glucoside (cynaroside), luteolin-7-O-sophoroside-5-O-arabinoside, luteolin-7-O-gentiobioside, luteolin-7,3′-di-O-glucoside, and luteolin-7-O-glucuronide). Capsaicin and related families of 'pungent' alkaloids (or 'pseudoalkaloids'), however, were not observed to be differentially expressed in the current study, a potential consequence of the maturity (age) of the fruit samples used for analyses. Overall, the information for the differentially expressed metabolites and their corresponding chemical classes reveal that pepper fruits can be good sources of these nutritional compounds in the diet.

The KEGG map annotations for a number of the differentially expressed metabolites were related to multiple metabolic pathways such as the biosynthesis of secondary metabolites (KO01110), including lutein (β,ε-carotene-3,3′-diol), which is the predominant metabolite in the 'NuMex LotaLutein' serrano pepper [23]. The production of lutein in plants generally takes place through the mevalonate pathway, in which isopentyl diphosphate (IPP) is converted to geranylgeranyl pyrophosphate (GPP), which is then synthesized into phytoene and lycopene, the precursor of lutein [47–49]. The most upregulated phenolic acid, chlorogenic acid (3-O-caffeoylquinic acid), had four different pathway annotations related to phenylpropanoid biosynthesis (pathway ID: KO00940); flavonoid biosynthesis (KO00941); stilbenoid, diarylheptanoid, and gingerol biosynthesis (KO00945); and the biosynthesis of secondary metabolites (KO01110). 3,7-Di-O-methylquercetin (flavonoid) and phenol (phenolic acid) had two pathway annotations; namely, K00944 and KO01110 (secondary metabolite biosynthesis), and KO00350 (Tyrosine metabolism) and KO01110 (secondary metabolite biosynthesis). Adenosine 5′-triphosphate (ATP), also an upregulated metabolite, has seven different KEGG pathway annotations, including oxidative phosphorylation (KO00190), photosynthesis (KO00195), nucleotide metabolism (KO01232), and the biosynthesis of cofactors (KO01240), among others. Altogether, these functional annotations indicate the relevance of these upregulated compounds in the production of a wide array of important nutritional metabolites in *Capsicum*.

4.3. Integrating Metabolomics with Genomics Can Direct Breeding and Selection Decisions for Nutritional Quality Trait Improvement in Chile Pepper

Metabolomics can render insights into the different metabolites which can be targeted further for breeding and selection using different genomics tools such as metabolite-quantitative trait loci (mQTL) analyses, metabolite-based genome-wide association studies (mGWAS), and metabolite-genome predictions [50]. Variation in the phenolic acid content in chile peppers is affected by the varietal type, location, and growing season, and, hence, can impose challenges on performing breeding and selection [32]. Given the complex nature of these phenotypes, caution must be observed when performing the collection of fruit samples for metabolomics analyses to minimize the errors and variation due to sampling. The high accuracy, reliability, and stability of the widely targeted metabolomics approach implemented in the current study, as indicated by the high values of R^2X, R^2Y, and model prediction ability, and the low values for CV, nonetheless, can mitigate the effects of sampling and environmental variation. It might also be necessary to survey the metabolite compositions of cultivated and wild relatives of the chile pepper for a more robust foundation for targeted genomic breeding strategies [39]. Altogether, 'multi-omics' tools (e.g., genomics + metabolomics) can help plant breeders perform informed breeding and selection decisions for the development of cultivars with improved nutritional quality traits [51].

5. Conclusions

A widely targeted metabolomics approach revealed differentially expressed metabolites in jalapeño ('NuMex Pumpkin Spice') and serrano ('NuMex LotaLutein') type peppers grown in New Mexico. Metabolomic profiles successfully distinguished the cultivars, indicating that metabolites can serve as efficient markers to differentiate chile pepper varieties. Upregulated metabolites comprised of phenolic compounds and flavonoids, which have potential implications in improving human health and nutrition when incorporated in the diet. The diversity of metabolites observed in the present work supports the nutritional value of chile peppers and can direct important breeding and selection decisions for designing more nutritious specialty vegetable crops in the future.

Supplementary Materials: The following supporting information can be downloaded at: https://www.mdpi.com/article/10.3390/metabo13020288/s1, Table S1: Metabolites identified for jalapeno- and serrano-type chile peppers, Table S2: Differentially expressed metabolites identified for jalapeno- and serrano-type chile peppers and their corresponding map annotations from the KEGG database.

Author Contributions: Conceptualization, D.N.L.; methodology, D.N.L. and S.R.P.; resources, D.N.L.; data curation, D.N.L.; writing—original draft preparation, D.N.L.; writing—review and editing, S.R.P. and F.O.H.; supervision, D.N.L.; project administration, D.N.L.; funding acquisition, D.N.L. All authors have read and agreed to the published version of the manuscript.

Funding: This research was funded by the US Department of Agriculture National Institute of Food and Agriculture (USDA-NIFA) Grant No. 2022-67014-37078.

Institutional Review Board Statement: Not applicable.

Informed Consent Statement: Not applicable.

Data Availability Statement: Data will be available from the authors upon reasonable request due to privacy or ethical restrictions.

Acknowledgments: The Authors would like to thank Seyed Shahabeddin Nourbakhsh for assistance in collecting fruit samples and Maryfrances Miller for editing the manuscript. The assistance of Jeffrey Chu (MetWare Bio) on the metabolomic data is also appreciated.

Conflicts of Interest: The authors declare no conflict of interest.

References

1. Martínez-Ispizua, E.; Martínez-Cuenca, M.-R.; Marsal, J.I.; Díez, M.J.; Soler, S.; Valcárcel, J.V.; Calatayud, Á. Bioactive Compounds and Antioxidant Capacity of Valencian Pepper Landraces. *Molecules* **2021**, *26*, 1031. [CrossRef]
2. Olatunji, T.L.; Afolayan, A.J. The suitability of chili pepper (*Capsicum annuum* L.) for alleviating human micronutrient dietary deficiencies: A review. *Food Sci. Nutr.* **2018**, *6*, 2239–2251. [CrossRef]
3. Clark, R.; Lee, S.-H. Anticancer properties of capsaicin against human cancer. *Anticancer Res.* **2016**, *36*, 837–843.
4. Batiha, G.E.-S.; Alqahtani, A.; Ojo, O.A.; Shaheen, H.M.; Wasef, L.; Elzeiny, M.; Ismail, M.; Shalaby, M.; Murata, T.; Zaragoza-Bastida, A. Biological properties, bioactive constituents, and pharmacokinetics of some *Capsicum* spp. and capsaicinoids. *Int. J. Mol. Sci.* **2020**, *21*, 5179. [CrossRef]
5. Materska, M.; Perucka, I. Antioxidant activity of the main phenolic compounds isolated from hot pepper fruit (*Capsicum annuum* L.). *J. Agric. Food Chem.* **2005**, *53*, 1750–1756. [CrossRef]
6. Sadef, Y.; Javed, T.; Javed, R.; Mahmood, A.; Alwahibi, M.S.; Elshikh, M.S.; AbdelGawwa, M.R.; Alhaji, J.H.; Rasheed, R.A. Nutritional status, antioxidant activity and total phenolic content of different fruits and vegetables' peels. *PLoS ONE* **2022**, *17*, e0265566. [CrossRef]
7. Lutz, M.; Hernández, J.; Henríquez, C. Phenolic content and antioxidant capacity in fresh and dry fruits and vegetables grown in Chile. *CyTA-J. Food* **2015**, *13*, 541–547.
8. Taofiq, O.; González-Paramás, A.M.; Barreiro, M.F.; Ferreira, I.C.F.R. Hydroxycinnamic acids and their derivatives: Cosmeceutical significance, challenges and future perspectives, a review. *Molecules* **2017**, *22*, 281. [CrossRef]
9. Kantar, M.B.; Anderson, J.E.; Lucht, S.A.; Mercer, K.; Bernau, V.; Case, K.A.; Le, N.C.; Frederiksen, M.K.; DeKeyser, H.C.; Wong, Z.-Z. Vitamin variation in *Capsicum* spp. provides opportunities to improve nutritional value of human diets. *PLoS ONE* **2016**, *11*, e0161464. [CrossRef]
10. Luo, J. Metabolite-based genome-wide association studies in plants. *Curr. Opin. Plant Biol.* **2015**, *24*, 31–38. [CrossRef]
11. Fiehn, O.; Kloska, S.; Altmann, T. Integrated studies on plant biology using multiparallel techniques. *Curr. Opin. Biotechnol.* **2001**, *12*, 82–86. [CrossRef]
12. Taiti, C.; Costa, C.; Migliori, C.A.; Comparini, D.; Figorilli, S.; Mancuso, S. Correlation between volatile compounds and spiciness in domesticated and wild fresh chili peppers. *Food Bioprocess Technol.* **2019**, *12*, 1366–1380. [CrossRef]
13. Barrajon-Catalan, E.; Álvarez-Martínez, F.J.; Borras, F.; Perez, D.; Herrero, N.; Ruiz, J.J.; Micol, V. Metabolomic analysis of the effects of a commercial complex biostimulant on pepper crops. *Food Chem.* **2020**, *310*, 125818. [CrossRef]
14. Osorio, S.; Alba, R.; Nikoloski, Z.; Kochevenko, A.; Fernie, A.R.; Giovannoni, J.J. Integrative comparative analyses of transcript and metabolite profiles from pepper and tomato ripening and development stages uncovers species-specific patterns of network regulatory behavior. *Plant Physiol.* **2012**, *159*, 1713–1729. [CrossRef]
15. Tang, Y.; Zhang, G.; Yang, T.; Yang, S.; Aisimutuola, P.; Wang, B.; Li, N.; Wang, J.; Yu, Q. Biochemical variances through metabolomic profile analysis of Jacq. during fruit development. *Folia Hortic.* **2021**, *33*, 17–26. [CrossRef]
16. Feng, X.; Yu, Q.; Li, B.; Kan, J. Comparative analysis of carotenoids and metabolite characteristics in discolored red pepper and normal red pepper based on non-targeted metabolomics. *LWT* **2022**, *153*, 112398. [CrossRef]

17. Cervantes-Hernández, F.; Ochoa-Alejo, N.; Martínez, O.; Ordaz-Ortiz, J.J. Metabolomic Analysis Identifies Differences Between Wild and Domesticated Chili Pepper Fruits During Development (*Capsicum annuum* L.). *Front. Plant Sci.* **2022**, *13*, 893055. [CrossRef]
18. Zhou, Y.; Shao, L.; Zhu, J.; Li, H.; Duan, H. Comparative analysis of tuberous root metabolites between cultivated and wild varieties of *Rehmannia glutinosa* by widely targeted metabolomics. *Sci. Rep.* **2021**, *11*, 11460. [CrossRef]
19. Adamski, J.; Suhre, K. Metabolomics platforms for genome wide association studies—Linking the genome to the metabolome. *Curr. Opin. Biotechnol.* **2013**, *24*, 39–47. [CrossRef]
20. Sawada, Y.; Akiyama, K.; Sakata, A.; Kuwahara, A.; Otsuki, H.; Sakurai, T.; Saito, K.; Hirai, M.Y. Widely targeted metabolomics based on large-scale MS/MS data for elucidating metabolite accumulation patterns in plants. *Plant Cell Physiol.* **2009**, *50*, 37–47. [CrossRef]
21. Chen, W.; Gong, L.; Guo, Z.; Wang, W.; Zhang, H.; Liu, X.; Yu, S.; Xiong, L.; Luo, J. A novel integrated method for large-scale detection, identification, and quantification of widely targeted metabolites: Application in the study of rice metabolomics. *Mol. Plant* **2013**, *6*, 1769–1780. [CrossRef]
22. Bosland, P.W.; Coon, D. 'NuMex Lemon Spice','NuMex Pumpkin Spice', and 'NuMex Orange Spice'Jalapenos. *HortScience* **2015**, *50*, 1104–1105. [CrossRef]
23. Guzman, I.; Coon, D.; Vargas, K.; Bosland, P.W. NuMex LotaLutein, a Lutein-rich Serrano Pepper. *HortScience* **2020**, *55*, 2052–2055. [CrossRef]
24. Fraga, C.G.; Clowers, B.H.; Moore, R.J.; Zink, E.M. Signature-discovery approach for sample matching of a nerve-agent precursor using liquid chromatography– mass spectrometry, XCMS, and chemometrics. *Anal. Chem.* **2010**, *82*, 4165–4173. [CrossRef]
25. Bujak, R.; Daghir-Wojtkowiak, E.; Kaliszan, R.; Markuszewski, M.J. PLS-based and regularization-based methods for the selection of relevant variables in non-targeted metabolomics data. *Front. Mol. Biosci.* **2016**, *3*, 35. [CrossRef]
26. Kanehisa, M.; Goto, S. KEGG: Kyoto encyclopedia of genes and genomes. *Nucleic Acids Res.* **2000**, *28*, 27–30. [CrossRef]
27. Singh, D.P.; Bisen, M.S.; Shukla, R.; Prabha, R.; Maurya, S.; Reddy, Y.S.; Singh, P.M.; Rai, N.; Chaubey, T.; Chaturvedi, K.K. Metabolomics-Driven Mining of Metabolite Resources: Applications and Prospects for Improving Vegetable Crops. *Int. J. Mol. Sci.* **2022**, *23*, 12062. [CrossRef]
28. Fiehn, O.; Kopka, J.; Dörmann, P.; Altmann, T.; Trethewey, R.N.; Willmitzer, L. Metabolite profiling for plant functional genomics. *Nat. Biotechnol.* **2000**, *18*, 1157–1161. [CrossRef]
29. Ward, J.L.; Harris, C.; Lewis, J.; Beale, M.H. Assessment of 1H NMR spectroscopy and multivariate analysis as a technique for metabolite fingerprinting of Arabidopsis thaliana. *Phytochemistry* **2003**, *62*, 949–957. [CrossRef]
30. Tohge, T.; Nishiyama, Y.; Hirai, M.Y.; Yano, M.; Nakajima, J.; Awazuhara, M.; Inoue, E.; Takahashi, H.; Goodenowe, D.B.; Kitayama, M. Functional genomics by integrated analysis of metabolome and transcriptome of Arabidopsis plants over-expressing an MYB transcription factor. *Plant J.* **2005**, *42*, 218–235. [CrossRef]
31. Litvinov, D.Y.; Karlov, G.I.; Divashuk, M.G. Metabolomics for Crop Breeding: General Considerations. *Genes* **2021**, *12*, 1602. [CrossRef]
32. Kim, E.-H.; Lee, K.M.; Lee, S.-Y.; Kil, M.; Kwon, O.-H.; Lee, S.-G.; Lee, S.-K.; Ryu, T.-H.; Oh, S.-W.; Park, S.-Y. Influence of genetic and environmental factors on the contents of carotenoids and phenolic acids in red pepper fruits (*Capsicum annuum* L.). *Appl. Biol. Chem.* **2021**, *64*, 1–11. [CrossRef]
33. Lozada, D.N.; Bhatta, M.; Coon, D.; Bosland, P.W. Single Nucleotide Polymorphisms Reveal Genetic Diversity in New Mexican Chile Peppers (*Capsicum* spp.). *BMC Genom.* **2021**, *22*, 356. [CrossRef]
34. Mi, S.; Yu, W.; Li, J.; Liu, M.; Sang, Y.; Wang, X. Characterization and discrimination of chilli peppers based on multi-element and non-targeted metabolomics analysis. *LWT* **2020**, *131*, 109742. [CrossRef]
35. Espichán, F.; Rojas, R.; Quispe, F.; Cabanac, G.; Marti, G. Metabolomic characterization of 5 native Peruvian chili peppers (*Capsicum* spp.) as a tool for species discrimination. *Food Chem.* **2022**, *386*, 132704. [CrossRef]
36. Florentino-Ramos, E.; Villa-Ruano, N.; Hidalgo-Martínez, D.; Ramírez-Meraz, M.; Méndez-Aguilar, R.; Velásquez-Valle, R.; Zepeda-Vallejo, L.G.; Pérez-Hernández, N.; Becerra-Martínez, E. 1H NMR-based fingerprinting of eleven Mexican *Capsicum annuum* cultivars. *Food Res. Int.* **2019**, *121*, 12–19. [CrossRef]
37. Villa-Ruano, N.; Ramírez-Meraz, M.; Méndez-Aguilar, R.; Zepeda-Vallejo, L.G.; Álvarez-Bravo, A.; Pérez-Hernández, N.; Becerra-Martínez, E. 1H NMR-based metabolomics profiling of ten new races from *Capsicum annuum* cv. serrano produced in Mexico. *Food Res. Int.* **2019**, *119*, 785–792. [CrossRef]
38. Lozada, D.N.; Coon, D.L.; Guzmán, I.; Bosland, P.W. Heat profiles of 'superhot'and New Mexican type chile peppers (*Capsicum* spp.). *Sci. Hortic.* **2021**, *283*, 110088. [CrossRef]
39. Carvalho Lemos, V.; Reimer, J.J.; Wormit, A. Color for life: Biosynthesis and distribution of phenolic compounds in pepper (*Capsicum annuum*). *Agriculture* **2019**, *9*, 81. [CrossRef]
40. Kumar, O.A.; Rao, S.A.; Tata, S.S. Phenolics quantification in some genotypes of *Capsicum annuum* L. *J. Phytol.* **2010**, *2*, 87–90.
41. Guzman, I.; Vargas, K.; Chacon, F.; McKenzie, C.; Bosland, P.W. Health-promoting carotenoids and phenolics in 31 Capsicum accessions. *HortScience* **2021**, *56*, 36–41. [CrossRef]
42. De Sa Mendes, N.; Santos, M.C.P.; Santos, M.C.B.; Cameron, L.C.; Ferreira, M.S.L.; Goncalves, E.C.B.A. Characterization of pepper (*Capsicum baccatum*)—A potential functional ingredient. *LWT* **2019**, *112*, 108209. [CrossRef]

43. Lee, Y.; Howard, L.R.; Villalon, B. Flavonoids and antioxidant activity of fresh pepper (*Capsicum annuum*) cultivars. *J. Food Sci.* **1995**, *60*, 473–476. [CrossRef]
44. Chen, L.; Kang, Y.-H. Anti-inflammatory and antioxidant activities of red pepper (*Capsicum annuum* L.) stalk extracts: Comparison of pericarp and placenta extracts. *J. Funct. Foods* **2013**, *5*, 1724–1731. [CrossRef]
45. Mokhtar, M.; Soukup, J.; Donato, P.; Cacciola, F.; Dugo, P.; Riazi, A.; Jandera, P.; Mondello, L. Determination of the polyphenolic content of a *Capsicum annuum* L. extract by liquid chromatography coupled to photodiode array and mass spectrometry detection and evaluation of its biological activity. *J. Sep. Sci.* **2015**, *38*, 171–178. [CrossRef]
46. Jeong, W.Y.; Jin, J.S.; Cho, Y.A.; Lee, J.H.; Park, S.; Jeong, S.W.; Kim, Y.; Lim, C.; El-Aty, A.M.A.; Kim, G. Determination of polyphenols in three *Capsicum annuum* L. (bell pepper) varieties using high-performance liquid chromatography-tandem mass spectrometry: Their contribution to overall antioxidant and anticancer activity. *J. Sep. Sci.* **2011**, *34*, 2967–2974. [CrossRef]
47. Rodrıguez-Concepción, M.; Boronat, A. Elucidation of the methylerythritol phosphate pathway for isoprenoid biosynthesis in bacteria and plastids. A metabolic milestone achieved through genomics. *Plant Physiol.* **2002**, *130*, 1079–1089. [CrossRef]
48. Lehmann, M.; Vamvaka, E.; Torrado, A.; Jahns, P.; Dann, M.; Rosenhammer, L.; Aziba, A.; Leister, D.; Rühle, T. Introduction of the carotenoid biosynthesis α-branch into Synechocystis sp. PCC 6803 for lutein production. *Front. Plant Sci.* **2021**, *12*, 699424. [CrossRef]
49. Zafar, J.; Aqeel, A.; Shah, F.I.; Ehsan, N.; Gohar, U.F.; Moga, M.A.; Festila, D.; Ciurea, C.; Irimie, M.; Chicea, R. Biochemical and Immunological implications of Lutein and Zeaxanthin. *Int. J. Mol. Sci.* **2021**, *22*, 10910. [CrossRef]
50. Sakurai, N. Recent applications of metabolomics in plant breeding. *Breed. Sci.* **2022**, *72*, 56–65. [CrossRef]
51. Lozada, D.N.; Bosland, P.; Barchenger, D.W.; Haghshenas-Jaryani, M.; Sanogo, S.; Walker, S. Chile Pepper (*Capsicum*) Breeding and Improvement in the "Multi-Omics" Era. *Front. Plant Sci.* **2022**, *13*, 879182. [CrossRef] [PubMed]

metabolites

Article

Cycloartane Saponins from *Astragalus glycyphyllos* and Their In Vitro Neuroprotective, Antioxidant, and hMAO-B-Inhibiting Effects

Ivan Stambolov [1], Aleksandar Shkondrov [1], Olaf Kunert [2], Franz Bucar [3], Magdalena Kondeva-Burdina [4] and Ilina Krasteva [1,*]

[1] Department of Pharmacognosy, Faculty of Pharmacy, Medical University of Sofia, 2 Dunav St., 1000 Sofia, Bulgaria; istambolov@pharmfac.mu-sofia.bg (I.S.); shkondrov@pharmfac.mu-sofia.bg (A.S.)
[2] Department of Pharmaceutical Chemistry, Institute of Pharmaceutical Sciences, University of Graz, Universitätsplatz 1, A-8010 Graz, Austria; olaf.kunert@uni-graz.at
[3] Department of Pharmacognosy, Institute of Pharmaceutical Sciences, University of Graz, Beethovenstrasse 8, A-8010 Graz, Austria; franz.bucar@uni-graz.at
[4] Laboratory of Drug Metabolism and Drug Toxicity, Department of Pharmacology, Pharmacotherapy and Toxicology, Faculty of Pharmacy, Medical University of Sofia, 2 Dunav St., 1000 Sofia, Bulgaria; mkondeva@pharmfac.mu-sofia.bg
* Correspondence: ikrasteva@pharmfac.mu-sofia.bg; Tel.: +359-2-9236-552

Abstract: *Astragalus glycyphyllos* (Fabaceae) is used in the traditional medicine of many countries against hepatic and cardiac disorders. The plant contains mainly flavonoids and saponins. From a defatted methanol extract from its overground parts, a new triterpenoid saponin, 3-*O*-[α-L-rhamnopyranosyl-(1$\rightarrow$2)]-β-D-xylopyranosyl]-24-*O*-α-L-arabinopyranosyl-3β,6α,16β,24(*R*),25-penta hydroxy-20*R*-cycloartane, together with the rare saponin astrachrysoside A, were isolated using various chromatography methods. The compounds were identified via extensive high resolution electrospray ionisation mass spectrometry (HRESIMS) and NMR analyses. Both saponins were examined for their possible antioxidant and neuroprotective activity in three different in vitro models. Rat brain synaptosomes, mitochondria, and microsomes were isolated via centrifugation using Percoll gradient. They were treated with the compounds in three different concentrations alone, and in combination with 6-hydroxydopamine or *tert*-butyl hydroperoxide as toxic agents. It was found that the compounds had statistically significant dose-dependent in vitro protective activity on the sub-cellular fractions. The compounds exhibited a weak inhibitory effect on the enzyme activity of human recombinant monoamine oxidase type B (hMAO-B), compared to selegiline.

Keywords: cycloartane saponins; isolation; structural elucidation; neuroprotection; antioxidant activity; hMAO-B-inhibition; brain synaptosomes; brain mitochondria; brain microsomes

Citation: Stambolov, I.; Shkondrov, A.; Kunert, O.; Bucar, F.; Kondeva-Burdina, M.; Krasteva, I. Cycloartane Saponins from *Astragalus glycyphyllos* and Their In Vitro Neuroprotective, Antioxidant, and hMAO-B-Inhibiting Effects. *Metabolites* **2023**, *13*, 857. https://doi.org/10.3390/metabo13070857

Academic Editors: Ramona Paltinean and Irina Ielciu

Received: 22 June 2023
Revised: 15 July 2023
Accepted: 17 July 2023
Published: 19 July 2023

1. Introduction

Astragalus glycyphyllos L., Fabaceae (Liquorice Milk-Vetch) is a perennial herbaceous plant with deep roots, distributed in the mountainous regions of Bulgaria on stony places in forest glades, forests, and scrub, up to 1750 m above sea level [1,2]. The species has been widely used in the folk medicine of the country as an anti-inflammatory, antihypertensive, and diuretic, etc. Its aerial parts can be used as an infusion for heart failure, kidney inflammation, and calculus, and as an adjuvant treatment for cancer diseases, tachycardia, and increased blood pressure, etc. [3]. Research on *A. glycyphyllos* has focused on its biologically active secondary metabolites such as saponins and flavonoids [4]. A phytochemical investigation of the overground parts of the species was initiated in Bulgaria 35 years ago and led to the isolation of some sapogenins (soyasapogenol B and 3β,22β,24-trihydroxyolean-12-en-19-one) after acid hydrolysis [5,6]. In later research, the saponins askendoside C and F were isolated from the roots of the plant [7]. Recently, from its aerial parts, an epoxycycloartane saponin 17(*R*),20(*R*)-3β,6α,16β-trihydroxycycloartanyl-23-carboxilic acid 16-lactone 3-*O*-β-D-glucopyranoside [8] was isolated.

A defatted extract obtained from the aerial parts of the species had in vivo antioxidant and hepatoprotective effects on Wistar rats with carbon tetrachloride-induced hepatotoxicity. The effects of the extract, containing mainly flavonoids and saponins, were commensurable to those of the classical hepatoprotector silymarin [9]. A purified saponin mixture (PSM) resulting from the same extract displayed antiproliferative and cytotoxic effects in vitro/in vivo on Graffi myeloid tumour cells and Graffi-tumour-bearing hamsters. Per oral treatment with PSM extended survival and reduced tumour growth. A statistically significant decrease in the proliferation and viability of the tumour cells was observed after this administration. Concentration- and time-dependent effects were proved. The antiproliferative effects were related to the induction of apoptosis by the saponins, which was demonstrated using fluorescence microscopy [10]. Several compounds obtained from the aerial parts, including the epoxycycloartane saponin, exhibited neuroprotective activity on 6-hydroxydopamine-induced neurotoxicity in isolated rat brain synaptosomes. The same saponin also displayed statistically significant hMAO-B-enzyme-inhibiting activity compared to selegiline [8]. Based on these data, the saponins from this species could be considered as perspective, possessing valuable pharmacological activities.

Neurodegenerative diseases, especially Parkinson's disease, are a major health concern. For the last two decades, they have established a new group of socially important ailments, which have a major impact on our modern life [11]. In pharmacological research, in vitro systems play an important role as the initial data collection tool. Their role in evaluating the ability of a compound to influence the mechanism of some disorders is invaluable. In vitro methods are fast and more acceptable than in vivo ones, especially in ethical terms. Unfortunately, the extrapolation of these results to a living organism is not always possible or adequate. For some pathogenetic mechanisms, their underlying factors are known. This could be reproduced in an in vitro model to accumulate results commensurable to those processes that are responsible for this effect in vivo [12,13]. Oxidative stress is considered to be the leading risk factor resulting in neurodegeneration. Thus, neuroprotection is in direct connection with antioxidant activity [11]. Sub-cellular fractions (synaptosomes, mitochondria, and microsomes) are a convenient object for the investigation of neuronal damage and its mechanisms, as included in [14]. The availability of specific enzymes (incl. human recombinant ones) capable of metabolizing neurotransmitters is another advantage of today's in vitro methods. It is known that human monoamine oxidase type B is responsible for metabolizing dopamine, so inhibiting it is a way of increasing the depleted dopamine in conditions such as Parkinson's [15].

In this continuation of our research, the aim was to isolate more polar and previously unidentified saponins from *Astragalus glycyphyllos* and to evaluate their possible antioxidant, neuroprotective, and hMAO-B-inhibiting effects in in vitro models.

2. Materials and Methods

2.1. General

AUTOPOL VI (Rudolph Research Analytical) was used to measure optical rotation. NMR spectra were recorded with a 700 MHz Bruker Avance II NMR spectrometer (Bruker, Rheinstetten, Germany; software MNova, Mestrelab Research, Santiago de Compostela, Spain) equipped with a cryo-probe. For both compounds, **S1** and **S2,** a data set consisting of 1D proton and carbon experiments and 2D COSY, HSQC, HMBC, and ROESY experiments was acquired at 298 K in methanol-d_4; in addition, a second data set for S2 was recorded at the same temperature in pyridine-d_5.

A UHPLC system (Dionex UltiMate 3000 RSLC, ThermoFisher Scientific, Bremen, Germany) connected to a Q Exactive Plus Orbitrap mass spectrometer with a heated electrospray ionisation (HESI) ion source (ThermoFisher Scientific, Bremen, Germany) was used. The parameters of the full scan MS were in the resolution of 70,000 (@200 m/z), $3e^6$AGC target, 100 ms max IT, and 250 to 1700 m/z scan range. The ion source operated at −2.5 or +3.5 kV and 320 °C (capillary and probe), 38 arbitrary units (a.u., as by the software Extactive Tune, ThermoFisher Scientific, Bremen, Germany) of sheath gas, and 12 a.u. of auxiliary gas (both N_2); an S-Lens RF level 50.0. A Kromasil C_{18} column (1.9 µm, 2.1 × 50 mm, Akzo Nobel, Bohus, Sweden) was used, maintained at 40 °C, and eluted with H_2O + 0.1% HCOOH (A) and MeCN + 0.1% HCOOH (B) (0.3 mL/min). The gradient program was for 0.5 min 10% B, for 7 min increasing to 30% B, for 1.5 min 30% B (isocratic), for 3.5 min increasing to 95% B, for 2 min with 95% B (isocratic), and for 0.1 min decreasing to 10% B. Diaion HP-20 resin was purchased from Supelco (Bellefonte, PA, USA). Silica gel cartridges (FlashPure® 40 µm, irregular, Buchi, Flawil, Switzerland) were used on a Reveleris X2 flash chromatograph (Buchi, Flawil, Switzerland). Sephadex LH-20 was obtained from Supelco (Bellefonte, PA, USA). TLC was performed on Kieselgel F_{254} sheets (Merck, Darmstadt, Germany), eluted with EtOAc/HCOOH/AcOH/H_2O (32/3/2/6). The saponins were visualised using p-anisaldehyde/conc. H_2SO_4 spraying and heating for 10 min at 104 °C were performed.

The determination of the absolute configuration of the monosaccharides forming the sugar chains was performed using a known GC-MS method [16]. Briefly, after acid hydrolysis, (2R)-2-butyl glycosides were prepared. Each compound (2 mg) was heated with 5.5 mL of H_2O, 3.5 mL of conc. AcOH, and 1 mL of conc. HCl on a water bath at 100 °C under a reflux condenser for 2 h. The hydrolysate was extracted with EtOAc and the aqueous residue was evaporated to dryness. To the residue, 0.45 mL of (2R)-2-BuOH and 0.1 mL of conc. HCl were added and heated at 100 °C for 15 h. The mixture was evaporated under N_2, and then 100 µL of Sigma-Sil-A (Sigma-Aldrich, Schnelldorf, Germany) was added to prepare TMS derivatives. The standard compounds (L-rhamnose, D-xylose, and L-arabinose, Sigma Aldrich, Schnelldorf, Germany) were treated in the same manner. The silated derivatives of the (2R)-2-butyl glycosides formed were analysed using GC-MS. For the GC-MS analysis, an Exactive Orbitrap GC-MS system (ThermoFisher Scientific, Bremen, Germany) was used. It was operated at 70 eV, with a 230 °C ion source, 280 °C interface, 270 °C injector temperature, and 1 µL injection volume (split, 20:1 ratio). A capillary column (fused silica, 5% phenyl/95% methyl polysiloxane, HP-5MS 30 m × 250 µm × 0.25 µm, Agilent, Santa Clara, CA, USA) was used. The temperature program was initially 100 °C at 3 °C/min to 270 °C. Helium 5.0 (1.5 mL/min) was the carrier gas. The data (50–450 u) were collected with Xcalibur v. 4 (ThermoFisher Scientific, Bremen, Germany).

All other chemicals and solvents were obtained from Sigma-Aldrich (Schnelldorf, Germany).

2.2. Extraction and Isolation of Compounds from Plant Material

The *A. glycyphyllos* overground parts were harvested in July 2020 from Vitosha Mt., Bulgaria. Prof. D. Pavlova identified the plant, and a specimen is kept in the Herbarium of the Faculty of Biology, Sofia University (S0 107 613). The air-dried plant material (200 g) was powdered (3 mm) and then extracted with dichloromethane (6 × 2 L) using percolation to remove the lipophilic constituents. The defatted plant substance was then aired and exhaustively extracted with 80% MeOH (24 × 3 L) using percolation. The obtained extract was filtered, concentrated under vacuum, and then lyophilized to produce a dry extract (42 g). The extract was separated over a Diaion HP-20 (4.7 × 45) column, eluting with H_2O: MeOH (0→100%). Seven main fractions were collected (I-VII). After the TLC analysis, fractions VI and VII were found to be rich in saponins. Fraction VI was chromatographed with CH_2Cl_2:MeOH:H_2O (step gradient 8:2:0.2→7:3:0.3) on a silica gel cartridge using flash chromatography to obtain 21 subfractions. Subfraction 13, which contained a main compound (TLC), was further separated with CH_2Cl_2:MeOH:H_2O (step gradient 9:1:0→5:6:1) on a silica gel cartridge using flash chromatography, affording compound **S1**. The saponin was further purified over Sephadex LH-20 (eluent MeOH) to obtain 33 mg of it. Fraction VII, containing another saponin (TLC analysis), was separated two subsequent times using flash chromatography on a silica gel cartridge with CH_2Cl_2:MeOH:H_2O (step gradient 8:2:0.1→5:5:0.5), to obtain compound **S2**. The final purification of the saponin over Sephadex LH-20 (eluent MeOH) provided 20 mg of S2.

*3-O-[α-L-rhamnopyranosyl-(1→2)]-β-D-xylopyranosyl]-24-O-α-L-arabinopyranosyl-3β,6α, 16β,24(R),25-pentahydroxy-20R-cycloartane (**S1**): a white amorphous powder (MeOH);* $C_{46}H_{78}O_{17}$*;* ${}^{1}H$ *NMR (methanol-d$_4$, 700 MHz),* see Table 1, Figures S1–S3; ${}^{13}C$ NMR (methanol-d_4, 175 MHz), see Table 1, Figure S4; HREIMS, m/z 901.5186 $[M-H]^-$ (calcd. for $C_{46}H_{77}O_{17}$, 901.5160), m/z 947.5242 $[M + HCOO]^-$ (calcd. for $C_{47}H_{79}O_{19}$, 947.5215), Figure S28; m/z 903.5323 $[M + H]^+$ (calcd. for $C_{46}H_{79}O_{17}$, 903.5317), m/z 925.5142 $[M + Na]^+$ (calcd. for $C_{46}H_{78}O_{17}Na$, 925.5136), Figure S29; $[\alpha]^{D}{}_{20} = -1.0546$ (c 0.1, MeOH).

Table 1. ${}^{1}H$ NMR spectroscopic data (700 MHz, J in Hz) and ${}^{13}C$ NMR spectroscopic data (175 MHz) of **S1** in methanol-d_4.

Position	δ_C [ppm], Type	δ_H [ppm], (J in Hz)
1	33.5, CH_2	1.54, t (13.0) 1.21
2	30.7, CH_2	1.91, m 1.66
3	89.3, CH	3.18, d (~13)
4	43.2, C	-
5	54.8, CH	1.35, d (9.9)
6	69.6, CH	3.44, m
7	38.8, CH_2	1.33 1.45
8	48.8, CH	1.79, dd (12.0, 4.2)
9	22.1, C	-
10	30.4, C	-
11	27.0, CH_2	1.191.97
12	33.9, CH_2	1.66 1.60, td (12.0, 3.2)
13	46.4, C	-
14	47.4, C	-
15	48.8, CH_2	1.99 1.38
16	73.1, CH	4.40, m
17	58.1, CH	1.66
18	19.2, CH_3	1.13 s
19	31.9, CH_2	0.37, d (4.0) 0.52, d (4.0)
20	32.5, CH	1.75
21	18.7, CH_3	0.93, d (6.5)
22	34.6, CH_2	2.08, brt (12.5) 0.96
23	30.4, CH_2	1.701.25
24	92.5, CH	3.30
25	74.9, C	-
26	26.4, CH_3	1.15, s
27	24.0, CH_3	1.18, s
28	16.8, CH_3	1.02, s
29	28.6, CH_3	1.28, s
30	20.4, CH_3	1.54, t (13.0) 1.21
Xyl-I		
1	106.2, CH	4.37, d (~6.9)
2	78.9, CH	3.42
3	78.8, CH	3.42
4	71.6, CH	3.47
5	66.5, CH_2	3.17, t (10.6) 3.84, dd (11.0, 5.3)
Rha-II		
1	102.1, CH	5.33, brs
2	72.2, CH	3.94, brs
3	72.2, CH	3.74, dd (9.3, 2.8)
4	74.0, CH	3.38, t (9.6)
5	70.1, CH	3.98, dq (9.4, 6.1)
6	18.1, CH_3	1.23, d (6.3)
Ara-III		
1	107.4, CH	4.38, d (7.5)
2	73.7, CH	3.58
3	75.0, CH	3.49
4	70.2, CH	3.78, brs
5	68.0, CH_2	3.56 3.87, dd (12.5, 1.8)

2.3. In Vitro Pharmacological Evaluation

Twenty Wistar rats (male, 200–250 g) were purchased from the National Breeding Centre of the Bulgarian Academy of Sciences in Slivnitsa, Bulgaria. They were kept in Plexiglas cells (3 in each) and a seven-day acclimatization period was allowed before the commencement of the experiment. A veterinary physician monitored the animals' health daily. The rats were given standardized pellets chew and drinking water ad libitum. The husbandry conditions in the Vivarium of the Faculty of Pharmacy at the Medical University of Sofia were checked by the Bulgarian Food Safety Agency at the Bulgarian Ministry of Foods and Agriculture. The experiment was allowed by permission № 200/2021 from the same agency. The University Ethical Commission (KENIMUS) gave ethical clearance for the experiment with the protocol 7338/11.2021.

For the isolation of the sub-cellular structures, the rats were decapitated, the cranial cavities were opened, and the brains were removed and stored over ice. The organs were pooled and homogenised using a Teflon pestle with the appropriate buffers, as described in the procedures below.

The synaptosomes and brain mitochondria were isolated using multiple differential centrifugations with the methods of [17,18]. Two buffers—A (HEPES 5 mM and Sucrose 0.32 M) and B (NaCl 290 mM, $MgCl_2.2H_2O$ 0.95 mM, KCl 10 mM, $CaCl_2.2H_2O$ 2.4 mM, NaH_2PO_4 2.1 mM, HEPES 44 mM, and D-Glucose 13 mM)—were prepared. Buffer A was used to prepare the brain homogenate. First, the homogenate was centrifuged twice at 1000× *g* for 10 min at 4 °C, after which, the supernatants of the two centrifuges were combined and centrifuged three times at 10,000× *g* for 20 min at 4 °C. The isolation of the synaptosomes and mitochondria was accomplished using Percoll. First, a 90% Percoll stock solution was prepared and, after that, 16% and 10% solutions were prepared. Amounts of 4 mL of 16% and 10% Percoll were carefully applied in layers. At the end, 4 mL of 7.5% Percoll was added to the precipitate from the last centrifugation. The tubes were centrifuged at 15,000× *g* for 20 min at 4 °C. After this centrifugation, three layers were formed: the lower, which contained mitochondria; the medium (between 16% and 10% Percoll), containing synaptosomes; and the top, lipids. Using a Pasteur pipette, we removed the middle and bottom layers. Each was centrifuged at 10,000× *g* for 20 min at 4 °C with buffer B. The resulting synaptosomes and mitochondria were diluted with buffer B to a protein content of 0.1 mg/mL. After incubation with the test substances (saponins) and 6-OHDA, the synaptosomes were centrifuged three times at 15,000× *g* for 1 min. The synaptosomal viability was determined using an MTT test with the method of [19]. After incubation with the substances (saponins), the synaptosomes were centrifuged at 400× *g* for 3 min. The precipitate was treated with 5% trichloroacetic acid, vortexed, and left on ice for 10 min, then centrifuged at 8000× *g* for 10 min. The supernatant was frozen at −20 °C. Immediately prior to their determination, each sample was neutralized with 5 M NaOH. The GSH levels were determined using Elmman's reagent (DTNB) spectrophotometrically at 412 nm with methods of [20,21]. For the isolation of the brain microsomes, the brain was homogenized in 9 volumes of 0.1 M Tris buffer containing 0.1 mM Dithiothreitol, 0.1 mM Phenylmethylsulfonyl fluoride, 0.2 mM EDTA, 1.15% KCl, and 20% (*v*/*v*) glycerol (pH 7.4). The resulting homogenate was centrifuged twice at 17,000× *g* for 30 min. The supernatants from the two centrifuges were combined and centrifuged twice at 100,000× *g* for 1 h. The pellet was frozen in 0.1 M Tris buffer at −20 °C [22]. The determination of MDA was performed spectrophotometrically at a wavelength of 535 nm using methods of [21,23]. A molar extinction coefficient of 1.56×10^5 M^{-1} cm^{-1} was used for the calculation [24].

The examined compounds were tested for their possible inhibitory activity on the hMAO-B enzyme (commercially available) using the Amplex UltraRed reagent fluorometric method [25] with small modifications [15].

Salts to prepare the buffer, 2,2′-dinitro-5,5′-dithiodibenzoic acid, reduced glutathione (GSH), oxidized glutathione (GSSG), human recombinant MAO type B enzyme, tyramine HCl, and horseradish peroxidase, as well as other reagents, were used for the pharmacological study and obtained from Sigma-Aldrich (Schnelldorf, Germany). The Amplex UltraRed kit was from Invitrogen (Thermo Scientific, Karlsruhe, Germany)

The statistical analysis of the results was performed using the statistical programme "Medcalc" v. 18 (MedCalc Software Ltd., Ostend, Belgium). Each experiment was performed in triplicate, and the values are represented as the mean of three (n = 3). A Mann–Whitney non-parametric test was used to examine the statistical significance of the results. When $p < 0.05$; $p < 0.01$; or $p < 0.001$, the differences were accepted as significant.

3. Results

3.1. Identification of the Compounds

Compound **S1** was isolated as a white amorphous powder (33 mg). In the negative HRESIMS spectrum (Figures S28 and S29) of the compound, an ion $[M - H]^-$ at m/z 901.5187 was observed, corresponding to the molecular formula $C_{46}H_{78}O_{17}$. The ^{1}H and ^{13}C NMR (Table 1, Table S1, Figures S1–S10) and HSQC spectrum pointed to a triterpene of the cycloartane-type with three sugars. The number of methine groups attached to the oxygen atoms (from –OH) was higher than expected for the three sugars, indicating the presence of at least five oxygenated positions in the aglycone. The complete resonance assignments (Table 1) led to 3,6,16,20,24,25-hexahydroxycycloartane as the aglycone. A list of all the essential HMBC correlations and ROESY cross-peaks are provided in the supporting information (Table S1). From the ROESY cross-peaks, it was deduced that H-3, H-5, H-30, H-16, and H-17 were on one face of the molecule, while H-6, H-8, H-18, H-19, and H-18 were on the opposite. The positions C-3, C-6, C-16, C-24, and C-25 were oxidized. Hence, the aglycone of compound **S1** was assigned as 3β,6α,16β,24,25-hexahydroxycycloartane. The aglycone was bound to three sugar moieties. After the complete resonance assignments, an analysis of the coupling patterns, and a comparison of their carbon resonances with the literature values, the three sugars were identified as β-xylose, α-rhamnose, and α-arabinose. Two of those, the xylose and the rhamnose, had identical resonance values and HMBC correlations as observed for astrachrysoside A (**S2**) in methanol-d_4 (Tables 1 and S2); therefore, both S1 and S2 had the same saccharide structure attached to C-3. A more detailed discussion was conducted for compound **S2** (see below). The remaining sugar moiety, an α-arabinose, was attached to the C-24 position of the aglycone according to the observed HMBC correlation for its anomeric proton.

NMR data from various *Astragalus* species were available for the same type of glycosylated side chain with a different configuration at the chiral carbon C-24, e.g., with a *24S*-configuration from *A. brachypterus* [25] and a 24R-configuration from *A. stereocalyx* [26]. A comparison of the ^{13}C-NMR shift values (Table 2) clearly indicated that, in compound **S1,** the position C-24 had an R-configuration. Additionally, a GC-MS analysis of the sugars after the preparation of the (2R)-butyl glycosides was performed to elucidate their absolute configuration. After the GC-MS analysis of the sililated (2*R*)-butylglycosides, it was found that S1 gave peaks at t_R = 24.97, 26.68 (L-Rha); 29.81, 31.42 (D-Xyl); and 24.62, 26.68 min (L-Ara), respectively (Figure S32) [16]. On basis of these spectroscopic data and the result of the sugar hydrolysis, the structure of S1 was established as 3-O-[α-L-rhamnopyranosyl-(1→2)]-β-D-xylopyranosyl]-24-O-α-L-arabinosyl-3β,6α,16β,24,25-pentahydroxy-20*R*,24*R*-cycloartane (Figure 1). To our knowledge, this is the first report on this compound.

Table 2. Comparison of ^{13}C chemical shifts of S1 with data of glycosylated sidechains which have different configuration at C-24. All data were recorded in methanol-d_4.

Atom	S1	24 S *A. brachypterus* [25]	24 R *A. stereocalyx* [26]
	δ_C [ppm]	δ_C [ppm]	δ_C [ppm]
C-21	18.7	17.5	18.2
C-20	32.5	30.9	32.4
C-22	34.6	33.0	34.2
C-23	30.4	29.4	30.1
C-24	92.5	89.7	91.8
C-25	74.9	73.5	75.0

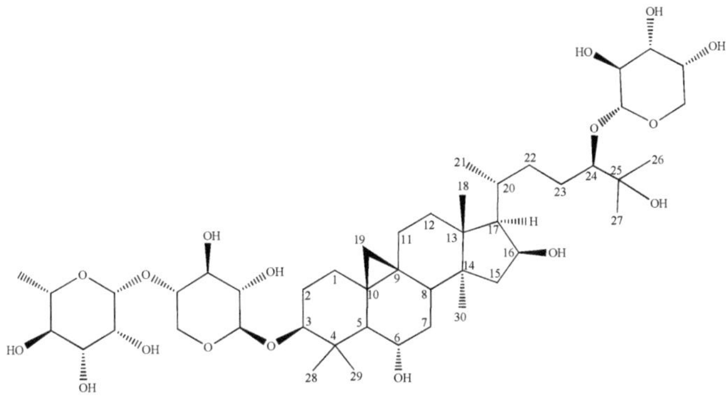

Figure 1. Structure of S1.

Compound **S2** was isolated as a white amorphous powder (20 mg) and had $[\alpha]^D{}_{20}$ −2.0698 (c 0.1, MeOH). It was identified via HRESIMS (Figures S30 and S31), 1D, 2D NMR experiments (Figures S11–S27), a GC-MS analysis of the absolute configuration of its sugar moieties (Figure S33) [16], and a comparison to the literature as 3-O-[α-L-rhamnopyranosyl-(1→2)]-β-D-xylopyranosyl]-cycloastragenol (astrachrysoside A) [27] (Figure 2). A major issue of NMR data recorded in methanol-d_4 is that the methine groups C-2 and C-3 of the xylose in compound **S2** have the same values for their carbon and proton resonances, meaning that the point of attachment of the rhamnose cannot be determined via HMBC or NOESY experiments. To circumvent this issue, a second NMR data set of S2 was recorded in pyridine-d_5 and all the resonances were assigned (Table S2). In pyridine, the resonances at C-2 and C-3 in the xylose were no longer identical, and the anomeric proton of the rhamnose gave an unambiguous HMBC correlation to C-2 of the xylose (Figure S18), allowing for a confirmation of the sugar substructure via NMR.

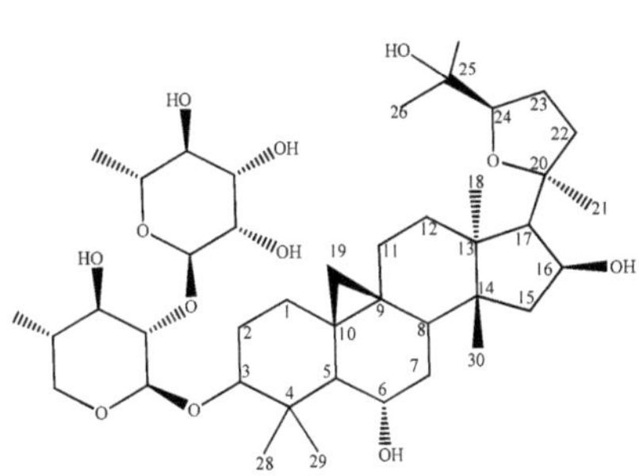

Figure 2. Structure of astrachrysoside A.

3.2. Pharmacological Investigation

Both compounds were tested in vitro for their possible antioxidant and neuroprotective activity in rat brain synaptosomes, mitochondria, and microsomes.

6-Hydroxydopamine (6-OHDA, 150 μM, administered alone) showed statistically significant neurotoxic effects. It reduced the synaptosomal viability and GSH levels by 55% and 50%, respectively, compared to the non-treated synaptosomes (control) (Figures 3 and 4).

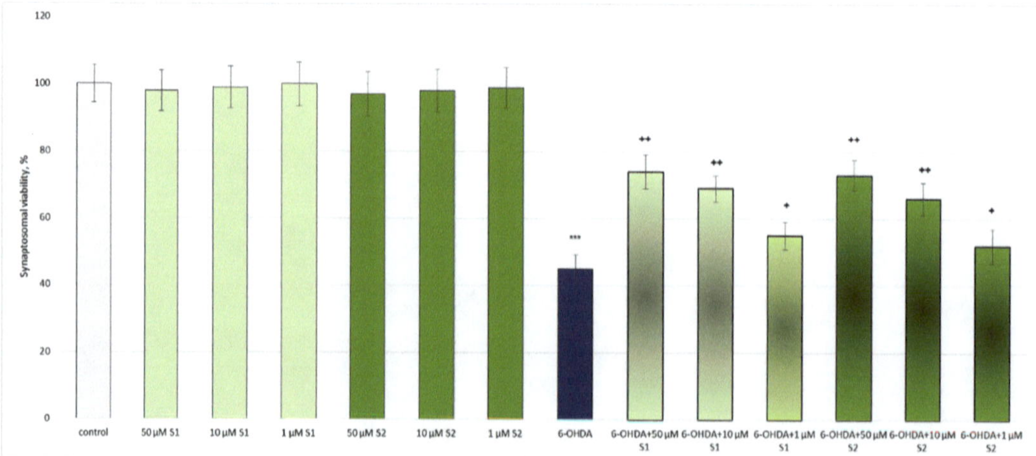

Figure 3. Effects of **S1** and **S2** on the synaptosomal viability, administered alone and in a model of 6-OHDA-induced oxidative stress; *** $p < 0.001$ vs. untreated synaptosomes (control); $^{+}$ $p < 0.05$; $^{++}$ $p < 0.01$ vs. 6-OHDA.

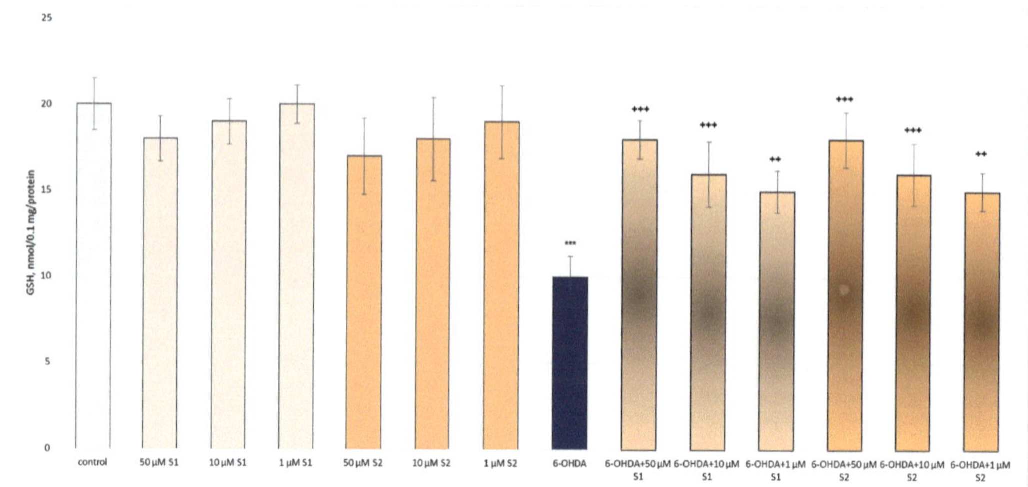

Figure 4. Effects of **S1** and **S2** on GSH level in isolated synaptosomes, administered alone and in a model of 6-OHDA-induced oxidative stress. *** $p < 0.001$ vs. untreated synaptosomes (control); $^{++}$ $p < 0.01$; $^{+++}$ $p < 0.001$ vs. 6-OHDA.

In the conditions of 6-OHDA-induced oxidative stress, **S1** and **S2** exhibited well-pronounced, concentration-dependent neuroprotective and antioxidant effects, preserving the synaptosomal viability and GSH levels in all the tested concentrations. At the highest tested concentration (50 μM), both **S1** and **S2** had the strongest effects on this model (Figures 3 and 4).

Administered alone, both saponins did not exhibit a statistically significant neurotoxic effect on the mitochondria. Tert-butyl hydroperoxide (t-BuOOH), applied alone, reduced the GSH levels by 50% and increased the MDA production by 152% compared to the untreated mitochondria (control), thus exhibiting a neurotoxic effect (Figures 5 and 6).

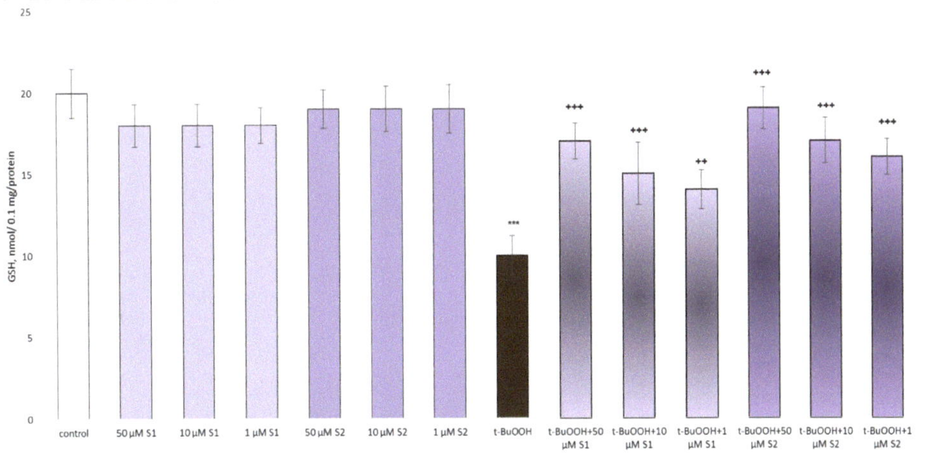

Figure 5. Effects of **S1** and **S2** on GSH level in isolated brain mitochondria, administered alone and in a model of t-BuOOH-induced oxidative stress; *** $p < 0.001$ vs. control (untreated mitochondria); ++ $p < 0.01$; +++ $p < 0.001$ vs. t-BuOOH.

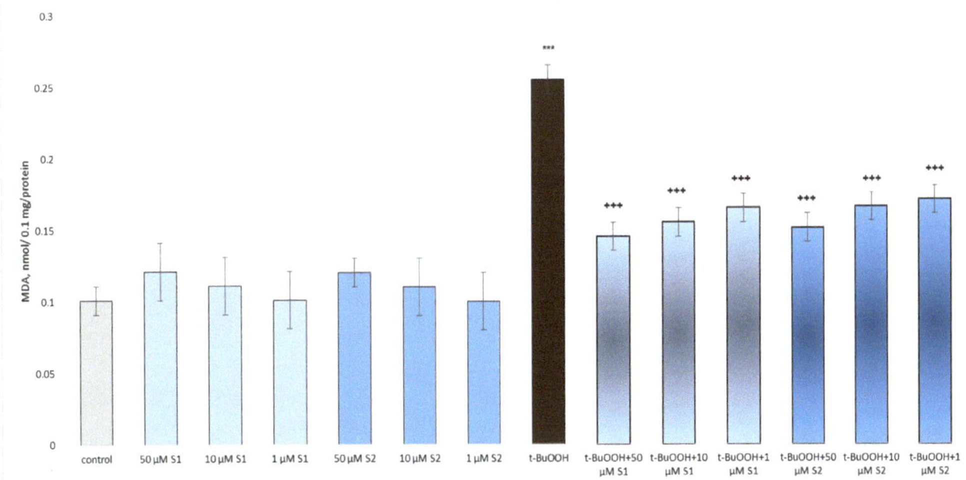

Figure 6. Effects of **S1** and **S2** on MDA production in isolated brain mitochondria, administered alone and in a model of t-BuOOH-induced oxidative stress; *** $p < 0.001$ vs. control (untreated mitochondria); +++ $p < 0.001$ vs. t-BuOOH.

Both saponins exhibited a pronounced, concentration-dependent neuroprotective and antioxidant effect on the isolated brain mitochondria in the tert-butyl hydroperoxide-induced oxidative stress model (Figures 5 and 6). The highest tested concentration (50 μM) was again the most effective. Also noteworthy was that the effect in this model on the level of GSH was more pronounced for **S2** than **S1**. For the MDA levels, this was the opposite.

Administered alone, the saponins did not exhibit a statistically significant pro-oxidant effect on the isolated rat brain microsomes. In the conditions of non-enzymatic lipid peroxidation and at the same concentrations (1 μM, 10 μM, and 50 μM), a significant decrease in the MDA production (by 124%) was observed compared to the non-treated microsomes (control) (Figure 7).

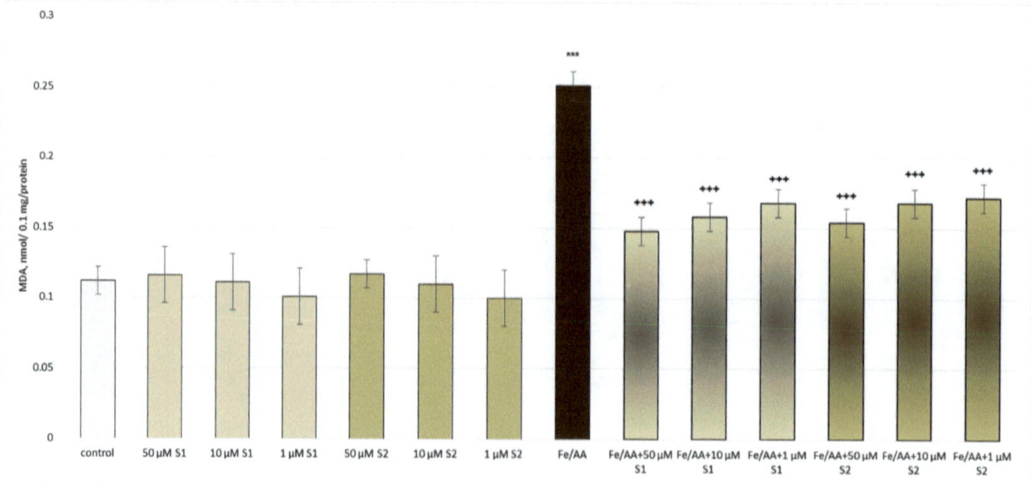

Figure 7. Effects of **S1** and **S2** on MDA production in isolated brain microsomes, administered alone and in a model of non-enzyme-induced lipid peroxidation; *** $p < 0.001$ vs. control (untreated microsomes); +++ $p < 0.01$ vs. Fe/AA.

On the activity of the human recombinant MAO-B enzyme (hMAO-B), the two saponins revealed a weak inhibition. **S1** inhibited the enzyme activity by 20%, and **S2** by 22%, while selegiline, a classical MAO-B inhibitor, decreased the enzyme activity by 55% compared to the control (pure enzyme) (Figure 8).

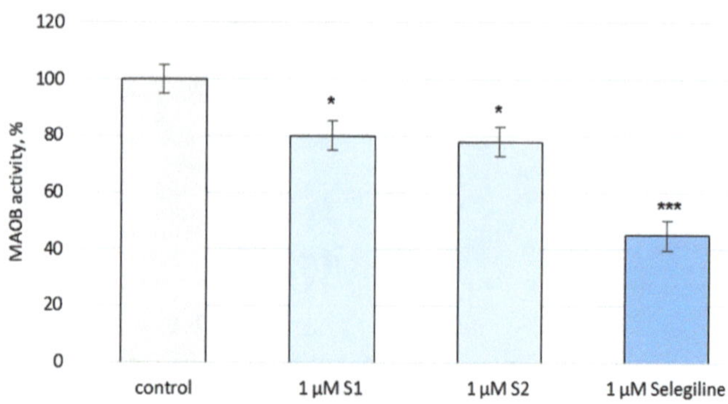

Figure 8. Effects of **S1** and **S2** on the hMAO-B enzyme activity, administered alone; * $p < 0.05$; *** $p < 0.001$ vs. control (untreated hMAO-B).

4. Discussion

Cycloartanes are considered to be the predominant type of saponins for Astragalus species, found in the Russian flora. In Bulgaria, these taxa produce mainly oleanane-type saponins [4]. Our studies on the phytochemistry of *A. glycyphyllos* grown in Bulgaria showed that this plant also contains cycloartane saponins. Two of them, isolated from the aerial parts of this species, are new natural compounds—17(R),20(R)-3β,6α,16β-trihydroxycycloartanyl-23-carboxilic acid 16-lactone 3-O-β-D-glucopyranoside [8] and the present 3-O-[α-rhamnopyranosyl-(1→2)]-β-xylopyranosyl]-24-O-α-arabinopyranosyl-3β,6α,16β,24(R),25-pentahydroxy-20R-cycloartane (**S1**), alongside the rare astrachrysoside A (S2). Saponin **S2** had previously been identified in *A. wiedemmanianus* [26], *A. trigonus* [27], and *A. chrysopterus* [28]. Nevertheless, this is the first instance of its isolation from a representative of the Bulgarian flora. Considering that the genus is represented by more than 3500 species worldwide [29], this saponin is rarely occurring. The significant structural difference between the novel saponin **S1** and astrachrysoside A is the sapogenin—cycloasgenin C of **S1** and cycloastragenol of **S2**. Compound **S1** has an aliphphatic side chain attached to C-20, unlike the substituent at C-20 of astrachrysoside A, which is a furan ring. In addition, the available –OH group at C-24 is glycosylated (L-Ara) in compound **S1**. The saponin **S1** has a structural similarity with the triperpenoids previously isolated from the roots of the species, i.e., askendosides C and F—the aglycone is the same and, like askendoside F, is a bisdesmoside [7]. The differences are the type of sugars and the attachment position.

In vitro systems have an important role in experimental toxicology for studying the biotransformation of substances and establishing the mechanisms by which oxidative stress develops. In addition, they are widely used to study the protective properties of natural compounds. The 6-hydroxydopamine model in isolated brain synaptosomes is a suitable in vitro subcellular system for studying the processes underlying the pathogenesis of degenerative diseases of the nervous system (including Parkinson's disease). The mechanism by which 6-OHDA neurotoxicity develops is a direct result of its metabolism in neuronal mitochondria. It consists of the production of reactive oxygen species, among other free radicals, and the destruction of nerve endings is thought to be primarily due to the oxidation of 6-OHDA to p-quinone, or free radical or superoxide anion production. It is these reactive intermediates that covalently interact with the nerve terminal, permanently inactivating it [30]. In the synaptosomes, prepared from rat brains with Percoll gradient, both saponins had statistically significant neuroprotective and antioxidant effects in a model of 6-OHDA-induced oxidative stress. This could be connected with their free-radical-scavenging potential, resulting in an elimination of the superoxide anion produced by the p-quinone [30].

There are other reports on the neuroprotective effects of triterpene saponins. Platycodin A, isolated from Platycodi radix, showed neuroprotective activities, increasing the cell viability by about 50% in a model of glutamate-induced toxicity [31]. Several tetracyclic triterpenoid saponins (at a concentration of 10 μM), isolated from Ginseng roots, showed neuroprotective effects on human neuroblastoma SH-SY5Y cells with H_2O_2-induced oxidative stress [32]. In vitro and in vivo studies have indicated that NgR1/RhoA/ROCK2 pathway expression regulation is the leading mechanism of action of these saponins [33]. The results obtained correlate with what is known in the literature about the protective effects of other saponins obtained from the species from the genus Astragalus. Other cycloartanes, such as cycloastragenol and astragaloside IV, have been proven as neuroprotectors [34,35]. It was found that astragaloside IV (100 μM) revealed neuroprotective effects in neuronal cell cultures treated with 6-OHDA [36], the same in vitro model of Parkinson's disease as that used in the present study. In a previous report, a cycloartane saponin, isolated from the aerial parts of *A. glycyphyllos*, displayed a strong neuroprotective effect on the same model at 100 μM [8]. A similar effect was proven for yet another oleanane-type saponin (100 μM) from *A. glycyphylloides* [37].

A suitable model of neuronal oxidative stress is t-BuOOH, which has a mitochondrial and microsomal metabolism, leading to free radicals formed through several steps. In microsomes, when NADPH is absent, the single-electron oxidation of peroxyl radicals occurs. Single-electron reduction to alkoxyl radicals happens in the presence of NADPH. In cells and isolated mitochondria, t-BuOOH forms, via β-cleavage, a methyl radical. All these radicals unlock the process of lipid peroxidation and reduce the level of reduced glutathione [38,39]. In the conditions of oxidative stress induced by t-BuOOH, both saponins tested exhibited statistically significant, concentration-dependent neuroprotective and antioxidant effects. These effects are most likely associated with the metabolism of t-BuOOH at the microsomal and mitochondrial levels, by preserving the reduced glutathione (cell protector) and possible free radical scavenging [38].

The human recombinant MAO-B enzyme is a convenient way of detecting possible inhibitors. The results observed for the isolated saponins were correlated with the data of a previous study on the inhibitory effect of an oleanane saponin from *A. glycyphylloides* [37] and a cycloartane saponin from *A. glycyphyllos* [8], showing again weak inhibitory effects on the hMAO-B enzyme activity.

These findings suggest that the protective effects on synaptosomes, brain mitochondria, and microsomes are not directly linked to direct hMAO-B inhibition. More likely, they are an indirect result of the weak radical-scavenging effects of the saponins.

5. Conclusions

From the aerial parts of *A. glycyphyllos*, two cycloartane-type triterpenpoid saponins, 3-O-[α-L-rhamnopyranosyl-(1→2)]-β-D-xylopyranosyl]-24-O-α-L-arabinopyranosyl-3β,6α,16β,24(R),25-pentahydroxycycloartane and astrachrysoside A, were isolated via various chromatographic techniques. They were structurally elucidated using extensive HRESIMS, NMR, and GC-MS experiments. Both compounds displayed significant in vitro neuroprotective and antioxidant effects in a 6-OHDA-induced neurotoxicity model on isolated brain synaptosomes, t-BuOOH-induced oxidative stress in brain mitochondria, and on isolated rat brain microsomes, in a model of lipid peroxidation (non-enzyme-induced). The observed effects were concentration-dependant. Both saponins revealed a weak inhibitory effect on the activity of hMAO-B in comparison with selegiline. These results suggest that *A. glycyphyllos* saponins could be considered as perspective for future research.

Supplementary Materials: The following supporting information can be downloaded at: https://www.mdpi.com/article/10.3390/metabo13070857/s1, Table S1. ^{1}H NMR spectroscopic data (700 MHz, J in Hz) and ^{13}C NMR spectroscopic data (175 MHz) of **S1** in methanol-d_4; Table S2. ^{1}H NMR spectroscopic data (700 MHz, J in Hz) and ^{13}C NMR spectroscopic data (175 MHz) of compound **S2** in methanol-d_4 and pyridine-d_5.; Figure S1: ^{1}H-NMR spectrum (700 MHz, methanol-d_4) of compound **S1**; Figure S2: Expansion of the ^{1}H-NMR spectrum of compound **S1** (sugar region); Figure S3: Expansion of the ^{1}H-NMR spectrum of compound **S1** (aglycone region); Figure S4: ^{13}C-NMR spectrum (175 MHz, methanol-d_4) of compound **S1**; Figure S5: COSY spectrum (700 MHz, methanol-d_4) of compound **S1**; Figure S6: HSQC spectrum (700/175 MHz, methanol-d_4) of compound **S1**; Figure S7: Expansion of the sugar region of the HSQC spectrum of compound **S1**; Figure S8: HMBC spectrum (700/175 MHz, methanol-d_4) of compound **S1**; Figure S9: Expansion of the sugar region of the HMBC spectrum of compound **S1**; Figure S10: ROESY-spectrum (700 MHz, methanol-d_4) of compound **S1**; Figure S11: ^{1}H-NMR spectrum (700 MHz, pyridine-d_5) of compound **S2**; Figure S12: Expansion of the ^{1}H-NMR (700 MHz, pyridine-d_5) spectrum of compound **S2**; Figure S13: Expansion of the ^{1}H-NMR (700 MHz, pyridine-d_5) spectrum of compound **S2**; Figure S14: ^{13}C-NMR spectrum (175 MHz, pyridine-d_5) of compound **S2**; Figure S15: COSY spectrum (700 MHz, pyridine-d_5) of compound **S2**; Figure S16: HSQC spectrum (700/175 MHz, pyridine-d_5) of compound **S2**; Figure S17: HMBC spectrum (700/175 MHz, pyridine-d_5) of compound **S2**; Figure S18: Expansion of the sugar region of the HMBC spectrum (700/175 MHz, pyridine-d_5) of compound **S2**; Figure S19: ROESY spectrum (700 MHz, pyridine-d_5) of compound **S2**; Figure S20: ^{1}H-NMR spectrum (700 MHz, methanol-d_4) of compound **S2**; Figure S21: Expansion of the ^{1}H-NMR (700 MHz, methanol-d_4) spectrum of compound **S2**; Figure S22: Expansion of the ^{1}H-NMR (700 MHz, methanol-d_4) spectrum of compound

S2; Figure S23: ^{13}C-NMR spectrum (175 MHz, methanol-d_4) of compound **S2**; Figure S24: COSY spectrum (700 MHz, methanol-d_4) of compound **S2**; Figure S25: HSQC spectrum (700/175 MHz, methanol-d_4) of compound **S2**; Figure S26: HMBC spectrum (700/175 MHz, methanol-d_4) of compound **S2**; Figure S27: ROESY spectrum (700 MHz, pyridine-d_5) of compound **S2**; Figure S28: HRESIMS of **S1** in negative mode; Figure S29: HRESIMS of **S1** in positive mode; Figure S30: HRESIMS of **S2** in negative mode; Figure S31: HRESIMS of **S2** in positive mode. Figure S32: GC chromatogram of the monosaccharides of S1. Figure S33: GC chromatogram of the monosaccharides of S2.

Author Contributions: Conceptualization, I.K., M.K.-B. and F.B.; methodology, I.S., A.S., O.K., F.B., M.K.-B. and I.K.; software, A.S., O.K. and F.B.; validation, A.S., O.K. and F.B.; formal analysis, A.S., O.K., F.B. and M.K.-B.; investigation, I.S., A.S., O.K., F.B., M.K.-B. and I.K.; resources, I.S., A.S., O.K., F.B., M.K.-B. and I.K.; data curation, A.S., O.K., F.B., M.K.-B. and I.K.; writing—original draft preparation, I.S., A.S., O.K., F.B. and I.K.; writing—review and editing, A.S., O.K., F.B. and I.K.; supervision, O.K., F.B. and I.K.; project administration, I.K.; funding acquisition, I.K. All authors have read and agreed to the published version of the manuscript.

Funding: This study is financed by the European Union-NextGenerationEU, through the National Recovery and Resilience Plan of the Republic of Bulgaria, project № BG-RRP-2.004-0004-C01.

Institutional Review Board Statement: The animal study protocol was approved by the Institutional Ethics Committee (KENIMUS) of Medical University of Sofia (7338/11.2021) and from the Bulgarian Food Safety Agency at the Ministry of Agriculture and Food (№ 200/2021).

Informed Consent Statement: Not applicable.

Data Availability Statement: Data connected with this study are freely available from the corresponding author, upon reasonable written request. The data are not publicly available due to the large volume of the files.

Conflicts of Interest: The authors declare no conflict of interest.

References

1. Asyov, B.; Petrova, A.; Dimitrov, D.; Vasilev, R. *Conspectus of the Bulgarian Vascular Flora. Distribution Maps and Floristic Elements*; Bulgarian Biodiversity Foundation: Sofia, Bulgaria, 2012.
2. Valev, S.A. *Astragalus* L. In *Flora Republicae Popularis Bulgaricae*; Yordanov, D., Ed.; Aedibus Academiae Scientiarum Bulgaricae: Sofia, Bulgaria, 1976; Volume 6, p. 167.
3. Nikolov, S.D. (Ed.) *Specialized Encyclopedia of Medicinal Plants in Bulgaria*; Labour Publishing House: Sofia, Bulgaria, 2006.
4. Krasteva, I.; Shkondrov, A.; Ionkova, I.; Zdraveva, P. Advances in Phytochemistry, Pharmacology and Biotechnology of Bulgarian *Astragalus* Species. *Phytochem. Rev.* **2016**, *15*, 567–590. [CrossRef]
5. Elenga, P.A.; Nikolov, S.; Panova, D. Triterpene Glycosides from *Astragalus glycyphyllos* L. a New Natural Compound of the Overgrowing Parts of *Astragalus glycyphyllos* L. *Pharmazie* **1987**, *42*, 422–423.
6. Elenga, P.A.; Nikolov, S.; Panova, D. Triterpene Glycosides and Sterols from *Astragalus glycyphyllos* L. *Pharmazie* **1986**, *41*, 41–42.
7. Linnek, J.; Mitaine-Offer, A.C.; Miyamoto, T.; Lacaille-Dubois, M.A. Two Cycloartane-Type Glycosides from the Roots of *Astragalus glycyphyllos*. *Planta Med.* **2008**, *74*, PB141. [CrossRef]
8. Shkondrov, A.; Krasteva, I.; Bucar, F.; Kunert, O.; Kondeva-Burdina, M.; Ionkova, I. A New Tetracyclic Saponin from *Astragalus glycyphyllos* L. and Its Neuroprotective and HMAO-B Inhibiting Activity. *Nat. Prod. Res.* **2020**, *34*, 511–517. [CrossRef]
9. Shkondrov, A.; Simeonova, R.; Kondeva-Burdina, M.; Vitcheva, V.; Krasteva, I. Study to Evaluate the Antioxidant Activity of *Astragalus glycyphyllos* Extract in Carbon Tetrachloride-Induced Oxidative Stress in Rats. *Eur. J. Med. Plants* **2015**, *7*, 59–66. [CrossRef]
10. Georgieva, A.; Popov, G.; Shkondrov, A.; Toshkova, R.; Krasteva, I.; Kondeva-Burdina, M.; Manov, V. Antiproliferative and Antitumour Activity of Saponins from *Astragalus glycyphyllos* on Myeloid Graffi Tumour. *J. Ethnopharmacol.* **2021**, *267*, 113519. [CrossRef]
11. Anastassova, N.; Aluani, D.; Hristova-Avakumova, N.; Tzankova, V.; Kondeva-Burdina, M.; Rangelov, M.; Todorova, N.; Yancheva, D. Study on the Neuroprotective, Radical-Scavenging and MAO-B Inhibiting Properties of New Benzimidazole Arylhydrazones as Potential Multi-Target Drugs for the Treatment of Parkinson's Disease. *Antioxidants* **2022**, *11*, 884. [CrossRef]
12. Wakefield, I.D.; Pollard, C.; Redfern, W.S.; Hammond, T.G.; Valentin, J.-P. The Application of in Vitro Methods to Safety Pharmacology. *Fundam. Clin. Pharmacol.* **2002**, *16*, 209–218. [CrossRef]
13. Salari, S.; Bagheri, M. In Vivo, in Vitro and Pharmacologic Models of Parkinson's Disease. *Physiol. Res.* **2019**, *68*, 17–24. [CrossRef]
14. Kondeva-Burdina, M.; Krasteva, I.; Popov, G.; Manov, V. Neuroprotective and Antioxidant Activities of Saponins' Mixture from *Astragalus glycyphylloides* in a Model of 6-Hydroxydopamine-Induced Oxidative Stress on Isolated Rat Brain Synaptosomes. *Pharmacia* **2019**, *66*, 233–236. [CrossRef]
15. Kasabova-Angelova, A.; Kondeva-Burdina, M.; Mitkov, J.; Georgieva, M.; Tzankova, V.; Zlatkov, A. Neuroprotective and MAOB Inhibitory Effects of a Series of Caffeine-8-Thioglycolic Acid Amides. *Braz. J. Pharm. Sci.* **2020**, *56*, e18255. [CrossRef]

16. Reznicek, G.; Susman, O.; Böhm, K. Bestimmung Der Reihenzugehörigkeit von Monosacchariden Aus Pflanzlichen Glykosiden Mittels GC—MS. *Sci. Pharm* **1993**, *61*, 35–45.
17. Taupin, P.; Zini, S.; Cesselin, F.; Ben-Ari, Y.; Roisin, M.-P. Subcellular Fractionation on Percoll Gradient of Mossy Fiber Synaptosomes: Morphological and Biochemical Characterization in Control and Degranulated Rat Hippocampus. *J. Neurochem.* **1994**, *62*, 1586–1595. [CrossRef] [PubMed]
18. Alpay, M.; Dulger, G.; Sahin, I.E.; Dulger, B. Evaluating Antimicrobial and Antioxidant Capacity of Endemic Phlomis Russeliana from Turkey and Its Antiproliferative Effect on Human Caco-2 Cell Lines. *An. Acad. Bras. Cienc.* **2019**, *91*, e20180404. [CrossRef]
19. Mungarro-Menchaca, X.; Ferrera, P.; Morán, J.; Arias, C. β-Amyloid Peptide Induces Ultrastructural Changes in Synaptosomes and Potentiates Mitochondrial Dysfunction in the Presence of Ryanodine. *J. Neurosci. Res.* **2002**, *68*, 89–96. [CrossRef] [PubMed]
20. Robyt, J.F.; Ackerman, R.J.; Chittenden, C.G. Reaction of Protein Disulfide Groups with Ellman's Reagent: A Case Study of the Number of Sulfhydryl and Disulfide Groups in Aspergillus Oryzae α-Amylase, Papain, and Lysozyme. *Arch. Biochem. Biophys.* **1971**, *147*, 262–269. [CrossRef]
21. Shirani, M.; Alizadeh, S.; Mahdavinia, M.; Dehghani, M.A. The Ameliorative Effect of Quercetin on Bisphenol A-Induced Toxicity in Mitochondria Isolated from Rats. *Environ. Sci. Pollut. Res.* **2019**, *26*, 7688–7696. [CrossRef]
22. Ravindranath, V.; Anandatheerthavarada, H.K. Preparation of Brain Microsomes with Cytochrome P450 Activity Using Calcium Aggregation Method. *Anal. Biochem.* **1990**, *187*, 310–313. [CrossRef]
23. Mansuy, D.; Sassi, A.; Dansette, P.M.; Plat, M. A New Potent Inhibitor of Lipid Peroxidation in Vitro and in Vivo, the Hepatoprotective Drug Anisyldithiolthione. *Biochem. Biophys. Res. Commun.* **1986**, *135*, 1015–1021. [CrossRef]
24. Cunha, M.P.; Lieberknecht, V.; Ramos-Hryb, A.B.; Olescowicz, G.; Ludka, F.K.; Tasca, C.I.; Gabilan, N.H.; Rodrigues, A.L.S. Creatine Affords Protection against Glutamate-Induced Nitrosative and Oxidative Stress. *Neurochem. Int.* **2016**, *95*, 4–14. [CrossRef] [PubMed]
25. Bautista-Aguilera, O.M.; Esteban, G.; Bolea, I.; Nikolic, K.; Agbaba, D.; Moraleda, I.; Iriepa, I.; Samadi, A.; Soriano, E.; Unzeta, M.; et al. Design, Synthesis, Pharmacological Evaluation, QSAR Analysis, Molecular Modeling and ADMET of Novel Donepezil—Indolyl Hybrids as Multipotent Cholinesterase/Monoamine Oxidase Inhibitors for the Potential Treatment of Alzheimer's Disease. *Eur. J. Med. Chem.* **2014**, *75*, 82–95. [CrossRef] [PubMed]
26. Polat, E.; Bedir, E.; Perrone, A.; Piacente, S.; Alankus-Caliskan, O. Triterpenoid Saponins from *Astragalus wiedemannianus* Fischer. *Phytochemistry* **2010**, *71*, 658–662. [CrossRef] [PubMed]
27. Gariboldi, P.; Pelizzoni, F.; Tatò, M.; Verotta, L.; El-Sebakhy, N.; Asaad, A.M.; Abdallah, R.M.; Toaima, S.M. Cycloartane Triterpene Glycosides from *Astragalus trigonus*. *Phytochemistry* **1995**, *40*, 1755–1760. [CrossRef] [PubMed]
28. Wang, H.K.; He, K.; Xu, H.X.; Zhang, Z.L.; Wang, Y.F.; Kikuchi, T.; Tezuka, Y. The structure of astrachrysosid A and the study of 2D-NMR on astrasieversianin XV and 7,2′-dihydroxy-3′,4′-dimethoxy-isoflavane-7-O-beta-D-glycoside. *Yao Xue Xue Bao* **1990**, *25*, 445–450. [PubMed]
29. Zarre, S.; Azani, N. Perspectives in Taxonomy and Phylogeny of the Genus *Astragalus* (Fabaceae): A Review. *Prog. Biol. Sci.* **2013**, *3*, 1–6.
30. Stokes, A.H.; Freeman, W.M.; Mitchell, S.G.; Burnette, T.A.; Hellmann, G.M.; Vrana, K.E. Induction of GADD45 and GADD153 in Neuroblastoma Cells by Dopamine-Induced Toxicity. *Neurotoxicology* **2002**, *23*, 675–684. [CrossRef]
31. Son, I.H.; Park, Y.H.; Lee, S.I.; Yang, H.D.; Moon, H.-I. Neuroprotective Activity of Triterpenoid Saponins from Platycodi Radix Against Glutamate-Induced Toxicity in Primary Cultured Rat Cortical Cells. *Molecules* **2007**, *12*, 1147–1152. [CrossRef]
32. Liu, X.-Y.; Wang, S.; Li, C.-J.; Ma, J.; Chen, F.-Y.; Peng, Y.; Wang, X.-L.; Zhang, D.-M. Dammarane-Type Saponins from the Leaves of Panax Notoginseng and Their Neuroprotective Effects on Damaged SH-SY5Y Cells. *Phytochemistry* **2018**, *145*, 10–17. [CrossRef]
33. Shi, X.; Yu, W.; Yang, T.; Liu, W.; Zhao, Y.; Sun, Y.; Chai, L.; Gao, Y.; Dong, B.; Zhu, L. Panax Notoginseng Saponins Provide Neuroprotection by Regulating NgR1/RhoA/ROCK2 Pathway Expression, in Vitro and in Vivo. *J. Ethnopharmacol.* **2016**, *190*, 301–312. [CrossRef]
34. Ikram, M.; Jo, M.H.; Choe, K.; Khan, A.; Ahmad, S.; Saeed, K.; Kim, M.W.; Kim, M.O. Cycloastragenol, a Triterpenoid Saponin, Regulates Oxidative Stress, Neurotrophic Dysfunctions, Neuroinflammation and Apoptotic Cell Death in Neurodegenerative Conditions. *Cells* **2021**, *10*, 2719. [CrossRef] [PubMed]
35. Liu, X.; Zhang, J.; Wang, S.; Qiu, J.; Yu, C. Astragaloside IV Attenuates the H_2O_2-Induced Apoptosis of Neuronal Cells by Inhibiting α-Synuclein Expression via the P38 MAPK Pathway. *Int. J. Mol. Med.* **2017**, *40*, 1772–1780. [CrossRef] [PubMed]
36. Chan, W.-S.; Durairajan, S.S.K.; Lu, J.-H.; Wang, Y.; Xie, L.-X.; Kum, W.-F.; Koo, I.; Yung, K.K.L.; Li, M. Neuroprotective Effects of Astragaloside IV in 6-Hydroxydopamine-Treated Primary Nigral Cell Culture. *Neurochem. Int.* **2009**, *55*, 414–422. [CrossRef] [PubMed]
37. Shkondrov, A.; Krasteva, I.; Bucar, F.; Kunert, O.; Kondeva-Burdina, M.; Ionkova, I. Flavonoids and Saponins from Two Bulgarian *Astragalus* Species and Their Neuroprotective Activity. *Phytochem. Lett.* **2018**, *26*, 44–49. [CrossRef]
38. Öllinger, K.; Brunk, U.T. Cellular Injury Induced by Oxidative Stress Is Mediated through Lysosomal Damage. *Free Radic. Biol. Med.* **1995**, *19*, 565–574. [CrossRef]
39. O'Donnell, V.; Burkitt, M.J. Mitochondrial Metabolism of a Hydroperoxide to Free Radicals in Human Endothelial Cells: An Electron Spin Resonance Spin-Trapping Investigation. *Biochem. J.* **1994**, *304*, 707–713. [CrossRef]

metabolites

MDPI

Article

Phragmanthera austroarabica A.G.Mill. and J.A.Nyberg Triggers Apoptosis in MDA-MB-231 Cells In Vitro and In Vivo Assays: Simultaneous Determination of Selected Constituents

Marwa S. Goda [1,2,†], Sameh S. Elhady [3,*,†], Mohamed S. Nafie [4], Hanin A. Bogari [5], Raina T. Malatani [5], Rawan H. Hareeri [6], Jihan M. Badr [1,*] and Marwa S. Donia [1]

1 Department of Pharmacognosy, Faculty of Pharmacy, Suez Canal University, Ismailia 41522, Egypt
2 Department of Pharmacognosy, Faculty of Pharmacy, Galala University, New Galala 43713, Egypt
3 Department of Natural Products, Faculty of Pharmacy, King Abdulaziz University, Jeddah 21589, Saudi Arabia
4 Department of Chemistry, Faculty of Science, Suez Canal University, Ismailia 41522, Egypt
5 Department of Pharmacy Practice, Faculty of Pharmacy, King Abdulaziz University, Jeddah 21589, Saudi Arabia
6 Department of Pharmacology and Toxicology, Faculty of Pharmacy, King Abdulaziz University, Jeddah 21589, Saudi Arabia
* Correspondence: ssahmed@kau.edu.sa (S.S.E.); gehan_ibrahim@pharm.suez.edu.eg (J.M.B.); Tel.: +966-544512552 (S.S.E.); +20-1091332451 (J.M.B.)
† These authors contributed equally to this work.

Citation: Goda, M.S.; Elhady, S.S.; Nafie, M.S.; Bogari, H.A.; Malatani, R.T.; Hareeri, R.H.; Badr, J.M.; Donia, M.S. *Phragmanthera austroarabica* A.G.Mill. and J.A.Nyberg Triggers Apoptosis in MDA-MB-231 Cells In Vitro and In Vivo Assays: Simultaneous Determination of Selected Constituents. *Metabolites* **2022**, *12*, 921. https://doi.org/10.3390/metabo12100921

Academic Editors: Ramona Paltinean and Irina Ielciu

Received: 11 September 2022
Accepted: 27 September 2022
Published: 29 September 2022

Abstract: *Phragmanthera austroarabica* (Loranthaceae), a semi-parasitic plant, is well known for its high content of polyphenols that are responsible for its antioxidant and anti-inflammatory activities. Gallic acid, catechin, and methyl gallate are bioactive metabolites of common occurrence in the family of Loranthaceae. Herein, the concentrations of these bioactive metabolites were assessed using high-performance thin layer chromatography (HPTLC). Methyl gallate, catechin, and gallic acid were scanned at 280 nm. Their concentrations were assessed as 14.5, 6.5 and 43.6 mg/g of plant dry extract, respectively. *Phragmanthera austroarabica* extract as well as the three pure compounds were evaluated regarding the cytotoxic activity. The plant extract exhibited promising cytotoxic activity against MDA-MB-231 breast cells with the IC_{50} value of 19.8 μg/mL while the tested pure compounds displayed IC_{50} values in the range of 21.26–29.6 μg/mL. For apoptosis investigation, *P. austroarabica* induced apoptotic cell death by 111-fold change and necrosis by 9.31-fold change. It also activated the proapoptotic genes markers and inhibited the antiapoptotic gene, validating the apoptosis mechanism. Moreover, in vivo studies revealed a significant reduction in the breast tumor volume and weight in solid Ehrlich carcinoma (SEC) mice. The treatment of SEC mice with *P. austroarabica* extract improved both hematological and biochemical parameters with amelioration in the liver and kidney histopathology to near normal. Taken together, *P. austroarabica* extract exhibited promising anti-cancer activity through an apoptosis-induction.

Keywords: sustainability of natural resources; *Phragmanthera austroarabica*; HPTLC; apoptosis; MDA-MB-231 cells; drug discovery

1. Introduction

Cancer is considered a major cause of death and a serious hindrance affecting life spans around the world. The remarkable causes of cancer deaths in males are liver, lung, and gastric cancer. On the other hand, breast, lung, and colorectal cancers are the principal causes among females [1,2]. In 2020, global estimations of new cancer cases and cancer deaths were 19.3 and 10 million, respectively [3–5]. The tendency of controlling cancer includes surgery, radiation therapy, or chemotherapy [6]. Natural products present a noteworthy chemical diversity and are considered as a golden mine regarding bioactive metabolites [7]. Between 1981 and 2019, about 25% of the newly approved anti-cancer drugs were naturally-derived [8,9]. The diversity of natural products is the most important factor that inspires the discovery of new anticancer drugs [10]. Among the widely used anticancer therapies are vincristine, etoposide, irinotecan, and paclitaxel. Obviously, camptothecin and taxol are considered the most felicitous examples [11–13]. Until now, their succession has endured as a series of new camptothecin derivatives with better properties [14,15]. Loranthaceae is one of the largest families that comprise about 70 genera and 1000 species. The family of Loranthaceae is largely distributed in a pantropical zone in the Southern Hemisphere and is widely used in complementary and alternative cancer therapy. Among the mistletoes belonging to the family of Loranthaceae is *Phragmanthera austroarabica.* It is broadly distributed in Arabian Peninsula, especially in Saudi Arabia and Yemen [16]. *Phragmanthera austroarabica* is a semi-parasitic plant. It is a perennial green mistletoe that directly binds itself to another plant through a haustorium. A haustorium is a very small modified root which composes a morphological and physiological connection between the parasite and the host [17]. Accordingly, the structure of the plant is mainly made of a stem and leaves. It is characterized by isobilateral mesophyll where the supporting collenchyma occurs below the vascular bundle. Stem axial parenchyma is of two types: paratracheal and diffused apotracheal [18]. For many years, *P. austroarabica* had been used in many regions of Africa and Asia to treat microbial and viral infections [19]. *P. austroarabica* is traditionally used for controlling hyperglycemia in Saudi Arabia [20]. There are no sufficient studies investigating the chemical composition and biological activities of *P. austroarabica.* From data reported in literature, it was found that *n*-butanol fraction of *P. austroarabica* showed a potent cytotoxic activity against cervical HeLa cell line while the methanolic extract exhibited significant cytotoxic activity against liver HepG2 and breast MCF-7 cell lines [21]. The methanolic crude extract of *P. austroarabica* ameliorated seizures via the enhancement of neurons' survival, suppression of necrotic pyknosis, the elevation of glutathione levels, and reduction in malondialdehyde levels in pentylenetetrazole-induced kindling in mice [22]. Our previous studies involved the isolation and identification of a number of anti-inflammatory and antioxidant compounds from *P. austroarabica* which are illustrated in Figure 1 [22–25].

Gallic acid, methyl gallate, and catechin are among the significant constituents of the family of Loranthaceae in general and are the major bioactive metabolites existing in *P. austroarabica* in particular. They are characterized by cytotoxic, antioxidant, and anti-inflammatory activities [24–28]. Many studies declared that there is a clear correlation between antioxidant and cytotoxic activity against proliferative cells [29–32]. Additionally, previous reports on natural products declared the possible correlation between anti-inflammatory potential and cytotoxic effect [33–36]. In parallel, methyl gallate was reported to be a potent antioxidant that inhibits oxidative stress in human adipocytes [35], attenuates doxorubicin-induced cardiotoxicity in rats [35], and suppresses the growth of different types of cancer [36,37]. Methyl gallate and gallic acid exhibit anti-inflammatory properties by blocking the activation of NF-κB [37–40]. Gallic acid has been reported to prevent the development and progression of various types of cancers inducing apoptosis through Janus kinase 2/signal transducer and activator of transcription 3 (JAK2/STAT3) signaling pathways [41,42] while catechin induced caspase-mediated apoptosis targeting phosphatidylinositol-3-kinase and the protein kinase B (PI3K/AKT) downstream pathway [43]. Catechins are strong anti-inflammatory agents that play a crucial role in the improvement of neurodegenerative and liver disease, metabolic, lung, and GIT disorders [44].

To date, there is no information about the effect of *P. austroarabica* extract on an animal model of breast cancer. Additionally, no previous studies focused on the quantification of its major bioactive metabolites. Accordingly, in the present work, we assessed the cytotoxic effect of the methanolic extract of *P. austroarabica* as well as the pure compounds gallic acid, methyl gallate, and catechin, against ovarian (A2780), prostate (PC-3), breast (MDA-MB-231), and lung (A549) cancer cell lines. This study was also oriented to examine the possible effect of the extract on the tumor mass, apoptotic and proapoptotic gene markers, hematological parameters, and histopathological examination in solid Ehrlich carcinoma (SEC) mice. Moreover, HPTLC quantification of the major bioactive metabolites, gallic acid, catechin, and methyl gallate was a topic of interest.

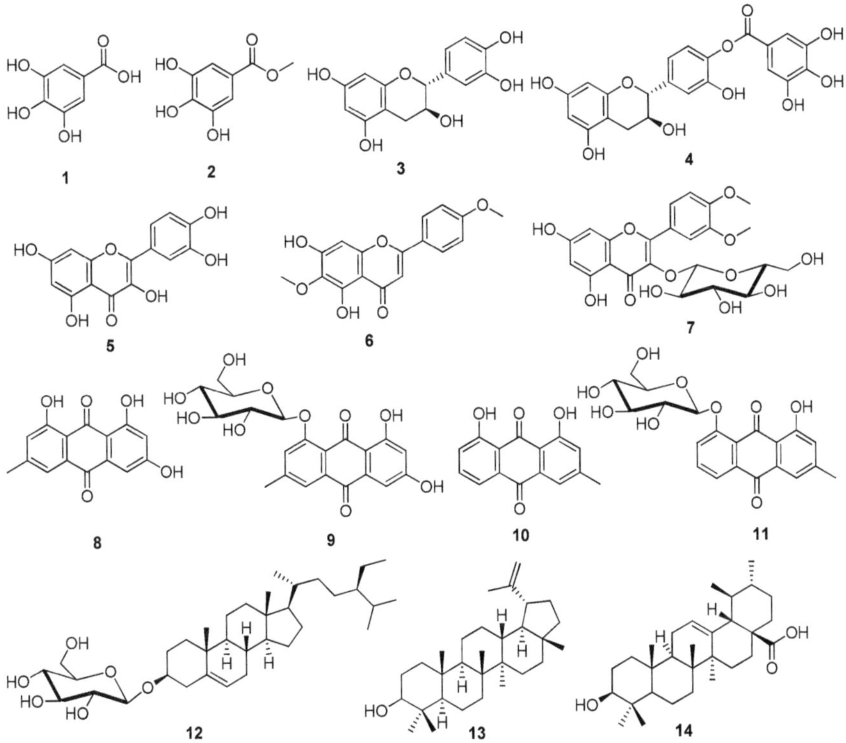

Figure 1. Chemical structures of previously isolated compounds from *P. austroarabica*; (**1**) gallic acid, (**2**) methyl gallate, (**3**) catechin, (**4**) catechin-4′-*O*-gallate, (**5**) quercetin, (**6**) pectolinarigenin, (**7**) dillenetin-3-*O*-glucoside, (**8**) emodin, (**9**) emodin-8-*O*-glucoside, (**10**) chrysophanic acid, (**11**) chrysophanic acid-8-*O*-glucoside, (**12**) β-sitosterol-3-*O*-glucoside, (**13**) lupeol, (**14**) ursolic acid.

2. Materials and Methods

2.1. Plant Material and Extraction Process

Phragmanthera austroarabica was previously collected in 2019 from South Abha, in the Southwestern part of Saudi Arabia. The plant was identified by Dr. Nahed M. Wally, Faculty of Science, King Abdulaziz University, and voucher samples were kept and given the code No. 2019-PU. For the extraction process, 50 g fresh weight of *P. austroarabica* was air-dried, finely ground, and then soaked with methanol (3 × 1000 mL) at 25 °C. The total methanolic extract was concentrated under a vacuum to obtain 3.27 g of crude extract of *P. austroarabica*.

2.2. Determination of Gallic Acid, Catechin, and Methyl Gallate in Methanolic Extract of P. austroarabica Using HPTLC Analysis

2.2.1. Preparation of Standard Solutions of Gallic Acid, Catechin, and Methyl Gallate

Analytical standards of methyl gallate, catechin, and gallic acid (≥98%; Merck™, Darmstadt, Germany) were used. An amount of 10 mg of each standard was used to prepare a methanolic mixture of standard solution at a concentration of 1 mg/mL. Then, the mixed standard solution was kept in a refrigerator till the construction of the calibration curve.

2.2.2. Analysis Conditions and Construction of Calibration Curves

According to the requirements of the International Council on Harmonization (ICH) guidelines [36], different concentrations of the methanolic mixture of gallic acid, catechin, and methyl gallate were applied using A CAMAG® (Muttenz, Switzerland) Linomat V controlled with CAMAG winCats™ software version 1.4.4. (CAMAG, Muttenz, Switzerland). A plate of silica gel 60 F_{254} (Merck™, Darmstadt, Germany) with the dimensions of 20 cm × 10 cm was used. The conditions were adjusted as follows: slit dimension of 6 mm in length and 0.1 mm in width, a distance between tracks of 10.5 mm, the distances from both x-axis and y-axis were 10 mm, and the rate of application used was 15 μL/second. The methanolic solutions were applied in triplicate. The plates were developed in a twin-trough chamber saturated with a solvent mixture consisting of chloroform–methanol–water (3:7:0.3) containing 1 % (*v*/*v*) glacial acetic acid. Development was performed after saturation for 20 min. After 15 min, the plates were air-dried, and then R_f was calculated. The spots were effectively separated with a suitable R_f at 0.53, 0.65, and 0.87 for gallic acid, catechin, and methyl gallate, respectively. Quantification was assessed using a CAMAG TLC Scanner III densitometer and CATS version 4 X software in the absorption mode using a deuterium source. Quantification was performed at different wavelengths of λ = 254, 280, and 366 nm. The wavelength corresponding to the best validation parameters was used for the construction of the calibration curve

2.2.3. Plant Sample Assay

For the determination of gallic acid, catechin, and methyl gallate concentration in the methanolic crude extract of *P. austroarabica*, 100 mg of the dry extract was dissolved, and transferred to a 100 mL volumetric flask. Methanol was added to complete the total volume. Different solutions of both sample and the methanolic mixture of standards were applied under the previously described conditions. Finally, different band areas were recorded.

2.3. Assessment of In Vitro Cytotoxic Activity of P. austroarabica

2.3.1. MTT Assay

Ovarian "A2780", prostate "PC-3", breast "MDA-MB-231", and lung "A549" cancer cell lines were collected from the "National Cancer Institute in Cairo and grown in RPMI-1640 medium L-Glutamine". The cells were grown in "10% fetal bovine serum (FBS) and 1% penicillin-streptomycin". On the second day, cells were cultured in triplicates on a 96-well plate at a density of 5 × 10^4 cells. This was followed by incubation with the crude extract of *P. austroarabica* and samples of pure compounds catechin, gallic acid, and methyl gallate at 0.1, 1, 10, 50, and 100 μg/mL, and cell viability was determined using MTT solution. An ELISA microplate reader (BIO-RAD microplate reader, model iMark, Japan) was used to measure the absorbance. Cell viability was measured by comparing the absorbance at each well to the control group, and the IC_{50} values were recorded [45].

2.3.2. Annexin V/PI Staining and Cell Cycle Flow Cytometry

MDA-MB-231 cells were incubated overnight in 6-well culture plates (3–5 × 10^5 cells/well) and then treated with crude extract of *P. austroarabica* (IC_{50} = 19.8 μg/mL, 48 h). After that, the cells were incubated in a 100 μL solution of Annexin binding buffer "25 mM $CaCl_2$, 1.4 M NaCl, and 0.1 M Hepes/NaOH, pH 7.4" in the dark for 30 min with "Annexin V-FITC solution (1:100) and propidium iodide (PI) at a concentration equivalent to 10 g/mL." The labelled cells were then extracted using the Cytoflex FACS machine (Beckman Coulter Inc., Brea, CA, USA). Data were analyzed using CytExpert software [46,47].

2.3.3. Gene Expression Analysis Using RT-PCR

To investigate the apoptosis-inducing activity in MDA-MB-231 cells, gene expression levels of P53, Bax, PUMA, Caspapses-3,8,9, and Bcl-2 were assessed. MDA-MB-231 cells were treated with crude extract of *P. austroarabica* (IC_{50} = 19.8 μg/mL, 48 h). Then, an RT-PCR reaction was performed following routine work, and the results were given in cycle thresholds (Ct) and ΔΔ Ct for calculating the relative quantities of each gene, as previously described [48].

2.3.4. Assessment of Caspase 3/7 Activity

Caspase 3/7 activity in untreated and treated MDA-MB-231 cells (with *P. austroarabica;* IC_{50} = 19.8 μg/mL, 48 h) were examined using the cell event Caspasepase-3/7 fluorescence method kit (No.10009135, Molecular probes, Eugene, OR, USA) following the detailed procedure in [49].

2.3.5. Autophagy Evaluation Using Acridine Orange Quantitative Assessment

Autophagic cell death in MDA-MB-231 cells treated with *P. austroarabica* (IC_{50} = 19.8 μg/mL, 48 h) was quantitatively assessed using acridine orange lysosomal stain coupled with flow-cytometric analysis following the previously mentioned procedure [50].

2.4. In Vivo Study

2.4.1. Animals

Forty male Swiss albino mice (body weight range: 21–28 g) were used. The mice were kept in a clean and hygienic environment with a normal day/night cycle. Prior to the experiment, mice were subjected to a ten-day period of adaptation to the study conditions. "The experimental protocol was permitted by the Research Ethics Committee (Approval number 202109RA2), Faculty of Pharmacy, Suez Canal University".

2.4.2. Experiment Design

Mice were equally and randomly divided into four groups: "normal control, SEC control, SEC+ *P. austroarabica*, SEC+5-FU". SEC cells (1×10^6 tumor cells/mouse) were implanted subcutaneously into the right thigh of the hind limb, and tumor masses were beginning to appear after ten days of tumor cell inoculation. During the experiment duration, seven doses (50 mg/Kg BW, IP) of the *P. austroarabica* and 5-FU (4.2 mg/Kg BW, IP) were used [46,51]. The weight and volume of the solid tumor masses were measured by a digital Vernier clipper (Tricle Brand, Shanghai, China) using the equation; $V = (L \times W \times W)/2$, where L is the length and W is the width of tumor mass. At the end of the procedure, animals of different groups were sacrificed and blood samples were collected. CBC parameters, liver enzymes ALT, AST, and kidney parameters of urea and creatinine were measured. Kidney and liver tissues were stained with Hematoxylin and Eosin. A light microscope was used for the histopathological examinations.

3. Results and Discussion

3.1. Simultaneous Determination of Gallic Acid, Catechin, and Methyl Gallate in a Methanolic Crude Extract of P. austroarabica Using High-Performance Thin Layer Chromatography (HPTLC)

The HPTLC quantification of the major bioactive metabolites, gallic acid, catechin, and methyl gallate, was performed. The spectra of different concentrations of a standard mixture of gallic acid, catechin, and methyl gallate are represented in Figure 2. Upon scanning the chromatographic plate at multi-wavelengths, it was found that λ = 280 nm is corresponding to the highest sensitivity; accordingly, the assessment method and validation parameters (e.g., linearity, precision, accuracy, stability, and limits of quantification and detections) were assessed at 280 nm.

3.1.1. Linearity

A linear relationship for each standard was detected over the concentration range recorded in Table 1. The correlation coefficients (R^2) were 0.99, and the linear regression equations for gallic acid, catechin, and methyl gallate are expressed in Table 1.

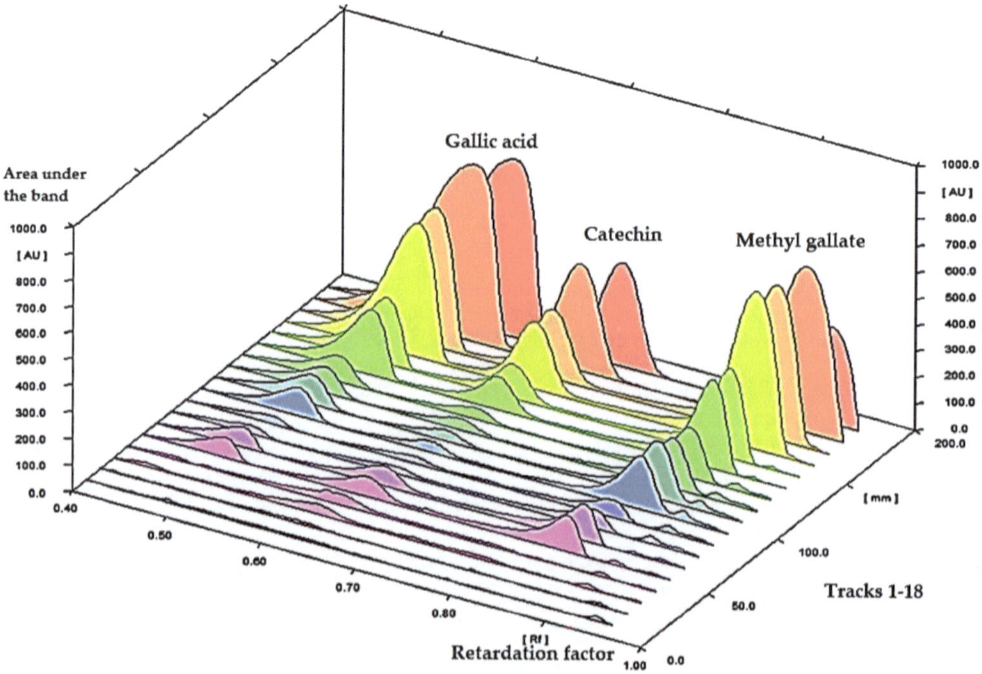

Figure 2. HPTLC spectrum of a standard mixture of gallic acid, catechin, and methyl gallate scanned at λ = 280 nm.

Table 1. Validation parameters of the method for the estimation of methyl gallate, catechin, and gallic acid using HPTLC with densitometer scanning.

Validation Parameters	Scanned at λ 280 nm		
	Methyl Gallate	Catechin	Gallic Acid
Linearity range (µg/band)	0.8–9	0.4–9	0.8–9
Correlation coefficient (R^2)	0.99	0.99	0.99
Regression equation	Y = 5926.9X + 1684.1	Y = 2175.9X + 1352.9	Y = 6831.9X − 1984.2
Limit of detection (µg/band)	0.1	0.09	0.11
Limit of quantification	0.31	0.29	0.34
System precision [%RSD]	3.72	3.27	2.74
Method precision [%RSD]	3.29	1.97	3.91
% Recovery	95.89	95.58	96.64
Conc. (mg/g extract)	14.5	6.5	43.6

Y: band area; X: concentrations of gallic acid, catechin, or methyl gallate; RSD: relative standard deviation.

3.1.2. System Precision

The system precision was verified by the determination of the band area corresponding to a standard mixture at a concentration of 1 mg/mL, applied in triplicate. The value of percent-relative standard deviation (%RSD) for each standard was calculated (Table 1).

3.1.3. Method Precision

The method precision was checked by the injection of the plant extract under the same procedure as described above. Five measurements were performed. The low value of % relative standard deviation (RSD) revealed the method's precision (Table 1).

3.1.4. Accuracy

The accuracy was assessed under the same conditions by scanning a fortified sample; the known sample was combined with a certain concentration of a standard mixture of gallic acid, catechin, and methyl gallate solution (Table 1).

3.1.5. Limits of Detection and Quantification

The limit of detection was computed through the equation of 3 σ/S while the limit of quantification was assessed through the equation of 10 σ/S. The σ refers to the standard deviation of the response and S refers to the slope of the calibration curve (Table 1).

3.1.6. Analytical Solution Stability

The stability of a standard mixture was assured by comparing the experimental results performed with those at ambient temperature for 2 days and after storage of them at 4 °C for 10 days.

3.1.7. Sample Analysis

The suggested method was utilized for the determination of the gallic acid, catechin, and methyl gallate in *P. austroarabica* which was applied to bands in triplicate. The bands were scanned at λ 280 nm (Figure 3). The concentrations were found to be 14.5, 6.5, and 43.6 mg/g of plant dry extract for methyl gallate, catechin, and gallic acid, respectively.

3.2. In Vitro Activities P. austroarabica Extract

3.2.1. Cytotoxicity of *P. austroarabica* against PC-3, MDA-MB-231, A2780, and A549 Cancer Cell Lines Using MTT Assay

The crude extract of *P. austroarabica,* as well as the pure compounds of gallic acid, methyl gallate, and catechin, were screened for their cytotoxicity against a panel of cancer cell lines, such as prostate "PC-3", breast "MDA-MB-231", ovarian "A2780", and lung "A549" cell lines using MTT assay. According to the US NCI (National Cancer Institute) plant screening program guidelines, a crude extract is considered to have in vitro cytotoxic activity if the IC_{50} value after incubation between 48 and 72 h is less than 20 µg/mL. While the pure compounds have in vitro cytotoxic activity if the IC_{50} value is less than 4 µg/mL [52,53]. Cytotoxicity results, as seen in Table 2, showed that *P. austroarabica* extract exhibited promising cytotoxicity against MDA-MB-231 cells with IC_{50} value of 19.8 µg/mL compared to the tested samples of pure compounds with IC_{50} values range (15.36–35.7 µg/mL). Additionally, *P. austroarabica* extract was non-toxic against normal breast cells MCF-10A with IC_{50} values of 47.26 µg/mL. These results agreed with the reported antioxidant activity of *P. austroarabica* extract, and this may elucidate the cytotoxic activity against cancer cells [24]. Additionally, the more prominent effects exhibited by *P. austroarabica* extract could be attributed to the combined activity of its major and other chemical constituents. The plant extract accumulates a number of active metabolites that were previously reported to possess antiproliferative effects against breast cancer cells. For example, β-sitosterol 3-*O*-glucoside was reported to suppress tumor growth through upregulating miR-10a expression as well as deactivating the PI3K-Akt-signaling pathway. Accordingly, it was recommended as a significant breast anticancer agent [54]. Lupeol inhibited the invasion of MDA-MB-231 cells through the suppression of the protein expression of COX-2, MMP-2, and MMP-9 [55]. Ursolic acid a showed proliferative inhibitory effect through downregulating Nrf2 via the Keap1/Nrf2 pathway and EGFR/Nrf2 pathway in MDA-MB-231 cells [56]. Both the anthraquinones emodin and chrysophanic acid were reported to inhibit invasion and metastasis of human breast cancer MDA-MB-231 cells [57,58]. The well-known flavonoid quercetin was reported to induce apoptosis in breast cancer cells by the suppression of Twist via the p38MAPK pathway [59], and in another study, it revealed inhibition of calcium-dependent urokinase activity and, hence, proved to be an effective antimetastatic agent [60]. Finally, catechin 4- gallate, when combined with 4-hydroxytamoxifen, disclosed synergistic cytotoxicity in MDA-MB-231 cells [61,62].

(a)

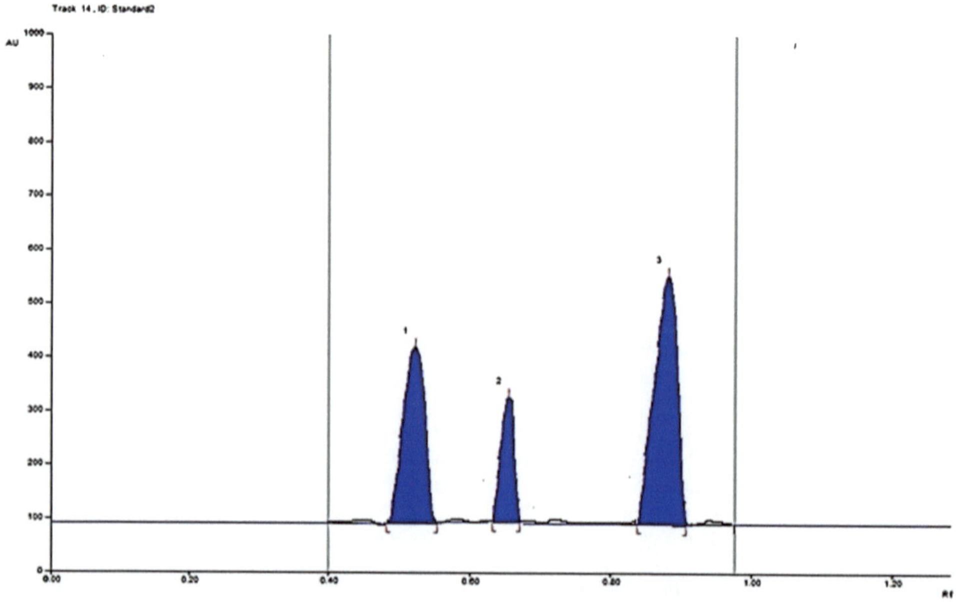

(b)

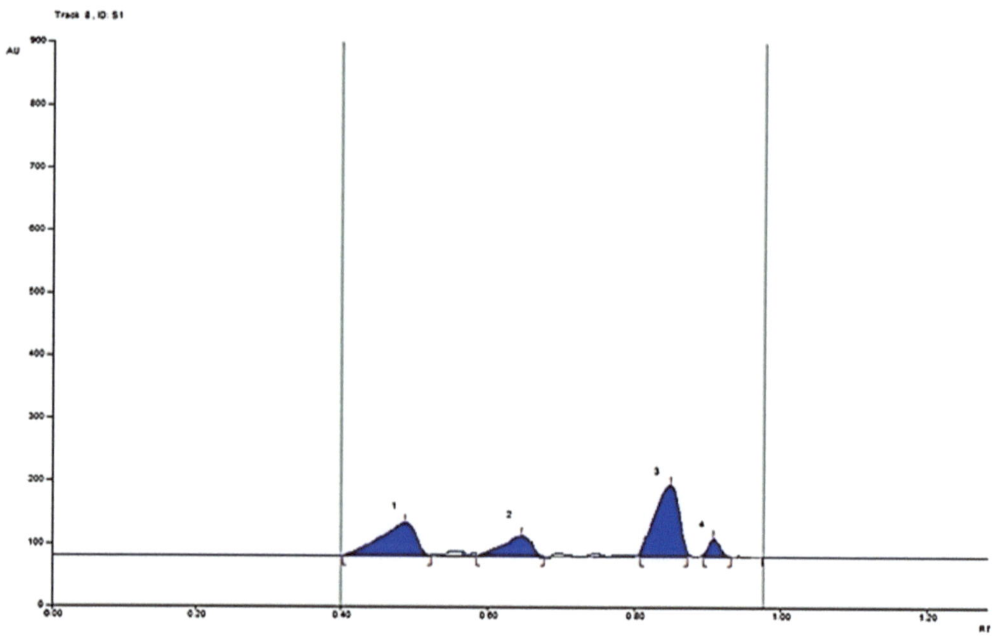

Figure 3. (**a**) HPTLC chromatogram of 1 µg/band of gallic acid, catechin, and methyl gallate scanned at λ = 280 nm; (**b**) HPTLC chromatogram of 10 µg per band of *P. austroarabica* extract scanned at λ = 280 nm.

Table 2. Cytotoxic activity of crude extract of *P. austroarabica*, gallic acid, methyl gallate, and catechin against panel of cancer cell lines using MTT assay.

Sample	IC_{50} * (µg/mL)				
	PC-3	**MDA-MB-231**	**A2780**	**A549**	**Normal MCF-10A**
P. austroarabica	36.9 ± 1.89	19.8 ± 0.76	38.6 ± 2.03	NA	47.26 ± 2.03
Gallic acid	22.15 ± 1.39	29.6 ± 0.96	33.41 ± 1.21	17.82 ± 0.92	≥50
Methyl gallate	15.36 ± 1.02	21.98 ± 0.95	35.7 ± 1.97	19.6 ± 2.12	≥50
Catechin	19.21 ± 0.78	21.26 ± 1.31	31.8 ± 2.19	26.39 ± 2.01	≥50
5-FU	2.64 ± 0.35	1.59 ± 0.36	3.64 ± 0.45	4.26 ± 0.64	49.65 ± 2.04

* IC_{50} are expressed as mean ± SD of three independent trials and were calculated using GraphPad prism.

3.2.2. *P. austroarabica* Treatment Induced Apoptosis in MDA-MB-231 Cells

Cytotoxic activity of *P. austroarabica* against MDA-MB-231 cells was investigated for its mechanism for apoptosis-induction using Annexin V/PI staining. As seen in Figure 4a, the extract induced total apoptotic cell death by 63.23% compared to 0.57% in the untreated control cells. Additionally, it induced necrotic cell death by 24.96% compared to 2.68%, these findings showed that the *P. austroarabica* treatment induced cell death by apoptosis by 111-fold change and necrosis by 9.31-fold change. DNA content-aided cell cycle analysis was carried out to determine the cell population at each phase. This step aimed to determine the cell cycle at which the cell proliferation was arrested. As seen in Figure 4b, *P. austroarabica* treatment significantly increased the cell population at G2/M and S-phases by 2.46-fold and 2.5-fold change compared to untreated control, so its treatment significantly induced the cell cycle arrest at G2/M and S-phases in MDA-MB-231 cells.

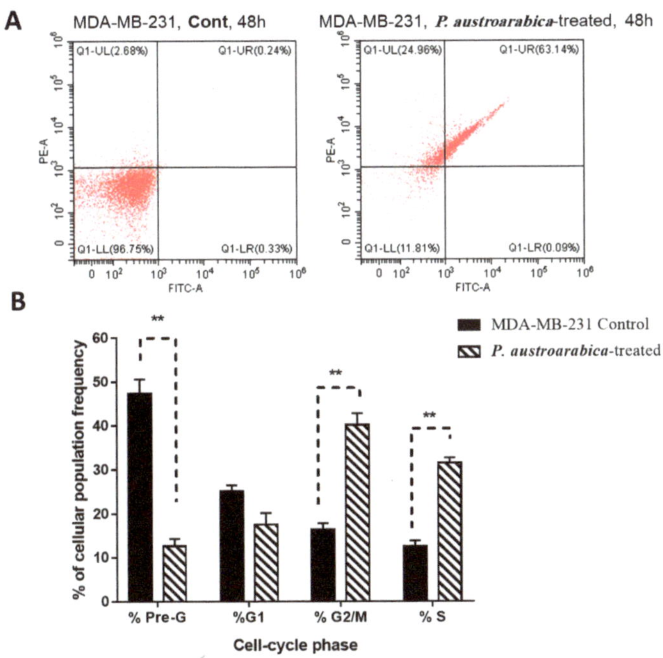

Figure 4. Flow cytometry analysis of MDA-MB-231 untreated and treated cells with *P. austroarabica* crude extract (IC_{50} = 19.8 µg/mL, 48 h). (**A**) Annexin V/PI staining for apoptosis/Necrosis assessment; (**B**) DNA content-cell cycle analysis. "Values are expressed as Mean ± SD of three independent values. ** ($p \leq 0.001$) significantly different between treated and control using unpaired t-test in GraphPad prism".

3.2.3. P. austroarabica Treatment Affected Gene Expression Analysis of Apoptosis-Related Genes

To further validate the apoptosis induction of *P. austroarabica* treatment in MDA-MB-231 cells, RT-PCR for the apoptosis-related genes was performed in the untreated and treated cells. As seen in Table 3, *P. austroarabica* treatment upregulated the P53 gene by 9.28-fold, the PUMA gene by 9.39-fold, the Bax gene by 7.39-fold, and caspases 3, 8, 9 by 10.36, 20.8, 12.39-fold, respectively. In contrast, it downregulated the Bcl-2 gene by 0.53-fold. This behavior of apoptosis-induction in MDA-MB-231 cells upon treatment agreed with the routine results [63,64] of proving apoptosis induction.

Table 3. Fold of change of apoptosis-related genes in untreated control and treated MDA-MB-231 with *P. austroarabica* crude extract (IC_{50} = 19.8 µg/mL, 48 h).

Sample	Gene Expression (Fold Change) *						
	Pro-Apoptotic Genes						Anti-Apoptotic Gene
	P53	PUMA	Bax	Casp-3	Casp-8	Casp-9	Bcl-2
Cont./ MDA-MB-231	1						
P. austroarabica MDA-MB-231	9.28 ± 1.38	9.39 ± 1.76	7.39 ± 1.38	10.36 ± 1.98	2.08 ± 0.98	12.39 ± 1.67	0.53 ± 0.01

* Fold of change is calculated by $2^{-\Delta\Delta CT}$, where ΔΔCT is the difference between mean values of genes CT values in the treated and control groups.

3.2.4. P. austroarabica Treatment Activated Caspase 3/7 Activity

Apoptosis is a form of cell death programmed by the family of cysteine protease. In response to the various cell death stimuli, a large irreversible proteolytic cascade is triggered and subsequently generated. The effector caspases activity has been evaluated to determine the apoptotic pathway initiated by the tested compounds, using the "Cell Event™ Caspase-3/7 Green Detection kit". As seen in Figure 5, the *P. austroarabica* treatment induced a higher percentage of MDA-MB-231 cancer cell death via apoptosis with a percentage of 13.04% than the untreated cells at 1.02%. Hence, apoptotic cell death was triggered by activation of effectors 3 and 7 caspases in mDA-MB-231 cells

3.2.5. P. austroarabica Induced MDA-MB-231 Cell Death through Autophagy

As seen in Figure 6, cell death in MDA-MB-231 cells through autophagy was assessed using acridine orange lysosomal stain. *P. austroarabica* induced autophagy by 1.5-fold, the cell population in the treated group was 63,456 cells compared to 41,775 cells in the untreated control.

3.3. In Vivo Study of P. austroarabica against Solid Ehrlich Carcinoma

Solid weight masses and volumes of solid Ehrlich carcinoma (SEC) were measured at the end of the experiment. As seen in Table 4, an increase in solid tumor weight of 203 mg was observed via tumor proliferation. Upon treatment with *P. austroarabica* extract and 5-FU, a significant reduction in the solid tumor mass of 96.8 mg and 78.3 mg, respectively, was noticed. Accordingly, treatments with *P. austroarabica* extract and Fluorouracil (5-FU) significantly inhibited tumor proliferation by 54% and 64%, respectively, by reducing the tumor volume from 357 mm^3 in untreated control to 169 mm^3 and 126 mm^3.

In SEC-bearing mice, hematological and biochemical parameters are summarized in Table 4. In the SEC control, all CBC parameters were altered, with Hb content and RBCs significantly reduced to 3.69 (g/dL) and 2.19 (10^6/L), respectively. While the WBC count elevated to 6.63 (10^3/L) compared to the normal control. Reduced hemoglobin and RBC levels, as well as an increase in WBC counts, are common side effects of tumor progression [49,65]. Treatment with *P. austroarabica* extract almost restored CBC levels to normal. It increased Hb (7.12 g/dL), RBC (5.01 × 10^6/L), and WBC (4.01 × 10^3/L) levels. Interestingly, our results support those of a previous study [51,66] which illustrated the ameliorative effect in the hematological parameters after treatment with the tested compound.

In terms of biochemical parameters, liver enzymes (ALT and AST) were significantly elevated to 66.51 and 92.5 (U/L), respectively, when compared to normal mice (43.6 and 46.5 U/L), respectively. Treatment with *P. austroarabica* substantially normalized liver enzymes to be 52.4 and 56.1 U/L, respectively, indicating a significant inhibition of hepatocellular toxicity caused by tumor proliferation. Additionally, kidney parameters, urea, and creatinine levels were deteriorated in cancer groups, while *P. austroarabica* treatment retained their levels to be 30.3 and 0.87 (mg/dL), respectively, nearly at normal control. In agreement with hematological and biochemical examinations, as seen in Figure 7, histopathological examinations of liver and kidney tissues exhibited improvement in their structure to be near normal with less hydropic degeneration and inflammation [66]. Taken together with improvement in the tumor volume and weight, treatment of SEC mice with *P. austroarabica* extract improved both hematological and biochemical parameters.

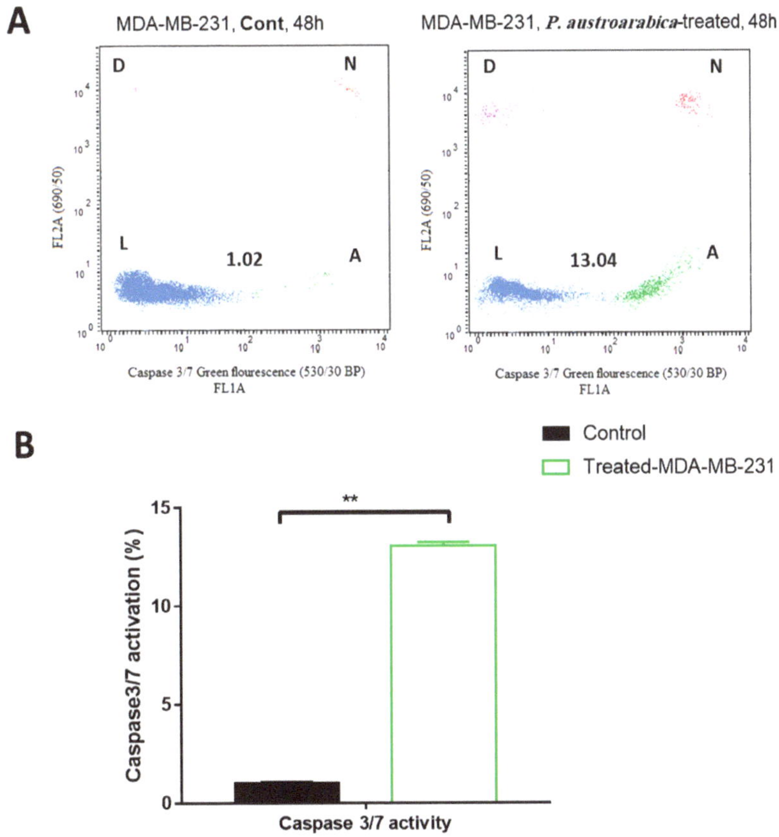

Figure 5. (**A**) Caspase 3/7 inhibitory activity in cancer MDA-MB-231 cells was treated with *P. austroarabica* crude extract (IC_{50} = 19.8 µg/mL, 48 h) using the "CellEvent® Caspase-3/7 Green Flow Cytometry kit", where "L, viable cells; A, apoptotic cells; N, necrotic cells; and D, dead cells"; (**B**) Bar presentation for comparison of apoptotic cancer cells due to active caspases 3/7 of the tested extract. The data are expressed as the mean ± SEM of three independent experiments in triplicate. ** ($p \leq 0.001$) significantly different between treated and control using unpaired t-test in GraphPad prism".

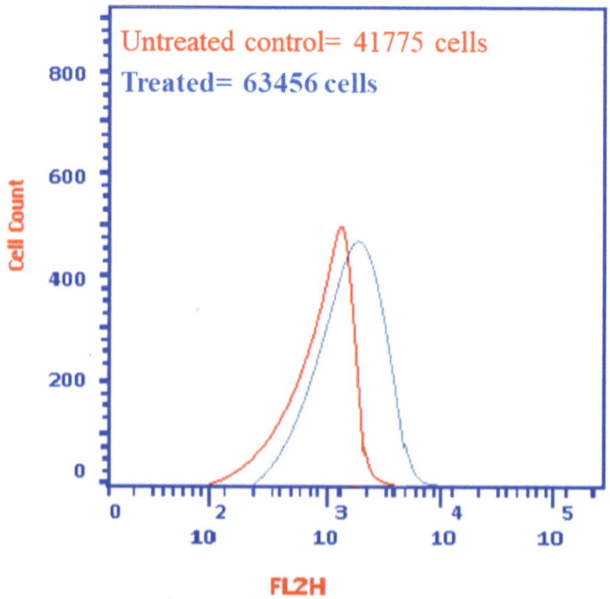

Figure 6. Cell death through autophagy in untreated and treated MDA-MB-231 cells with *P. austroarabica* crude extract (IC_{50} = 19.8 μg/mL, 48 h) using the acridine orange lysosomal stain coupled with the flow cytometric analysis. Red: Negative control (untreated), Blue: treated cells.

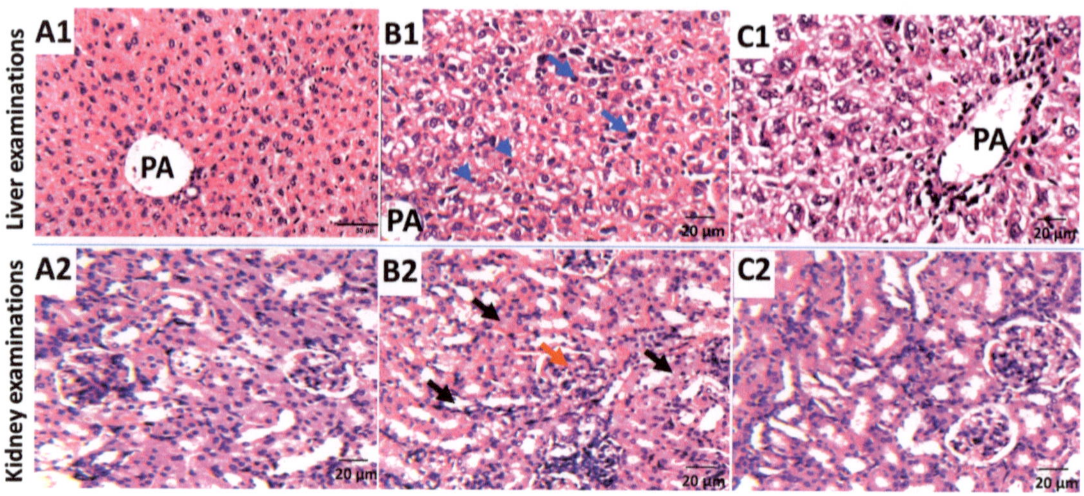

Figure 7. Histopathological examinations of liver and kidney tissues in (**A1**,**A2**) Normal control; (**B1**,**B2**) SEC control; (**C1**,**C2**) *P. austroarabica*-treated SEC group. Portal area (PA); chronic inflammation (blue arrow); hydropic degeneration (Arrow heads). Normal glomeruli (Black arrows), urinary spaces, and tubules (Red arrows).

Table 4. Hematological and biochemical parameters in the normal, untreated, and treated SEC-bearing mice.

	Parameters	Normal Control	SEC Control	SEC + *P. austroarabica*	SEC + 5-FU
Anti-tumor potentiality	Tumor weight (mg)	–	203.6 ± 4.26	96.8 ± 2.34	78.3 ± 2.38
	Tumor volume (mm^3)	–	356.9 ± 22.3	168.8 ± 19.8	126.2 ± 18.6
	Tumor inhibition ratio (TIR%)	–	–	54.26 ± 1.36	64.7 ± 1.65
Hematological parameters	Hb (g/dL)	8.16 ± 0.67	3.69 * ± 0.6	7.12 # ± 0.64	7.89 # ± 0.54
	RBC's count ($\times 10^6/\mu L$)	5.98 ± 0.56	2.19 * ± 0.54	5.01 # ± 0.56	5.21 # ± 0.44
	WBC's count ($\times 10^3/\mu L$)	3.27 ± 0.34	6.63 * ± 0.41	4.01 # ± 0.55	3.69 # ± 0.69
Liver and kidney parameters	ALT (I/U)	43.3 ± 1.23	66.5 * ± 1.99	52.4 # ± 1.4	48.5 # ± 1.7
	AST (I/U)	46.5 ± 0.78	92.6 * ± 1.45	56.1 # ± 2.0	50.8 # ± 1.5
	Urea (mg/dL)	23.2 ± 1.06	41.3 * ± 1.01	30.3 # ± 1.36	30.3 # ± 1.01
	Creatinine (mg/dL)	0.76 ± 0.02	1.01 ± 0.17	0.87 ± 0.01	0.64 ± 0.06

Mean ± SEM values of mice in each group (n = 6). * Values are significantly different ($p \leq 0.05$) between SEC control and normal group, while # values are significantly different ($p \leq 0.05$) between Treated SCE and untreated SEC mice using un-paired test in GraphPad prism.

4. Conclusions

In conclusion, the present study handled the HPTLC quantification of gallic acid, catechin, and methyl gallate. It is a platform for cytotoxic activity of *P. austroarabica* extract against MDA-MB-231 cells with IC_{50} value of 19.8 µg/mL compared to the tested samples of pure compounds with potent cytotoxic activities with IC_{50} values range (15.36–35.7 µg/mL). For apoptosis-investigation, *P. austroarabica* treatment induced cell death by apoptosis by 111-fold change and necrosis by 9.31-fold change, and it activated the proapoptotic genes markers, while it inhibited the antiapoptotic gene. Moreover, in vivo results exhibited inhibition in the tumor volume and weight, and the treatment of SEC mice with *P. austroarabica* extract improved both hematological and biochemical parameters with amelioration in the liver and kidney histopathology being near normal. Taken together, *P. austroarabica* extract exhibited promising anti-cancer activity through apoptosis-induction. Hence, further future work of the formulation of crude extract and purified compounds will be handled and correlated with their biological activity. In addition, semisynthetic derivatives of methyl gallate, catechin, and gallic acid will be developed and assessed as target-oriented chemotherapeutic agents against breast cancer.

Author Contributions: Conceptualization, J.M.B. and S.S.E.; methodology, M.S.G., S.S.E. and M.S.N.; software, M.S.G., H.A.B., R.T.M., R.H.H. and M.S.N.; validation, M.S.G., R.H.H., H.A.B., R.T.M. and M.S.N.; data curation, M.S.G., S.S.E., M.S.N. and M.S.D.; writing—original draft preparation, M.S.G., M.S.N., J.M.B. and M.S.D.; writing—review and editing, M.S.G., M.S.N., S.S.E., R.T.M., J.M.B. and M.S.D.; resources, S.S.E., H.A.B., R.T.M., R.H.H. and J.M.B.; supervision, S.S.E. and J.M.B.; project administration, H.A.B. and S.S.E.; funding acquisition, H.A.B., R.T.M., R.H.H. and S.S.E. All authors have read and agreed to the published version of the manuscript.

Funding: This research was funded by the Deanship of Scientific Research (DSR) at King Abdulaziz University (KAU), Jeddah, Saudi Arabia, under grant number (RG-29-166-43).

Institutional Review Board Statement: Experimental protocol was permitted by the Research Ethics Committee (Approval number 202109RA2), Faculty of Pharmacy, Suez Canal University.

Data Availability Statement: The data presented in this study are available in the article.

Informed Consent Statement: Not applicable.

Acknowledgments: The Deanship of Scientific Research (DSR) at King Abdulaziz University (KAU), Jeddah, Saudi Arabia, has funded this project under grant no. (RG-29-166-43). Therefore, all the authors acknowledge, with thanks, DSR for technical and financial support. The authors thank Nahed Morad Wally, Faculty of Science, King Abdulaziz University, for her taxonomical identification of *Phragmanthera austroarabica*.

Conflicts of Interest: The authors declare no conflict of interest.

References

1. Lowy, D.R.; Collins, F.S. Aiming high-changing the trajectory for cancer. *N. Engl. J. Med.* **2016**, *374*, 1901–1904. [CrossRef] [PubMed]
2. Nagai, H.; Kim, Y.H. Cancer prevention from the perspective of global cancer burden patterns. *J. Thorac. Dis.* **2017**, *9*, 448–451. [CrossRef] [PubMed]
3. Sung, H.; Ferlay, J.; Siegel, R.L.; Laversanne, M.; Soerjomataram, I.; Jemal, A.; Bray, F. Global cancer statistics 2020: GLOBOCAN estimates of incidence and mortality worldwide for 36 cancers in 185 countries. *CA Cancer J. Clin.* **2020**, *71*, 209–249. [CrossRef] [PubMed]
4. Merriel, S.W.D.; Ingle, S.M.; May, M.T.; Martin, R.M. Retrospective cohort study evaluating clinical, biochemical, and pharmacological prognostic factors for prostate cancer progression using primary care data. *BMJ* **2021**, *11*, e044420. [CrossRef] [PubMed]
5. Ganesh, K.; Massagué, J. Targeting metastatic cancer. *Nat. Med.* **2021**, *27*, 34–44. [CrossRef] [PubMed]
6. Roy, A.; Li, S.D. Modifying the tumor microenvironment using nanoparticle therapeutics. *Wiley Interdiscip. Rev. Nanomed. Nanobiotechnol.* **2016**, *8*, 891–908. [CrossRef] [PubMed]
7. Huang, M.; Lu, J.-J.; Ding, J. Natural products in cancer therapy: Past, present and future. *Nat. Prod. Bioprospect.* **2021**, *11*, 5–13. [CrossRef] [PubMed]
8. Huang, M.Y.; Zhang, L.L.; Ding, J.; Lu, J.J. Anticancer drug discovery from Chinese medicinal herbs. *Chin. Med.* **2018**, *13*, 35. [CrossRef] [PubMed]
9. Newman, D.J.; Cragg, G.M. Natural products as sources of new drugs over the nearly four decades from 01/1981 to 09/2019. *J. Nat. Prod.* **2020**, *83*, 770–803. [CrossRef] [PubMed]
10. Demain, A.L.; Vaishnav, P. Natural products for cancer chemotherapy. *Microb. Biotechnol.* **2011**, *4*, 687–699. [CrossRef] [PubMed]
11. Wall, M.E.; Wani, M.C. Camptothecin and taxol: Discovery to clinic–thirteenth Bruce, F. Cain Memorial Award Lecture. *Cancer Res.* **1995**, *55*, 753–760. [PubMed]
12. Wall, M.E. Camptothecin and taxol: Discovery to clinic. *Med. Res. Rev.* **1998**, *18*, 299–314. [CrossRef]
13. Oberlies, N.H.; Kroll, D.J. Camptothecin and taxol: Historic achievements in natural products research. *J. Nat. Prod.* **2004**, *67*, 129–135. [CrossRef] [PubMed]
14. Pommier, Y. Topoisomerase I inhibitors: Camptothecins and beyond. *Nat. Rev. Cancer* **2006**, *6*, 789–802. [CrossRef] [PubMed]
15. Huang, M.; Gao, H.; Chen, Y.; Zhu, H.; Cai, Y.; Zhang, X.; Miao, Z.; Jiang, H.; Zhang, J.; Shen, H.; et al. Chimmitecan, a novel 9-substituted camptothecin, with improved anticancer pharmacologic profiles in vitro and in vivo. *Clin. Cancer Res.* **2007**, *13*, 1298–1307. [CrossRef] [PubMed]
16. Alahdal, F.A.M.; Qashqoosh, M.T.A.; Manea, Y.K.; Salem, M.A.S.; Khan, A.M.T.; Naqvi, S. Eco-friendly synthesis of zinc oxide nanoparticles as nanosensor, nanocatalyst and antioxidant agent using leaf extract of *P. austroarabica*. *OpenNano* **2022**, *8*, 100067. [CrossRef]
17. Tawfik, M.K.; Badr, J.M. Evaluation of hepatoprotective activity of *Plicosepalus acacia* and *Phragmanthera austroarabica* extracts on paracetamol-induced hepatotoxicity in rats. *Wulfenia* **2012**, *19*, 325–337.
18. Nahed, M.; Waly, N.M. Anatomical and statistical analysis of six parasitic Loranthaceae species. *Am. J. Res. Commun.* **2013**, *1*, 317–332.
19. Ibrahim, H.M.; Ajlan, A.A.; Alriany, Y.H.; Al-Gifri, A.N. Correction in *Phragmanthera tiegh.* (Loranthaceae) in the Flora of AlHujariyah—Taiz Governorate, Yemen. *Univ. Aden J. Nat. Appl. Sci.* **2014**, *18*, 449–459.
20. Hanafy, A.; Badr, J.M. Anti-hyperglycaemic effect of *Phragmenthera austroarabica* A.G. Mill.& J.A. Nyberg extract in streptozotocin induced diabetes in rats. *Nat. Prod. Res.* **2014**, *28*, 2351–2354. [CrossRef]
21. Almehdar, H.; Abdallah, H.M.; Osman, A.-M.M.; Abdel-Sattar, E.A. In vitro cytotoxic screening of selected Saudi medicinal plants. *J. Nat. Med.* **2012**, *66*, 406–412. [CrossRef] [PubMed]
22. Aldawsari, H.M.; Eid, B.G.; Neamatallah, T.; Zaitone, S.A.; Badr, J.M. Anticonvulsant and neuroprotective activities of *Phragmanthera austroarabica* extract in pentylenetetrazole-kindled mice. *Evid-Based Complement. Altern. Med.* **2017**, *2017*, 5148219. [CrossRef] [PubMed]
23. Bamane, F.H.; Badr, J.M.; Amin, O.R.M. Antioxidant activities and flavonoid contents of selected plants belonging to Family Loranthaceae. *Afr. J. Biotechnol.* **2012**, *11*, 14380–14385. [CrossRef]
24. Badr, J.M. Chemical constituents of *Phragmanthera austroarabica* A. G. Mill and J. A. Nyberg with potent antioxidant activity. *Pharmacogn. Res.* **2014**, *7*, 335–340. [CrossRef]

25. Badr, J.M.; Shaala, L.A.; Youssef, D.T.A. Loranthin: A new polyhydroxylated flavanocoumarin from *Plicosepalus acacia* with significant free radical scavenging and antimicrobial activity. *Phytochem. Lett.* **2013**, *6*, 113–117. [CrossRef]
26. Correa, L.B.; Pádua, T.A.; Seito, L.N.; Costa, T.E.M.M.; Silva, M.A.; Candéa, A.L.P.; Rosas, E.C.; Henriques, M.G. Anti-inflammatory effect of methyl gallate on experimental arthritis: Inhibition of neutrophil recruitment, production of inflammatory mediators, and activation of macrophages. *J. Nat. Prod.* **2016**, *79*, 1554–1566. [CrossRef]
27. BenSaad, L.A.; Kim, K.H.; Quah, C.C.; Kim, W.R.; Shahimi, M. Anti-inflammatory potential of ellagic acid, gallic acid and punicalagin A&B isolated from *Punica granatum*. *BMC Complement. Altern. Med.* **2017**, *17*, 47. [CrossRef]
28. Fan, F.Y.; Sang, L.X.; Min Jiang, M. Catechins and their therapeutic benefits to inflammatory bowel disease. *Molecules* **2017**, *22*, 484. [CrossRef]
29. Mates, J.M. Effects of antioxidant enzymes in the molecular control of reactive oxygen species toxicology. *Toxicology* **2000**, *153*, 83–104. [CrossRef]
30. Rao, Y.K.; Geethangili, M.; Fang, S.-H.; Tzeng, Y.-M. Antioxidant and cytotoxic activities of naturally occurring phenolic and related compounds: A comparative study. *Food Chem. Toxicol.* **2007**, *45*, 1770–1776. [CrossRef]
31. Pandey, K.B.; Rizvi, S.I. Plant polyphenols as dietary antioxidants in human health and disease. *Oxid. Med. Cell. Longev.* **2009**, *2*, 270–278. [CrossRef] [PubMed]
32. Valko, M.; Leibfritz, D.; Moncol, J.; Cronin, M.T.D.; Mazur, M.; Telser, J. Free radicals and antioxidants in normal physiological functions and human disease. *Int. J. Biochem. Cell. Biol.* **2007**, *39*, 44–84. [CrossRef] [PubMed]
33. Harput, U.S.; Saracoglu, I.; Inoue, M.; Ogihara, Y. Anti-inflammatory and cytotoxic activities of five *Veronica* species. *Biol. Pharm. Bull.* **2002**, *25*, 483–486. [CrossRef] [PubMed]
34. Nemudzivhadi, V.; Masoko, M. In vitro assessment of cytotoxicity, antioxidant, and anti-Inflammatory activities of *Ricinus communis* (Euphorbiaceae) leaf extracts. *Evid. Based Complement. Alternat. Med.* **2014**, *2014*, 625961. [CrossRef] [PubMed]
35. Kurek, J.; Myszkowski, K.; Okulicz-Kozaryn, I.; Kurant, A.; Kamińska, E.; Szulc, M.; Rubiś, B.; Kaczmarek, M.; Mikołajczak, P.; Murias, M. Cytotoxic, analgesic and anti-inflammatory activity of colchicine and its C-10 sulfur containing derivatives. *Sci. Rep.* **2021**, *11*, 9034. [CrossRef]
36. Elhady, S.S.; Abdelhameed, R.F.A.; Mehanna, E.T.; Wahba, A.S.; Elfaky, M.A.; Koshak, A.E.; Noor, A.O.; Bogari, H.A.; Malatani, R.T.; Goda, M.S. Metabolic profiling, chemical composition, antioxidant capacity, and in vivo hepato- and nephroprotective effects of *Sonchus cornutus* in mice exposed to cisplatin. *Antioxidants* **2022**, *11*, 819. [CrossRef]
37. Abdel-Hamed, A.R.; Mehanna, E.T.; Hazem, R.M.; Badr, J.M.; Abo-Elmatty, D.M.; Abdel-Kader, M.S.; Goda, M.S. Plicosepalus acacia extract and its major constituents, methyl gallate and quercetin, potentiate therapeutic angiogenesis in diabetic hind limb ischemia: HPTLC quantification and LC-MS/MS metabolic profiling. *Antioxidants* **2021**, *10*, 1701. [CrossRef]
38. Huang, C.-Y.; Chang, Y.-J.; Wei, P.-L.; Hung, C.-S.; Wang, W. Methyl gallate, gallic acid-derived compound, inhibit cell proliferation through increasing ROS production and apoptosis in hepatocellular carcinoma cells. *PLoS ONE* **2021**, *16*, e0248521. [CrossRef]
39. Lee, H.; Lee, H.; Kwon, Y.; Lee, J.-H.; Kim, J.; Shin, M.-K.; Kim, S.-H.; Bae, H. Methyl gallate exhibits potent antitumor activities by inhibiting tumor infiltration of CD4+CD25+ regulatory T cells. *J. Immunol.* **2010**, *185*, 6698–6705. [CrossRef]
40. Subramanian, A.P.; John, A.A.; Vellayappan, M.V.; Balaji, A.; Jaganathan, S.K.; Supriyanto, E.; Yusof, M. Gallic acid: Prospects and molecular mechanisms of its anticancer activity. *RSC Adv.* **2015**, *5*, 35608–35621. [CrossRef]
41. Tsai, C.-L.; Chiu, Y.-M.; Ho, T.-Y.; Hsieh, C.-T.; Shieh, D.-C.; Lee, Y.-J.; Tsay, G.J.; Wu, Y.-Y. Gallic acid induces apoptosis in human gastric adenocarcinoma cells. *Anticancer Res.* **2018**, *38*, 2057–2067. [PubMed]
42. Zhang, T.; Ma, L.; Wu, P.; Li, W.; Li, T.; Gu, R.; Dan, X.; Li, Z.; Fan, X.; Xiao, Z. Gallic acid has anticancer activity and enhances the anticancer effects of cisplatin in non-small cell lung cancer A549 cells via the JAK/STAT3 signaling pathway. *Oncol. Rep.* **2019**, *41*, 1779–1788. [CrossRef] [PubMed]
43. Sun, H.; Yin, M.; Hao, D.; Shen, Y. Anti-cancer activity of catechin against A549 lung carcinoma cells by induction of cyclin kinase inhibitor P21 and suppression of cyclin E1 and P–AKT. *Appl. Sci.* **2020**, *10*, 2065. [CrossRef]
44. Kim, J.M.; Heo, H.J. The roles of catechins in regulation of systemic inflammation. *Food Sci. Biotechnol.* **2022**, *31*, 957–970. [CrossRef]
45. Mosmann, T. Rapid colorimetric assay for cellular growth and survival: Application to proliferation and cytotoxicity assays. *J. Immunol. Methods* **1983**, *65*, 55–63. [CrossRef]
46. Nafie, M.S.; Boraei, A.T.A. Exploration of novel VEGFR2 tyrosine kinase inhibitors via design and synthesis of new alkylated indolyl-triazole Schiff bases for targeting breast cancer. *Bioorg. Chem.* **2022**, *122*, 105708. [CrossRef]
47. Nafie, M.S.; Kishk, S.M.; Mahgoub, S.; Amer, A.M. Quinoline-based thiazolidinone derivatives as potent cytotoxic and apoptosis-inducing agents through EGFR inhibition. *Chem. Biol. Drug Des.* **2022**, *99*, 547–560. [CrossRef]
48. Abdelhameed, R.F.A.; Elhady, S.S.; Sirwi, A.; Samir, H.; Ibrahim, E.A.; Thomford, A.K.; El Gindy, A.; Hadad, G.M.; Badr, J.M.; Nafie, M.S. *Thonningia sanguinea* extract: Antioxidant and cytotoxic activities supported by chemical composition and molecular docking simulations. *Plants* **2021**, *10*, 2156. [CrossRef]
49. Gad, E.M.; Nafie, M.S.; Eltamany, E.H.; Hammad, M.S.A.G.; Barakat, A.; Boraei, A.T.A. Discovery of new apoptosis-inducing agents for breast cancer based on ethyl 2- amino-4,5,6,7-tetra hydrobenzo[b]thiophene-3-carboxylate: Synthesis, in vitro, and in vivo activity evaluation. *Molecules* **2020**, *25*, 2523. [CrossRef]

50. ElZahabi, H.S.A.; Nafie, M.S.; Osman, D.; Elghazawy, N.H.; Soliman, D.H.; EL-Helby, A.A.H.; Arafa, R.K. Design, synthesis, and evaluation of new quinazolin-4-one derivativesas apoptotic enhancers and autophagy inhibitors with potent antitumor activity. *Eur. J. Med. Chem.* **2021**, *222*, 113609. [CrossRef]
51. Goda, M.S.; Nafie, M.S.; Awad, B.M.; Abdel-Kader, M.S.; Ibrahim, A.K.; Badr, J.M.; Eltamany, E.E. In vitro and in vivo studies of anti-lung cancer activity of *Artemesia judaica* L. crude extract combined with LC-MS/MS metabolic profiling, docking simulation and HPLC-DAD quantification. *Antioxidants* **2021**, *11*, 17. [CrossRef] [PubMed]
52. Ibrahim, R.S.; Seif El-Din, A.A.; Abu-Serie, M.; Abd El Rahman, N.M.; El-Demellawy, M.; Metwally, A.M. Investigation of in vitro cytotoxic and potential anticancer activities of flavonoidal aglycones from Egyptian Propolis. *Rec. Pharm. Biomed. Sci.* **2017**, *2*, 13–20. [CrossRef]
53. Abdelhameed, R.F.A.; Habib, E.S.; Goda, M.S.; Fahim, J.R.; Hassanean, H.A.; Eltamany, E.E.; Ibrahim, A.K.; AboulMagd, A.M.; Fayez, S.; El-kader, A.M.A.; et al. Thalassosterol, a new cytotoxic aromatase inhibitor ergosterol derivative from the Red Sea seagrass *Thalassodendron ciliatum*. *Mar. Drugs* **2020**, *18*, 354. [CrossRef]
54. Xu, H.; Li, Y.; Han, B.; Li, Z.; Wang, B.; Jiang, P.; Zhang, J.; Ma, W.; Zhou, D.; Li, X.; et al. Anti-breast-cancer activity exerted by β-sitosterol-d-glucoside from sweet potato via upregulation of microRNA-10a and via the PI3K-Akt signaling pathway. *J. Agric. Food Chem.* **2018**, *66*, 9704–9718. [CrossRef]
55. Wang, M.; Cui, H.-X.; Sun, C.; Li, G.; Wang, H.-I.; Xia, C.-H.; Wang, Y.-C.; Liu, J.-C. Effect of lupeol on migration and invasion of human breast cancer MDA-MB-231 cells and its mechanism. *Acta Pharm. Sin.* **2016**, *51*, 558–562. [CrossRef]
56. Zhang, X.; TLi, T.; Gong, F.S.; Liu, R.H. Antiproliferative activity of ursolic acid in MDA-MB-231 human breast cancer cells through Nrf2 pathway regulation. *J. Agric. Food Chem.* **2020**, *68*, 7404–7415. [CrossRef]
57. Sun, Y.; Wang, X.; Zhou, Q.; Lu, Y.; Zhang, H.; Chen, Q.; Zhao, M.; Su, S. Inhibitory effect of emodin on migration, invasion and metastasis of human breast cancer MDA-MB-231 cells in vitro and in vivo. *Oncol. Rep.* **2014**, *33*, 338–346. [CrossRef]
58. Ren, L.; Li, Z.; Dai, C.; Zhao, D.; Wang, Y.; Ma, C.; Liu, C. Chrysophanol inhibits proliferation and induces apoptosis through NF-κB/cyclin D1 and NF-κB/Bcl-2 signaling cascade in breast cancer cell lines. *Mol. Med. Rep.* **2018**, *17*, 4376–4382. [CrossRef]
59. Ranganathan, S.; Halagowder, D.; Devaraj, N.; Sivasithambaram, N.D. Quercetin suppresses twist to induce apoptosis in MCF-7 breast cancer cells. *PLoS ONE* **2015**, *22*, e0141370. [CrossRef]
60. Devipriya, S.; Vani, G.; Ramamurthy, N.; Shyamaladevi, C.S. Regulation of intracellular calcium levels and urokinase activity in MDA MB 231 cells by quercetin. *Chemotherapy* **2006**, *52*, 60–65. [CrossRef]
61. Chisholm, K.; Bray, B.J.; Rosengren, R.J. Tamoxifen and epigallocatechin gallate are synergistically cytotoxic to MDA-MB-231 human breast cancer cells. *Anti-Cancer Drugs* **2004**, *15*, 889–897. [CrossRef]
62. Schroder, L.; Marahrens, P.; Koch, J.G.; Heidegger, H.; Vilsmeier, T.; Phan-Brehm, T.; Hofmann, S.; Mahner, S.; Jeschke, U.; Richter, D.U. Effects of green tea, matcha tea and their components epigallocatechin gallate and quercetin on MCF 7 and MDA-MB-231 breast carcinoma cell. *Oncol. Rep.* **2019**, *41*, 387–396. [CrossRef] [PubMed]
63. Eltamany, E.E.; Elhady, S.S.; Ahmed, H.A.; Badr, J.M.; Noor, A.O.; Ahmed, S.A.; Nafie, M.S. Chemical profiling, antioxidant, cytotoxic activities and molecular docking simulation of *Carrichtera annua* DC. (Cruciferae). *Antioxidants* **2020**, *9*, 1286. [CrossRef] [PubMed]
64. Abdelhameed, R.F.A.; Habib, E.S.; Ibrahim, A.K.; Yamada, K.; Abdel-Kader, M.S.; Ibrahim, A.K.; Ahmed, S.A.; Badr, J.M.; Nafie, M.S. Chemical profiling, cytotoxic activities through apoptosis induction in MCF-7 cells and molecular docking of *Phyllostachys heterocycla* bark nonpolar extract. *J. Biomol. Struct. Dyn.* **2021**, *2021*, 1–12. [CrossRef] [PubMed]
65. Nafie, M.S.; Arafa, K.; Sedky, N.K.; Alakhdar, A.A.; Arafa, R.K. Triaryl dicationic DNA minor-groove binders with antioxidant activity display cytotoxicity and induce apoptosis in breast cancer. *Chem.-Biol. Interact.* **2020**, *324*, 109087. [CrossRef]
66. Khalifa, M.M.; Al-Karmalawy, A.A.; Elkaeed, E.B.; Nafie, M.S.; Tantawy, M.A.; Eissa, I.H.; Mahdy, H.A. Topo II Inhibition and DNA Intercalation by New Phthalazine-Based Derivatives as Potent Anticancer Agents: Design, Synthesis, Anti-Proliferative, Docking, and in Vivo Studies. *J. Enzym. Inhib. Med. Chem.* **2022**, *37*, 299–314. [CrossRef]

metabolites

Article

Cytotoxic Metabolites from *Calophyllum tacamahaca* Willd.: Isolation and Detection through Feature-Based Molecular Networking

Elise Gerometta [1], Gaëtan Herbette [2], Elnur Garayev [3], Arnaud Marvilliers [1], Jean-Valère Naubron [2], Carole Di Giorgio [4], Pierre-Eric Campos [1,5], Patricia Clerc [1], Allison Ledoux [6], Michel Frederich [6], Béatrice Baghdikian [3], Isabelle Grondin [1] and Anne Gauvin-Bialecki [1,*]

1 Laboratoire de Chimie et de Biotechnologie des Produits Naturels, Faculté des Sciences et Technologies, Université de La Réunion, Campus du Moufia, 97744 St Denis, France; elise.gerometta@univ-reunion.fr (E.G.); arnaud.marvilliers@univ-reunion.fr (A.M.); pierre-eric.campos@univ-orleans.fr (P.-E.C.); patricia.clerc@univ-reunion.fr (P.C.); isabelle.grondin@univ-reunion.fr (I.G.)
2 Spectropole, FSCM, Centrale Marseille, CNRS, Aix-Marseille Université, Campus de St Jérôme–Service 511, 13397 Marseille, France; gaetan.herbette@univ-amu.fr (G.H.); jean-valere.naubron@univ-amu.fr (J.-V.N.)
3 IMBE, CNRS, IRD, Aix Marseille Université, Faculté de Pharmacie, Service de Pharmacognosie, 13331 Marseille, France; elnur.garayev@univ-amu.fr (E.G.); beatrice.baghdikian@univ-amu.fr (B.B.)
4 IMBE, CNRS, IRD, Aix-Marseille Université, Faculté de Pharmacie, Service de Mutagénèse Environnementale, 13385 Marseille, France; carole.di-giorgio@univ-amu.fr
5 Institut de Chimie Organique et Analytique, UMR 6759, Université d'Orléans–CNRS, Pôle de Chimie, Rue de Chartres, BP6759, CEDEX 2, 45067 Orléans, France
6 Laboratoire de Pharmacognosie, Centre Interfacultaire de Recherche sur le Médicament (CIRM), Université de Liège, Département de Pharmacie, Campus du Sart-Tilman, Quartier Hôpital, B-4000 Liège, Belgium; allison.ledoux@uliege.be (A.L.); m.frederich@uliege.be (M.F.)
* Correspondence: anne.bialecki@univ-reunion.fr

Citation: Gerometta, E.; Herbette, G.; Garayev, E.; Marvilliers, A.; Naubron, J.-V.; Di Giorgio, C.; Campos, P.-E.; Clerc, P.; Ledoux, A.; Frederich, M.; et al. Cytotoxic Metabolites from *Calophyllum tacamahaca* Willd.: Isolation and Detection through Feature-Based Molecular Networking. *Metabolites* **2023**, *13*, 582. https://doi.org/10.3390/metabo13050582

Academic Editor: Marijana Zovko Končić

Received: 29 March 2023
Revised: 14 April 2023
Accepted: 18 April 2023
Published: 23 April 2023

Abstract: Isocaloteysmannic acid (**1**), a new chromanone, was isolated from the leaf extract of the medicinal species *Calophyllum tacamahaca* Willd. along with 13 known metabolites belonging to the families of biflavonoids (**2**), xanthones (**3–5**, **10**), coumarins (**6–8**) and triterpenes (**9**, **11–14**). The structure of the new compound was characterized based on nuclear magnetic resonance (NMR), high-resolution electrospray mass spectrometry (HRESIMS), ultraviolet (UV) and infrared (IR) data. Its absolute configuration was assigned through electronic circular dichroism (ECD) measurements. Compound (**1**) showed a moderate cytotoxicity against HepG2 and HT29 cell lines, with IC_{50} values of 19.65 and 25.68 µg/mL, respectively, according to the Red Dye method. Compounds **7**, **8** and **10–13** exhibited a potent cytotoxic activity, with IC_{50} values ranging from 2.44 to 15.38 µg/mL, against one or both cell lines. A feature-based molecular networking (FBMN) approach led to the detection of a large amount of xanthones in the leaves extract, and particularly analogues of the cytotoxic isolated xanthone pyranojacareubin (**10**).

Keywords: *Calophyllum tacamahaca*; xanthones; triterpenes; cytotoxicity; feature-based molecular networking

1. Introduction

The genus *Calophyllum* (Calophyllaceae) includes approximately 200 species, distributed across all tropical regions. They are traditionally used against many ailments, including ulcers, malaria, tumor, infections, eye diseases, pain, inflammation and rheumatism [1,2]. This genus is an important source of bioactive natural products, including coumarins, xanthones, chromanones and triterpenes [3,4]. Xanthones and coumarins from *Calophyllum* species are known to possess cytotoxic, antiviral, antimicrobial, antiparasite, analgesic, anti-inflammatory and chemopreventive properties [5,6]. (+)-Calanolide A, a pyranocoumarin isolated from *C. lanigerum*, reached phase II of a clinical trial for its potent inhibitory activity of HIV-1 reverse transcriptase [4].

The species *Calophyllum tacamahaca* Willd., commonly known as "Takamaka des Hauts", is an endemic tree to Mauritius and Reunion Island. The leaf species is registered in the List of plants used in traditional medicine of French pharmacopoeia since April 2022, for eye diseases, fever, headaches and as veinotonic. This species is also traditionally employed to treat skin diseases, memory disorders, rheumatism and blood circulation troubles [7]. Previous investigations showed that leaf extract possesses hypotensive [8], antiplasmodial [9], antimicrobial [7], antiviral [10] and anti-inflammatory [11] activities. Nevertheless, the chemical composition of the species has never been studied and so bioactive compounds of the species have never been isolated nor identified so far.

Thus, the ethyl acetate (EtOAc) leaf extract of *C. tacamahaca* was subjected to a bio-guided chemical investigation in order to identify bioactive metabolites. Herein, we report the isolation, structure characterization and in vitro cytotoxic activity of one new chromanone (**1**), along with 13 known compounds (**2–14**). A feature-based molecular networking (FBMN) approach was performed in order to detect analogues of the bioactive compounds, and the obtained results are discussed below.

2. Materials and Methods

2.1. General Experimental Procedures

Optical rotations were determined using an Anton Paar MCP200 polarimeter (589 nm, 25 °C), and UV spectra were acquired on a Thermo Scientific DAD spectrophotometer. IR spectra were recorded on a Vertex 70 (Bruker) ATR-FTIR spectrometer. For compound (**1**), UV–vis and experimental ECD spectra were recorded on a JASCO J-815 spectrometer equipped with a JASCO Peltier cell holder PTC-423 to maintain the temperature at 20.0 °C. The handedness of the circular polarized light was modulated at 50 kHz with a quartz photoelastic modulator set at 1/4 retardation. A quartz cell of 1 mm of optical path length was used. Sample was prepared in dry methanol at a concentration of 0.0005 mol. L^{-1}. ECD spectra were recorded using CD_3OD as a reference and are presented without smoothing and further data processing. NMR spectra were acquired in CD_3OD (δ_{1H} 3.31 ppm, δ_{13C} 49.00 ppm) on a Bruker Avance II$^+$ 600 MHz (TCI cryoprobe) spectrometer at 300 K. NMR spectra were analyzed with the TopSpin (v 4.1.1) software. Structural assignments were based on ^{1}H NMR, ^{13}C NMR, COSY, HSQC and HMBC spectra. The chemical shifts δ are provided in ppm and coupling constants J in Hz. UHPLC-HRESIMS and UHPLC-HRESIMS/MS analyses were performed on an Impact II Bruker Daltonics Qq-TOF spectrometer with an ESI source using a 2.1 × 150 mm 1.6 µm RP-C_{18} column (Luna Omega C18, Phenomenex, Torrance, CA, USA) and an elution gradient of H_2O-CH_3CN with 0.1% HCO_2H (98:2 to 0:100). Solid reverse-phase extraction was performed over 10 g SPE Strata 55 µm C18 tubes (Phenomenex), with three elution steps (H_2O/CH_3CN v/v). Preparative HPLCs were performed on a Waters 2545 system using MassLynx software and a 21.2 × 150 mm 5 µm RP-C_{18} column (Gemini C18, Phenomenex), with an appropriate elution gradient of H_2O-CH_3CN with 0.1% HCO_2H at a flow rate of 20 mL/min. Semi-preparative HPLC was performed on a Dionex Ultimate 3000 system (Thermo Scientific, Waltham, MA, USA) using Chromeleon software and a 10 × 250 mm 5 µm RP-C_{18} column (Gemini C18, Phenomenex) with an appropriate elution gradient of H_2O-CH_3CN with 0.1% HCO_2H at a flow rate of 4.5 mL/min. Analytical HPLC was performed on a Dionex Ultimate 3000 system (Thermo Scientific) using Chromeleon software and a 4.6 × 150 mm 3 µm RP-C_{18} column (Gemini C18, Phenomenex) with an appropriate elution gradient of H_2O-CH_3CN with 0.1% HCO_2H at a flow rate of 0.8 mL/min.

2.2. Plant Material

Leaves of *Calophyllum tacamahaca* were collected in March 2019 on Reunion Island (Saint Denis). The taxonomic identification of the plant species was performed by Mr. H. Thomas (Parc National de La Réunion). A voucher specimen was deposited in the Herbarium of the University of La Réunion for confirmation of identification, with the following accession number: REU024075.

2.3. Extraction and Isolation

Leaves of C. *tacamahaca* were dried at 40 °C for 48 h and powdered. An accelerated solvent extractor (ASE 300 Dionex) was used to exhaustively extract the ground material (237.0 g). Four successive extractions were performed at 40 °C with EtOAc. The extract was evaporated under reduced pressure at 38 °C to obtain 20.3 g of crude extract.

A total of 3.32 g of crude extract were fractionated by solid reverse-phase extraction using combinations of H_2O/CH_3CN (v/v) of decreasing polarity. Three fractions (F1–F3) were obtained and evaluated for their cytotoxic activity against cancer cell lines. Fraction F2 (240.1 mg) was then subjected to preparative HPLC using an elution gradient of H_2O-CH_3CN with 0.1% HCO_2H (55:45 over 5 min, 55:45 to 20:80 over 35 min) at a flow rate of 20 mL/min (UV 260 nm). The purification of fraction F2 afforded the pure compounds isocaloteysmannic acid (**1**, 13.1 mg), amentoflavone (**2**, 12.0 mg), 6-(4-hydroxy-3-methylbutyl)-1,5-dihydroxyxanthone (**3**, 1.7 mg), scriblitifolic acid (**4**, 1.7 mg), pancixanthone B (**5**, 1.9 mg) and isocalophyllic acid (**7**, 29.4 mg). Subfraction F2–7 (11.4 mg) was subjected to semi-preparative HPLC using an isocratic elution of H_2O-CH_3CN with 0.1% HCO_2H (35:65) for 18 min at a flow rate of 4.5 mL/min (UV 320 nm) and afforded the pure compounds isocalophyllic acid (**7**, 1.0 mg) and inophyllum E (**8**, 2.8 mg). The purification of subfraction F2–8 (13.7 mg) was performed by semi-preparative HPLC using an isocratic elution of H_2O-CH_3CN with 0.1% HCO_2H (35:65) for 20 min at a flow rate of 4.5 mL/min (UV 280 nm) and yielded the pure compound calophyllic acid (**6**, 2.5 mg). Fraction F3 (673.3 mg) was subjected to preparative HPLC using an elution gradient of H_2O-CH_3CN with 0.1% HCO_2H (30:70 over 2 min, 30:70 to 0:100 over 10 min, 0:100 over 9 min) at a flow rate of 20 mL/min (ELSD). The purification of fraction F3 afforded the pure compounds isocalophyllic acid (**7**, 35.8 mg), calophyllic acid (**6**, 15.9 mg), canophyllalic acid (**11**, 22.8 mg), canophyllol (**12**, 21.9 mg) and canophyllic acid (**13**, 25.2 mg). The purification of subfraction F3–6 (5.7 mg) was performed by analytical HPLC using an isocratic elution of H_2O-CH_3CN with 0.1% HCO_2H (30:70) for 16 min at a flow rate of 0.8 mL/min (CAD) and yielded the pure compound 27-hydroxyacetate-canophyllic acid (**9**, 0.5 mg). Subfraction F3–7 (5.6 mg) was subjected to analytical HPLC using an isocratic elution of H_2O-CH_3CN with 0.1% HCO_2H (22:78) for 13 min at a flow rate of 0.8 mL/min (UV 280 nm) and yielded the pure compound pyranojacareubin (**10**, 0.9 mg). The purification of subfraction F3–19 was performed by semi-preparative HPLC using an isocratic elution of H_2O-CH_3CN with 0.1% HCO_2H (10:90) for 32 min at a flow rate of 4.5 mL/min (ELSD) and yielded the pure compound canophyllal (**14**, 0.3 mg).

Isocaloteysmannic acid (1): yellow–green powder, $[\alpha]_D^{25}$ −31.7 (*c* 0.1, MeOH); UV (MeOH) λ_{max} 200, 264–274, 299–312, 368 nm; IR ν_{max} 3087, 2977, 2926, 2855, 1709, 1627, 1300, 1000 cm^{-1}; for ^1H and ^{13}C NMR spectroscopic data, see Table 1; HRESIMS m/z 423.1791 $[M+H]^+$ (calculated for $C_{25}H_{27}O_6^+$, 423.1802).

Amentoflavone (2): yellow powder, UV (MeOH) λ_{max} 220, 272, 332 nm; ^1H NMR (CD_3OD, 600 MHz) 7.96 (1H, brs), 7.84 (1H, brd, J = 7.8 Hz), 7.50 (2H, d, J = 7.8 Hz), 7.08 (1H, brd, J = 7.8), 6.69 (2H, d, J = 7.8), 6.56 (1H, s), 6.55 (1H, s), 6.37 (1H, brs), 6.34 (1H, s), 6.16 (1H, brs); ^{13}C NMR (CD_3OD, 150 MHz) 184.2 (C=O), 183.8 (C=O), 166.2 (CO), 166.0 (COH, CO), 164.3 (COH), 163.2 (COH), 162.5 (COH), 161.2 (COH), 159.4 (CO), 156.5 (CO), 132.8 (CH), 129.4 (CH), 128.8 (CH), 123.3 (C), 123.2 (C), 121.9 (C), 117.7 (CH), 116.8 (CH), 105.6 (C), 105.3 (C), 104.0 (CH), 103.4 (CH), 100.4 (CH), 100.2 (CH), 95.2 (CH); HRESIMS m/z 539.0941 $[M+H]^+$ (calculated for $C_{30}H_{19}O_{10}^+$, 539.0973).

Table 1. ^{1}H and ^{13}C NMR data of isocaloteysmannic acid (**1**) in CD_3OD (600 MHz for ^{1}H and 150 MHz for ^{13}C).

Position	δ_H m (*J* in Hz)	δ_C
2	-	177.2
3	3.07, dd (15.2, 7.2) 3.27, dd (15.2, 8.2)	38.2
4	5.0, 7 brt [a] (7.7)	36.3
4a	-	113.0
4b	-	160.9
6	-	79.2
7	5.48, d (10.0)	127.3
8	6.49, d (10.0)	116.7
8a	-	102.9
8b	-	156.8
10	4.18, dq (11.3, 6.2)	80.3
11	2.61, dq (11.3, 6.9)	46.9
12	-	200.7
12a	-	102.6
12b	-	162.2
13	1.01, s	27.5
14	1.41, s	28.5
15	1.49, d (6.2)	19.8
16	1.19, d (6.9)	10.3
1′	-	145.2
2′, 6′	7.33, d (7.6)	128.8
3′, 5′	7.20, brt (7.5)	128.8
4′	7.10, brt (7.3)	126.7

[a] br: broad.

6-(4-Hydroxy-3-methylbutyl)-1,5-dihydroxyxanthone (**3**): yellow powder, UV (MeOH) λ_{max} 250, 316, 370 nm; ^{1}H NMR (CD_3OD, 600 MHz) 7.65 (1H, brt, *J* = 8.2 Hz), 7.65 (1H, d, *J* = 8.1 Hz), 7.20 (1H, d, *J* = 8.1 Hz), 7.08 (1H, d, *J* = 8.2 Hz), 6.76 (1H, d, *J* = 8.2 Hz), 3.48 (1H, dd, *J* = 10.7, 5.9 Hz), 3.41 (1H, dd, *J* = 10.7, 6.5 Hz), 2.89 (1H, ddd, *J* = 13.3, 10.0, 5.5 Hz), 2.80 (1H, ddd, *J* = 13.3, 9.0, 6.2 Hz), 1.81 (1H, m), 1.66 (1H, m), 1.46 (1H, m), 1.01 (1H, d, *J* = 6.8 Hz); ^{13}C NMR (CD_3OD, 150 MHz) 183.8 (C=O), 163.1 (COH), 157.5 (CO), 146.8 (CO), 144.6 (COH), 138.6 (C), 137.9 (CH), 126.5 (CH), 120.3 (C), 116.2 (CH), 111.2 (CH), 109.6 (C), 108.2 (CH), 68.3 (CH_2OH), 36.8 (CH), 34.4 (CH_2), 29.0 (CH_2), 17.0 (CH_3); HRESIMS *m*/*z* 315.1221 $[M+H]^+$ (calculated for $C_{18}H_{19}O_5{}^+$, 315.1227).

Scriblitifolic acid (**4**): yellow–beige powder, UV (MeOH) λ_{max} 237, 249, 298, 366 nm; ^{1}H NMR (CD_3OD, 600 MHz) 7.84 (1H, d, *J* = 8.0 Hz), 7.63 (1H, t, *J* = 8.2 Hz), 7.25 (1H, d, *J* = 8.0 Hz), 7.02 (1H, d, *J* = 8.2 Hz), 6.75 (1H, d, *J* = 8.2 Hz), 4.03 (3H, s), 2.81 (2H, t, *J* = 7.1 Hz), 2.45 (1H, m), 1.98 (2H, m), 1.72 (1H, m), 1.21 (1H, d, *J* = 6.5 Hz); ^{13}C NMR (CD_3OD, 150 MHz) 183.4 (C=O), 182.4 (COOH), 162.9 (COH), 157.3 (CO), 151.1 (CO), 147.1 (CO), 144.8 (C), 138.1 (CH), 126.5 (CH), 121.3 (C), 121.0 (CH), 111.4 (CH), 109.6 (C), 108.3 (CH), 62.2 (OCH_3), 42.0 (CH), 36.0 (CH_2), 29.3 (CH_2), 18.1 (CH_3); HRESIMS *m*/*z* 343.1166 $[M+H]^+$ (calculated for $C_{19}H_{19}O_6{}^+$, 343.1176).

Pancixanthone B (**5**): beige powder, UV (MeOH) λ_{max} 219, 248, 325, 363 nm; ^{1}H NMR (CD_3OD, 600 MHz) 7.61 (1H, d, *J* = 7.8 Hz), 7.24 (1H, brs), 7.18 (1H, t, *J* = 7.8 Hz), 6.15 (1H, s), 4.55 (1H, q, *J* = 6.6 Hz), 1.61 (3H, s), 1.41 (3H, d, *J* = 6.6 Hz), 1.33 (3H, s); ^{13}C NMR (CD_3OD, 150 MHz) 182.2 (C=O), 167.8 (CO), 165.3 (COH), 154.1 (CO), 148.0 (COH), 146.6 (CO), 124.9 (CH), 122.7 (C), 121.2 (CH), 116.2 (CH), 114.5 (C), 104.6 (C), 94.3 (CH), 92.5 (CH), 45.0 (C), 25.9 (CH_3), 21.4 (CH_3), 14.6 (CH_3); HRESIMS *m*/*z* 313.1075 $[M+H]^+$ (calculated for $C_{18}H_{17}O_5{}^+$, 313.1071).

Calophyllic acid (**6**): dark green powder, UV (MeOH) λ_{max} 200, 270, 320, 366 nm; ^{1}H NMR ($CDCl_3$, 600 MHz) 12.55 (1H, s), 7.38 (2H, m), 7.32 (1H, m), 7.30 (2H, m), 6.53 (1H, d, *J* = 9.5 Hz), 6.44 (1H, s), 5.42 (1H, d, *J* = 9.5 Hz), 4.27 (1H, dq, *J* = 11.5, 5.9 Hz), 2.63 (1H, dq, *J* = 11.5, 6.9 Hz), 1.54 (3H, d, *J* = 5.9 Hz), 1.26 (3H, s), 1.22 (3H, d, *J* = 6.9 Hz), 1.06 (3H, s); ^{13}C

NMR ($CDCl_3$, 150 MHz) 198.7 (C=O), 170.2 (COOH), 160.5 (COH), 158.7 (CO), 156.7 (CO), 149.7 (C), 140.8 (C), 129.3 (CH), 128.5 (CH), 127.3 (CH), 126.3 (CH), 120.1 (CH), 115.6 (CH), 108.0 (C), 101.7 (C), 101.4 (C), 79.1 (CH), 78.4 (C), 45.8 (CH), 28.4 (CH_3), 28.2 (CH_3), 19.9 (CH_3), 10.1 (CH_3); HRESIMS *m/z* 421.1651 $[M+H]^+$ (calculated for $C_{25}H_{25}O_6^+$, 421.1646).

Isocalophyllic acid (7): dark green powder, UV (MeOH) λ_{max} 200, 270, 320, 366 nm; ^1H NMR (CD_3OD, 600 MHz) 7.35 (2H, m), 7.31 (3H, m), 6.56 (1H, d, *J* = 10.1 Hz), 6.43 (1H, s), 5.49 (1H, d, *J* = 10.1 Hz), 4.68 (1H, qd, *J* = 6.9, 3.8 Hz), 2.65 (1H, qd, *J* = 7.3, 3.8 Hz), 1.44 (3H, d, *J* = 6.9 Hz), 1.29 (3H, s), 1.19 (3H, d, *J* = 7.3 Hz), 0.97 (3H, s); ^{13}C NMR (CD_3OD, 150 MHz) 202.8 (C=O), 169.9 (COOH), 162.0 (COH), 160.0 (CO), 157.7 (CO), 148.6 (C), 142.4 (C), 129.8 (CH), 129.3 (CH), 128.1 (CH), 127.4 (CH), 122.6 (CH), 116.5 (CH), 109.8 (C), 102.7 (C), 102.1 (C), 79.4 (C), 78.0 (CH), 45.6 (CH), 28.7 (CH_3), 28.2 (CH_3), 16.7 (CH_3), 9.7 (CH_3); HRESIMS *m/z* 421.1639 $[M+H]^+$ (calculated for $C_{25}H_{25}O_6^+$, 421.1646).

Inophyllum E (8): yellow powder, UV (MeOH) λ_{max} 200, 270, 310, 366 nm; ^1H NMR (CD_3OD, 600 MHz) 7.28 (2H, m), 7.23 (2H, m), 7.22 (1H, m), 6.49 (1H, d, *J* = 10.0 Hz), 6.02 (1H, s), 5.46 (1H, d, *J* = 10.0 Hz), 4.65 (1H, qd, *J* = 6.5, 3.4 Hz), 2.64 (1H, qd, *J* = 7.4, 3.4 Hz), 1.41 (3H, d, *J* = 6.5 Hz), 1.18 (3H, d, *J* = 7.4 Hz), 1.04 (3H, s), 0.99 (3H, s); ^{13}C NMR (CD_3OD, 150 MHz) 202.7 (C=O), 162.0 (OC=O), 160.5 (CO), 157.7 (CO), 141.8 (C), 129.9 (CH), 128.4 (CH), 128.3 (CH), 127.7 (CH), 125.2 (CH), 146.4 (C), 116.3 (CH), 113.3 (C), 102.9 (C), 101.9 (C), 79.3 (C), 77.9 (CH), 45.5 (CH), 27.9 (CH_3), 27.7 (CH_3), 16.5 (CH_3), 9.7 (CH_3); HRESIMS *m/z* 403.1529 $[M+H]^+$ (calculated for $C_{25}H_{23}O_5^+$, 403.1540).

27-Hydroxyacetate-canophyllic acid (9): yellow powder, ^1H NMR ($CDCl_3$, 600 MHz) 4.43 (1H, d, *J* = 12.2 Hz), 4.34 (1H, d, *J* = 12.2Hz), 3.73 (1H, m), 2.43 (1H, m), 2.21 (1H, m), 2.07 (1H, s), 1.96 (1H, m), 1.72 (1H, m), 1.55 (1H, m), 1.51 (2H, m), 1.50 (1H, m), 1.39 (1H, m), 1.38 (1H, m), 1.36 (1H, m), 1.34 (1H, m), 1.32 (1H, m), 1.31 (1H, m), 1.30 (1H, brs), 1.26 (1H, m), 1.24 (2H, m), 1.14 (1H, m), 1.11 (1H, m), 1.02 (3H, s), 0.96 (3H, s), 0.92 (3H, d, *J* = 7.3 Hz), 0.90 (3H, s), 0.89 (3H, s), 0.85 (1H, brs), 0.85 (3H, s); ^{13}C NMR ($CDCl_3$, 150 MHz) 182.9 (COOH), 171.7 (OC=O), 72.7 (COH), 65.4 (CH_2), 61.3 (CH), 53.2 (CH), 49.0 (CH), 44.6 (C), 42.2 (C), 41.3 (CH_2), 38.4 (CH), 38.1 (C), 37.8 (C), 37.5 (C), 36.3 (CH_2), 36.1 (CH_2), 35.7 (CH_2), 34.7 (CH_3), 32.6 (CH_2), 31.9 (CH_2), 29.7 (CH_2), 29.6 (CH_3), 28.3 (C), 25.1 (CH_2), 21.6 (CH_3), 21.4 (CH_3), 18.6 (CH_2), 17.9 (CH_3), 16.5 (CH_3), 15.9 (CH_2), 11.9 (CH_3).

Pyranojacareubin (10): yellow powder, UV (MeOH) λ_{max} 200, 290–300, 350 nm; ^1H NMR ($CDCl_3$, 600 MHz) 13.30 (1H, s), 7.47 (1H, s), 6.72 (1H, d, *J* = 10.3 Hz), 6.43 (1H, s), 6.43 (1H, d, *J* = 10.5 Hz), 5.73 (1H, d, *J* = 10.5 Hz), 5.59 (1H, d, *J* = 10.3 Hz), 1.53 (6H, s), 1.47 (6H, s); ^{13}C NMR ($CDCl_3$, 150 MHz) 180.0 (C=O), 160.4 (CO), 157.8 (COH), 157.2 (CO), 145.1 (CO), 132.1 (COH), 131.2 (CH), 127.7 (CH), 121.5 (CH), 117.8 (C), 115.6 (CH), 113.7 (CH), 104.8 (C), 103.3 (C), 95.4 (CH), 79.1 (C), 78.2 (C), 28.6 (CH_3), 28.5 (CH_3).

Canophyllalic acid (11): green powder, ^1H NMR ($CDCl_3$, 600 MHz) 2.40 (1H, brdd, *J* = 13.6, 4.3 Hz), 2.39 (1H, m), 2.34 (1H, m), 2.28 (1H, m), 2.23 (1H, q, *J* = 6.9 Hz), 1.95 (1H, m), 1.74 (1H, m); 1.68 (1H, m), 1.67 (1H, m), 1.52 (1H, brdd, *J* = 12.7, 2.7 Hz), 1.51 (1H, m), 1.49 (1H, m), 1.476 (1H, m), 1.472 (1H, m), 1.44 (1H, m), 1.42 (1H, m), 1.41 (1H, brdd, *J* = 10.1, 2.2), 1.39 (1H, m), 1.35 (1H, brt, *J* = 13.6 Hz), 1.29 (1H, m), 1.27 (1H, m), 1.25 (1H, m), 1.20 (1H, m), 1.195 (1H, m), 1.17 (1H, brdd, *J* = 13.6, 4.3 Hz), 1.04 (3H, s), 1.03 (3H, s), 0.94 (3H, s), 0.87 (3H, d, *J* = 6.9 Hz), 0.86 (3H, s), 0.81 (3H, s), 0.71 (3H, s); ^{13}C NMR ($CDCl_3$, 150 MHz) 213.3 (C=O), 184.9 (COOH), 59.4 (CH), 58.4 (CH), 53.2 (CH), 45.0 (C), 42.2 (C), 41.6 (CH_2), 41.3 (CH_2), 39.1 (C), 38.0 (CH), 37.9 (C), 37.8 (C), 36.1 (CH_2), 35.6 (CH_2), 35.0 (CH_2), 34.7 (CH_3), 32.8 (CH_2), 32.6 (CH_2), 31.2 (CH_2), 29.9 (CH_3), 29.6 (CH_2), 28.6 (C), 22.4 (CH_2), 20.8 (CH_3), 18.7 (CH_3), 18.3 (CH_2), 17.7 (CH_3), 14.8 (CH_3), 7.0 (CH_3).

Canophyllol (12): green powder, ^1H NMR ($CDCl_3$, 600 MHz) 3.64 (1H, d, *J* = 11.9 Hz), 3.61 (1H, d, *J* = 11, 9 Hz), 2.38 (1H, m), 2.28 (1H, m), 2.24 (1H, q, *J* = 6.6 Hz), 1.96 (1H, m), 1.84 (1H, m), 1.75 (1H, m), 1.68 (1H, m), 1.53 (1H, brdd, *J* = 12.3, 2.2 Hz), 1.48 (1H, m), 1.47 (1H, m), 1.46 (2H, m), 1.41 (2H, m), 1.35 (1H, m), 1.32 (2H, m), 1.31 (1H, m), 1.30 (1H, m), 1.29 (2H, m), 1.27 (1H, m), 1.26 (1H, m), 1.12 (3H, s), 0.99 (3H, s), 0.97 (3H, s), 0.91 (3H, s), 0.87 (3H, d, *J* = 6.6 Hz), 0.86 (3H, s), 0.71 (3H, s); ^{13}C NMR ($CDCl_3$, 150 MHz) 213.3 (C=O), 68.2 (COH), 59.6 (CH), 58.4 (CH), 52.6 (CH), 42.2 (C), 41.6 (CH_2), 41.4 (CH_2), 39.6 (CH), 39.5

(C), 38.3 (C), 37.6 (C), 35.6 (CH_2), 35.3 (C), 34.6 (CH_2), 34.4 (CH_3), 33.5 (CH_2), 33.0 (CH_3), 31.5 (CH_2), 31.4 (CH_2), 30.2 (CH_2), 29.3 (CH_2), 28.3 (C), 22.4 (CH_2), 19.3 (CH_3), 19.2 (CH_3), 18.4 (CH_2), 18.2 (CH_3), 14.8 (CH_3), 7.0 (CH_3).

Canophyllic acid (13): orange powder, ^{1}H NMR ($CDCl_3$, 600 MHz) 3.73 (1H, m), 2.38 (1H, brdd, *J* = 13.3, 4.0 Hz), 1.89 (1H, m), 1.73 (1H, m), 1.66 (1H, m), 1.55 (1H, m), 1.54 (1H, m), 1.50 (1H, m), 1.45 (1H, m), 1.43 (1H, m), 1.42 (1H, m), 1.35 (1H, m), 1.34 (1H, m), 1.33 (1H, m), 1.29 (1H, m), 1.25 (1H, m), 1.245 (1H, m), 1.23 (2H, m), 1.17 (1H, m), 1.13 (1H, m), 1.03 (3H, s), 1.0 (3H, s), 0.97 (1H, m), 0.96 (3H, s), 0.934 (3H, s), 0.93 (3H, d, *J* = 7.0 Hz), 0.89 (1H, m), 0.85 (3H, s), 0.80 (3H, s); ^{13}C NMR ($CDCl_3$, 150 MHz) 184.0 (COOH), 72.9 (COH), 61.3 (CH), 53.3 (CH), 49.3 (CH), 44.9 (CH), 41.7 (CH_2), 39.1 (C), 38.1 (CH), 38.0 (C), 37.9 (C), 37.5 (C), 36.1 (CH_2), 35.6 (CH_2), 35.3 (CH_2), 35.0 (CH_2), 34.7 (CH_3), 32.8 (CH_2), 32.7 (CH_2), 31.4 (CH_2), 29.9 (CH_3), 29.7 (CH_2), 28.6 (C), 20.7 (CH_3), 18.7 (CH_3), 18.0 (CH_3), 17.6 (CH_2), 16.5 (CH_3), 16.0 (CH_2), 11.8 (CH_3).

Canophyllal (14): off-white powder, ^{1}H NMR ($CDCl_3$, 600 MHz) 9.47 (1H, s), 2.39 (1H, ddd, *J* = 13.4, 5.1, 2.0), 2.28 (1H, tdd, *J* = 13.4, 7.4, 0.9), 2.23 (1H, q, *J* = 7.7 Hz), 2.18 (1H, dd, *J* = 13.4, 4.4 Hz), 2.01 (1H, m), 1.99 (1H, m), 1.95 (1H, m), 1.74 (1H, dt, *J* = 12.5, 3.0 Hz), 1.52 (1H, dd, *J* = 12.3, 3.0 Hz), 1.50 (1H, m), 1.46 (1H, m), 1.43 (1H, m), 1.38 (1H, m), 1.37 (1H, m), 1.25 (2H, m), 1.07 (3H, s), 0.98 (3H, s), 0.95 (3H, s), 0.87 (3H, d, *J* = 7.7 Hz), 0.84 (3H, s), 0.71 (3H, s), 0.67 (3H, s); ^{13}C NMR ($CDCl_3$, 150 MHz) 213.3 (C=O), 209.4 (HC=O), 59.4 (CH), 58.3 (CH), 53.0 (CH), 48.0 (C), 42.1 (C), 41.6 (CH_2), 41.5 (CH_2), 38.5 (C), 38.0 (C), 37.7 (C), 36.5 (C), 35.5 (CH_2), 34.9 (CH_2), 34.6 (CH_3), 32.6 (CH_2), 32.5 (CH_2), 30.7 (CH_2), 29.5 (CH_3), 28.2 (C), 27.3 (CH_2), 22.5 (CH_2), 19.9 (CH_3), 18.8 (CH_3), 18.2 (CH_2), 17.3 (CH_3), 14.6 (CH_3), 6.8 (CH_3).

2.4. *Molecular Modeling*

2.4.1. Calculation of Averaged NMR Spectra

The GAUSSIAN 09 program [12] using the hybrid B3LYP exchange–correlation functional [13,14] and the 6-31+G(d,p) basis set was used to carry out all DFT calculations. Tight convergence criteria were used for geometry optimization. All stationary points were confirmed as true minima via vibrational frequency calculations. Frequencies calculated in the harmonic approximation were multiplied by 0.98. Density functional theory (DFT) was used to perform the quantum chemical calculations. The molecular geometries were optimized by the DFT/B3LYP/6-31+G(d,p) method. Gauge including atomic orbitals (GIAO) NMR chemical shifts were calculated for the obtained geometries using the polarizable continuum model, PCM, with methanol as solvent, mPW1PW91 DFT functional and 6-31+G(d,p) basis sets to be in agreement with the DP4+ probability calculation. Averaged NMR chemical shifts were calculated from the unscaled chemical shifts of individual conformers according to their contribution calculated by Boltzmann weighting and using TMS as reference standard.

2.4.2. Conformational Study for UV–ECD Calculations

Conformational analysis was performed by stochastic exploration of the potential energy surface (PES) using the simulated annealing algorithm proposed by the Ampac11 software and combined with semi-empirical levels RM1 [15]. For the annealing, a geometry optimized at GD3BJ-B3LYP/6-311G(d,p) level was used as a starting structure. The GD3BJ term stands for empirical dispersion which was added with the D3 version of Grimme's dispersion with Becke–Johnson damping (GD3BJ) [16]. During each annealing, only the dihedral angles of this initial geometry were allowed to relax, the bond lengths and the valence angles were kept constant. A set of 24 geometries (the conformations with energy lower than 3 kcal mol^{-1} compared to the lower energy conformation) were selected for each diastereomer from the structures generated by 4 simulated annealing algorithms, each performed either with an initial geometry with some dihedral angles modified or with a different annealing temperature. Then, these geometries were fully optimized (i.e., all internal coordinates released) using GD3BJ-B3LYP/6-311G(d,p) level.

2.4.3. Calculation of Averaged UV and ECD Spectra

Based on the GD3BJ-B3LYP/6-311G(d,p) optimized geometries, the UV and ECD spectra were calculated using time-dependent density functional theory (TDSCF-DFT) with CAM-B3LYP functional and 6-31++G(d,p) basis set and with the SMD(CH_3OH) solvation model. SMD indicates the implicit solvent model used which is a dielectric continuum model that simulates the average effects of the solvent [17]. Calculations were performed for vertical 1A singlet excitation for 50 states. For a comparison between theoretical results and the experimental values, the calculated UV and ECD spectra have been modeled with a gaussian function using a half-width of 0.33 eV. Due to the approximations of the theoretical model used, an almost constant offset was observed between measured and calculated wavenumbers. Using UV spectra, all frequencies were calibrated by a factor of 1.05. Gaussian 16 package [18] was used to perform all calculations. It should be noted that similar calculations were performed using the LC-whPBE functional instead of CAM-B3LYP (SMD(CH_3OH)/LC-whPBE/6-31++G(d,p)//GD3BJ-B3LYP/6-311G(d,p)) and led to a similar result, which is not presented here.

2.5. *In Vitro Cytotoxic Assay*

HepG2 (human liver cancer) and HT29 (human colon and colorectal adenocarcinoma) cell lines were used to assess the toxicity of samples. In the performed assay, cytotoxicity was expressed as a concentration-dependent reduction in the uptake of the vital dye Neutral Red (NR) when measured 24 h after treatment. NR is a weak cationic dye that readily penetrates cell membranes by non-diffusion and accumulates intracellularly in lysosomes. Alterations of the sensitive lysosomal membrane lead to lysosomal fragility and other changes that gradually become irreversible. This results in a decreased uptake and binding of NR in non-viable cells. HT29 (ATCC® HTB-38™) and HepG2 (ATCC® HB-8065™), low passage number (<50), were cultivated into DMEM (Dulbecco's Minimum Essential Medium, PAN BIOTECH. lot 1874561) supplemented with penicillin 100 IU/mL and streptomycin 100 µg/mL (PAN BIOTECH, Lot 945514), and 10% of inactivated calf serum (PAN BIOTECH, Lot P56314), pH 7.2, freshly prepared, stored no longer than 1 week. Cells were seeded into 96-well tissue culture plates (0.1 mL per well) at a concentration of 1.10^5 cells/mL and incubated at 37 °C (5% CO_2) until semi-confluent. The test material was diluted into sterile DMSO (stock solutions 0.1, 1 and 10 mg/mL) at final concentrations ranging from 0.1 to 250 µg/mL. The culture medium was decanted and replaced by 100 µL of fresh medium containing the various concentrations of the test material; then, cells were incubated for 24 h at 37 °C (5% CO_2). At the end of the incubation period, cells were placed into Neutral Red medium (50 µg/mL NR in complete medium) and incubated for 3 h at 37 °C, 5% CO_2. Then, the medium was removed, and cells were washed three times with 0.2 mL of HBSS to remove excessive dye. The Neutral Red medium was removed and the distaining solution (50% ethanol, 1% acetic acid, 49% distilled water; 50 µL per well) was added into the wells. Then, the plates were shaken for 15–20 min at room temperature in the dark. The test samples and controls were run in triplicates in three independent experiments. A fluorescence–luminescence reader Infinite M200 Pro (TECAN) was used to measure the degree of membrane damage (i.e., the increase in released NR). For each well, the Optical Density (OD) was read at 540 nm. The results obtained for test material wells were compared to those of untreated control wells (HBSS, 100% viability) and converted to percentage values. The concentrations of the test material causing a 50% release of the preloaded NR (IC_{50}) compared to the control culture were calculated using software Phototox Version 2.0. The mean OD value of blank wells (only NR desorbed solution) was subtracted from the mean OD value of three test/untreated wells.

2.6. Feature-Based Molecular Networking

The leaf crude extract of *C. tacamahaca* as well as the isolated metabolites were profiled by UHPLC-QqTOF-MS/MS in a mass range from *m/z* 50 to 1200 using positive (+) mode for the ESI source. The following parameters were used: end plate offset at 500 V; nebulizer gas pressure at 3.5 bar; dry gas flow at 12 L/min; drying temperature at 200 °C; acquisition rate at 4.0 Hz. The capillary voltage was set at 4500 V, with a fragmentation energy of 20–40 eV. The UHPLC conditions were as follows: sample concentrations: 5 mg/mL (crude extract), 0.2 mg/mL (isolated compounds) in 100% MeOH, injection volume: 2 μL, column temperature: 40 °C, elution gradient of H_2O-CH_3CN with 0.1% HCO_2H (98:02 over 2 min, 98:02 to 0:100 over 12 min, 0:100 over 3 min) at a flow rate of 0.5 mL/min. Raw data obtained from the crude extract analysis were converted into open format .mzXML using software Bruker Compass DataAnalysis Version 4.2 and processed using software MZmine Version 2.53 [19–21]. Then, a feature-based molecular network (FBMN) was created on the GNPS platform [22], and it is available via the following link https://gnps.ucsd.edu/ProteoSAFe/status.jsp?task=f0c193d2141d463ba34af46df7bfe57c (accessed on 29 March 2022). The Mzmine MS/MS data processing comprises .mzXML file import, MS peak detection, ADAP chromatogram builder, chromatogram deconvolution, isotopic peaks grouper, alignment, filtering, fragment search, adduct search and spectra normalization. Setting parameters were as follows: positive ionization mode, centroid detection, MS1 peak detection limit: 1^E3, MS2 peak detection limit: 1^E2, *m/z* tolerance: 10 ppm, peak/top edge ratio: 2, peak duration range: 0.03–1 min, *m/z* range for MS2 pairing: 0.02 Da, RT range for MS2 pairing: 0.1 min, representative isotope: most intense, alignment weight for *m/z*: 75, weight for RT: 25, filtering RT tolerance: 0.1 min, filtering *m/z* tolerance: 0.001 *m/z*, adduct search $[M+Na]^+$, $[M+NH_4]^+$, spectra normalization type: average intensity. Processed files including an mgf and a csv file were uploaded to the GNPS platform. An FBMN was then developed using the Advanced Analysis Tools—Feature Networking workflow [23]. Advanced Network Options parameters were as follows: min pair cos: 0.7, minimum matched fragment ions: 6, network topK: 10, maximum connected component size: 100, mass tolerance for precursor and fragment ions: 0.02 Da. The output was imported into Cytoscape Version 3.8.2 in order to visualize the network. Node annotations were performed manually for isolated compounds and with GNPS spectral databases (score threshold: 0.7) and In Silico MS/MS DataBase ISDB (score threshold: 0.2) [24].

3. Results and Discussion

3.1. Isolation of Compounds **1–14**

C. tacamahaca leaf EtOAc extract was subjected to a solid reverse-phase extraction and yielded three fractions (F1–F3). Fractions F2 and F3 were further purified by preparative, semi-preparative and analytical reverse-phase HPLC, resulting in the isolation of one new chromanone (**1**) and 13 known compounds (**2–14**) (Figure 1). The latter were identified by comparison with previously reported spectroscopic data as amentoflavone (**2**) [25], scriblitifolic acid (**4**) [26], pancixanthone B (**5**) [27], calophyllic acid (**6**) [28] isocalophyllic acid (**7**) [28], inophyllum E (**8**) [28,29], 27-hydroxyacetate-canophyllic acid (**9**) [30], pyranojacareubin (**10**) [31], canophyllalic acid (**11**) [32], canophyllol (**12**) [32], canophyllic acid (**13**) [32] and canophyllal (**14**) [33]. Spectroscopic data of the known metabolite **3**, identified as 6-(4-hydroxy-3-methylbutyl)-1,5-dihydroxyxanthone [34], have not been published so far and are provided here (Section 2 and Figures S11–S15). The structure of the new compound **1** was established based on 1D and 2D NMR, IR and UV spectroscopic and HRESIMS spectrometric data.

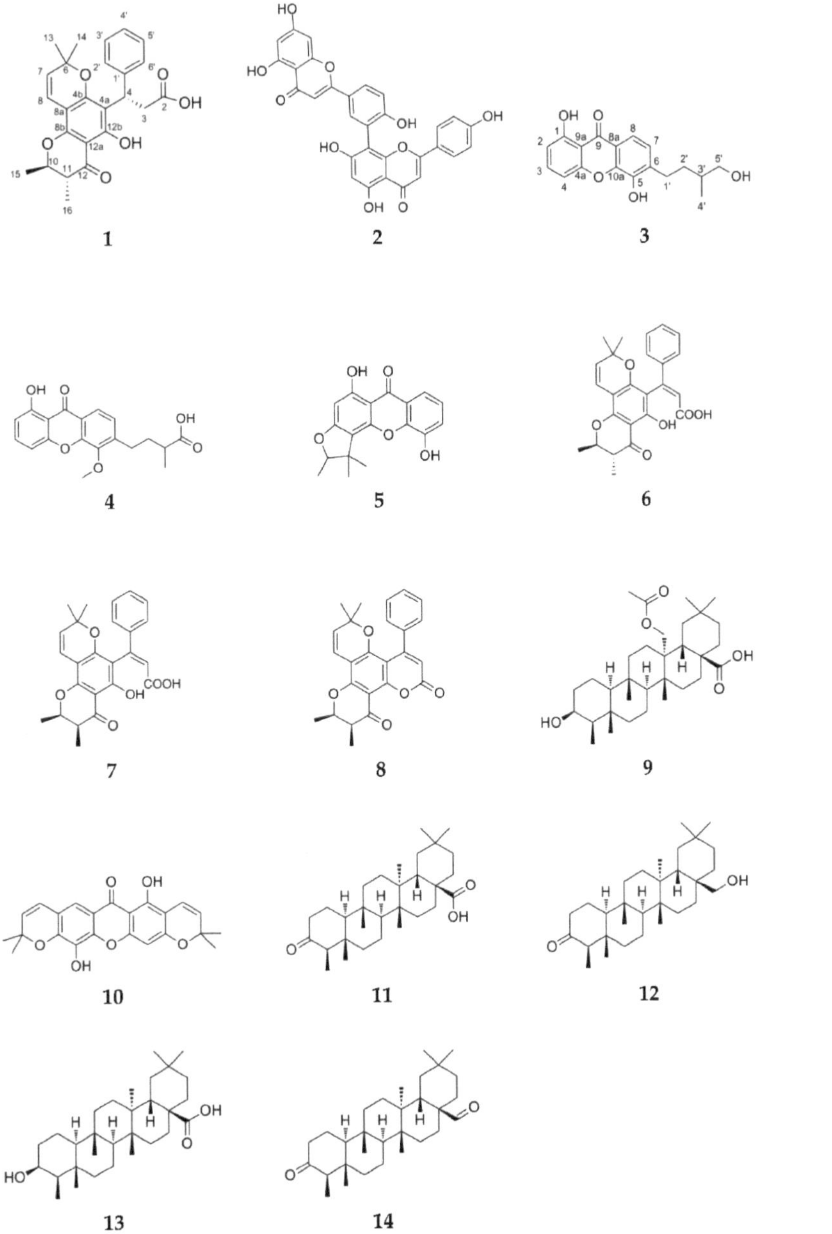

Figure 1. Structures of compounds **1–14** isolated from *C. tacamahaca*.

*3.2. Structure Elucidation of Isocaloteysmannic Acid (**1**)*

Isocaloteysmannic acid (**1**), $[\alpha]_D^{25}$ −31.7 (c 0.1, MeOH), was isolated as a yellow-green powder. The molecular formula $C_{25}H_{26}O_6$ was established from HRESIMS data showing a molecular ion peak at *m*/*z* 423.1791 $[M+H]^+$ (calculated for $C_{25}H_{27}O_6{}^+$, 423.1802), suggesting the occurrence of 13 degrees of insaturation. The UV spectrum exhibited absorption maxima at 200, 264–274, 299–312 and 368 nm, characteristic of a pyranochromanone moiety [35]. The IR spectrum exhibited characteristic bands of sp^3 type CH (2926 cm^{-1}), sp^2 type CH (3087 cm^{-1}), carboxylic acid function (1709 cm^{-1}), aromatic rings (1627 cm^{-1}) and ether function (1000 and 1300 cm^{-1}). The ^1H and ^{13}C NMR data of (**1**) (Table 1 and Figures S2 and S3) are similar to those of caloteysmannic acid [35]. The ^1H and ^{13}C NMR spectra showed aromatic signals at $\delta_{H/C}$ 7.33 (H-2′, H-6′, doublet)/128.8 (C-2′, C-6′), $\delta_{H/C}$ 7.20 (H-3′, H-5′, borad triplet)/128.8 (C-3′, C-5′) and $\delta_{H/C}$ 7.10 (H-4′, triplet)/126.7 (C-4′), consistent with the phenyl group of the chromanone. The COSY spectrum (Figure S4) showed correlations consistent with the spin system H-2′–H-3′–H-4′–H-5′–H-6′. Signals observed at $\delta_{H/C}$ 5.48 (H-7, doublet)/127.3 (C-7) and $\delta_{H/C}$ 6.49 (H-8, doublet)/116.7 (C-8) correspond to the spin-pair of two sp^2 methine protons. Characteristic signals of protons H-10 and H-11 are observed at $\delta_{H/C}$ 4.18 (H-10, doublet of quadruplet)/80.3 (C-10) and $\delta_{H/C}$ 2.61 (H-11, doublet of quadruplet)/46.9 (C-11). The two signals observed at $\delta_{H/C}$ 3.07 (H-3a, doublet of doublet)/38.2 (C-3) and 3.27 (H-3b, doublet of doublet)/38.2 (C-3) correspond to diastereotopic protons. The ^1H and ^{13}C NMR spectra show a deshielded signal at $\delta_{H/C}$ 5.07 (H-4, broad triplet)/36.3 (C-4), corresponding to the alkane proton in beta position of the acid carboxylic function. These positions were confirmed with the COSY spectrum (Figure S4) showing a correlation between H-3 (δ_H 3.07, 3.27) and H-4 (δ_H 5.07). Four signals corresponding to methyl groups are observed at $\delta_{H/C}$ 1.01 (H-13, singlet)/27.5 (C-13), $\delta_{H/C}$ 1.41 (H-14, singlet)/28.5 (C-14), $\delta_{H/C}$ 1.49 (H-15, doublet)/19.8 (C-15) and $\delta_{H/C}$ 1.19 (H-16, doublet)/10.3 (C-16). The COSY spectrum (Figure S4) shows correlations between H-15 (δ_H 1.49) and H-10 (δ_H 4.18), and between H-16 (δ_H 1.19) and H-11 (δ_H 2.61). Finally, the characteristic signals of the acid carboxylic and the ketone functions are observed on the ^{13}C NMR spectrum at δ_C 177.2 (C-2) and δ_C 200.7 (C-12), respectively. The linkage and the substitution pattern of (**1**) is determined from HMBC correlations (Figure 2 and Figure S6). The HMBC correlations of H-4 (δ_H 5.07) to C-2′ (δ_C 128.8) and C-6′ (δ_C 128.8) and those of H-3 (δ_H 3.07, 3.27) to C-1′ (δ_C 145.2) indicate the substitution of C-4 (δ_C 36.3) by the phenyl group. The carboxylic acid function position in C-2 (δ_C 177.2) is confirmed by the $^2J_{HC}$ correlation of H-3 (δ_H 3.07, 3.27) to C-2 (δ_C 177.2). The HMBC correlations of methyl protons H-13 (δ_H 1.01) and H-14 (δ_H 1.41) to C-6 (δ_C 79.2) indicate these two methyl groups are borne by the same carbon C-6 (δ_C 79.2). The HMBC correlations of H-7 (δ_H 5.48) to C-14 (δ_C 28.5) and C-8a (δ_C 102.9), and of H-8 (δ_H 6.49) to C-4b (δ_C 160.9), C-6 (δ_C 79.2) and C-8b (δ_C 156.8) confirmed the A and C rings linkage. The HMBC correlations of H-4 (δ_H 5.07) to C-4b (δ_C 160.9) and C-12b (δ_C 162.2), and of H-3 (δ_H 3.07, 3.27) to C-4a (δ_C 113.0) indicate the substitution of C-4a (δ_C 113.0) by the phenyl-bearing saturated chain. Finally, the HMBC correlations of H-10 (δ_H 4.18) and H-11(δ_H 2.61) to C-16 (δ_C 10.3) and C-15 (δ_C 19.8), respectively, of H-10 (δ_H 4.18) to C-12 (δ_C 200.7) and C-8b (δ_C 156.8), and those of H-16 (δ_H 1.19) to C-12 (δ_C 200.6) confirm the D ring configuration. Based on NMR data, the $^3J_{H\text{-}10/11}$ coupling constant (11.3 Hz) between the vicinal protons H-10 and H-11 indicate a dihedral angle consistent with an axial–axial coupling constant [36]. In a previous work, Patil et al. showed that the only possible configuration for these trans-diaxial H-10 and H-11 vicinal protons is a configuration of C-10 and C-11 carbons 10R, 11R [28]. Consequently, two potential diastereoisomers were conceivable for compound **1**: (4*R*,10*R*,11*R*) or (4*S*,10*R*,11*R*) (Figure 3).

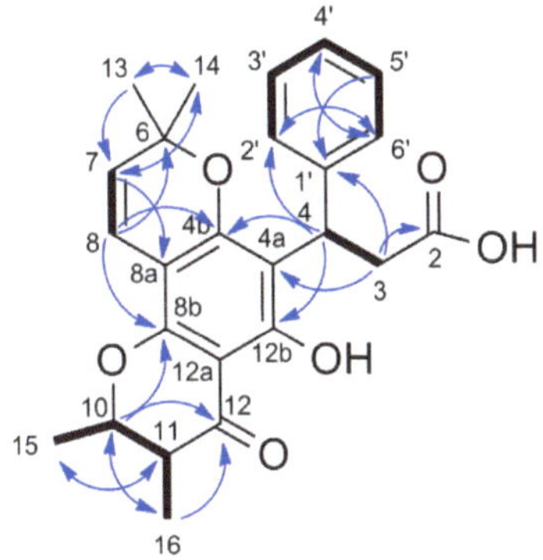

Figure 2. Key ^{1}H-^{1}H COSY (bold) and ^{1}H-^{13}C HMBC (blue arrows) correlations of (**1**).

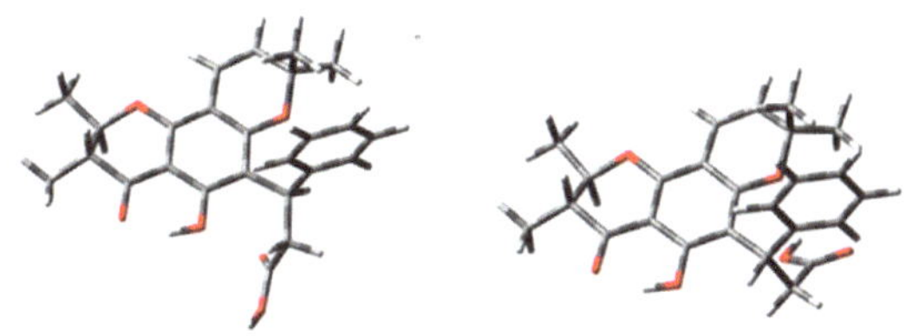

4*R*,10*R*,11*R* 4*S*,10*R*,11*R*

Figure 3. The two potential enantiomers for (**1**).

3.3. Absolute Configuration of Isocaloteysmannic Acid (1)

First, to confirm the configuration of C-10 and C-11 carbons and to determine the configuration of C-4 carbon, experimental chemical shifts (^{1}H and ^{13}C) of compound (**1**) were compared with calculated chemical shifts of four isomers (4*S*,10*S*,11*S*), (4*S*,10*R*,11*R*), (4*S*,10*R*,11*S*), (4*S*,10*S*,11*R*). For these four isomers, the equilibrium population of each conformer was calculated from its relative free energy using Boltzmann statistics, 36 conformers for (4*S*,10*S*,11*S*), 37 conformers for (4*S*,10*R*,11*R*), 39 conformers for (4*S*,10*R*,11*S*), and 48 conformers for (4*S*,10*S*,11*R*) (Tables S1–S4). NMR chemical shifts have been calculated with the GIAO method at the PCM/mPW1PW91/6-31+G(d,p) level allowing to use the DP4+ probability [37]. Experimental chemical shifts have been compared to theorical chemical shifts of each isomer individually by linear regressions of $\delta^1H_{theorical} = f(\delta^1H_{experimental})$ and $\delta^{13}C_{theorical} = f(\delta^{13}C_{experimental})$ and all together with the DP4+ probability (Table S5). Assignment by ^{1}H-DP4+ and ^{13}C-DP4+ did not converge to the same isomer, and when including all the data, probabilities were shared between two isomers (4*S*,10*S*,11*S*) (41.07%) and (4*S*,10*R*,11*R*) (58.93%) (Figure 4). Therefore, the results of these comparisons did not allow unambiguous determination of the absolute configuration of compound **1** but did confirm the trans-configuration of C-10 and C-11.

The absolute configuration of (**1**) was established by ECD by comparing the measured spectra with those calculated using DFT and TD-DFT for diastereomers (4*S*,10*R*,11*R*) and (4*R*,10*R*,11*R*) according to the previous NMR analysis (Figure 3).

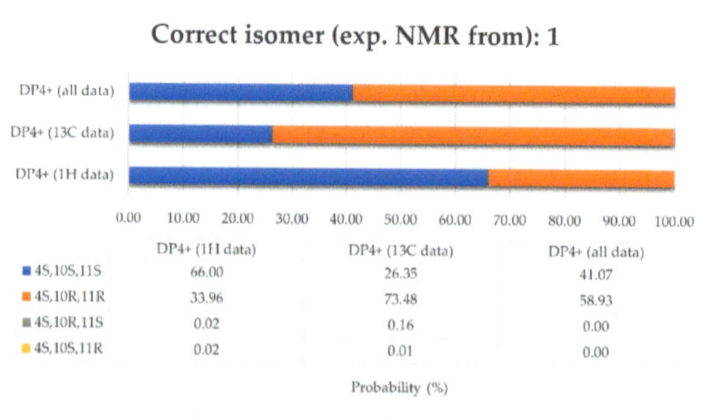

Figure 4. Graph of ^{1}H-DP4+, ^{13}C-DP4+, and DP4+ (PCM/mPW1PW91/6-31+Gdp) probabilities obtained by correlating the experimental NMR of compound **1** with the calculated data of the four isomers (4*S*,10*S*,11*S*), (4*S*,10*R*,11*R*), (4*S*,10*R*,11*S*), (4*S*,10*S*,11*R*).

The UV and ECD spectra of (4*S*,10*R*,11*R*) and (4*R*,10*R*,11*R*) were built, respectively, from the individual spectra of the A_{1-6} and B_{1-6} conformations weighed by their Boltzmann population (Appendix A). The comparison of the calculated UV spectra for the two diastereomers showed a good agreement with the measured spectrum, without allowing to establish the absolute configuration of the C-4 atom. Furthermore, the calculated ECD spectra showed a clear sign difference around 215 nm: positive bands for (4*S*,10*R*,11*R*) and negative bands for (4*R*,10*R*,11*R*) (Figure 5A–D).

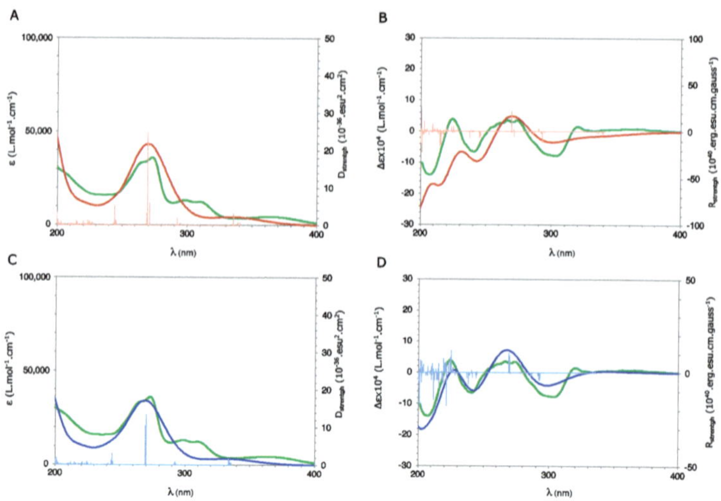

Figure 5. UV (**left-A,C**) and ECD (**right-B,D**) spectra measured in CD_3OD for (**1**) (green) and calculated using SMD(CH_3OH)/CAM-B3LYP/6-31++G(d,p)//GD3BJ-B3LYP/6-311G(d,p) level for (4*R*,10*R*,11*R*) (red) and (4*S*,10*R*,11*R*) (blue).

Comparison with the corresponding measured spectrum showed excellent agreement with that calculated for the (4*S*,10*R*,11*R*) configuration (Figure 5A–D). In particular, the band around 215 nm is positive as in the measured spectrum. This ECD analysis therefore confirmed the *R*-configuration of the C-10 and C-11 atoms, but also unambiguously established that the C-4 atom is of absolute configuration *S*. Consequently, compound **1** has the absolute configuration (4*S*,10*R*,11*R*).

Compound **1** is a trans-epimer of caloteysmannic acid, a chromanone with (4*S*) configuration and cis-configuration of vicinal protons H-10, H-11 (10*S*,11*R*) previously isolated from *Calophyllum teysmannii* [35]. Therefore, (**1**) was named isocaloteysmannic acid.

3.4. Cytotoxic Activity of the Isolated Compounds

Ten isolated compounds were evaluated for their cytotoxic properties against the two cancer cell lines HepG2 and HT29. Due to their paucity, compounds **3–5** and **14** were not evaluated. Compounds **7**, **8**, **10**, **11**, **12** and **13** showed a potent activity against one or both cell lines, with IC_{50} values ranging from 2.44 to 15.38 μg/mL (Table 2). The new compound **1**, as well as compound **6**, exhibited a moderate activity against both cell lines with IC_{50} values ranging from 15.98 to 25.68 μg/mL.

Table 2. Cytotoxic activity of the isolated compounds.

Compound	IC_{50} (μg/mL) [a]	
	HepG2	HT29
1	19.65 ± 2.34	25.68 ± 2.08
2	39.03 ± 3.23	41.97 ± 2.54
6	15.98 ± 3.65	18.97 ± 2.94
7	2.44 ± 0.67	4.24 ± 0.67
8	7.03 ± 1.56	5.94 ± 0.07
9	45.09 ± 2.09	56.98 ± 3.76
10	9.54 ± 1.22	10.46 ± 2.08
11	3.34 ± 0.94	5.97 ± 0.99
12	15.38 ± 2.07	10.26 ± 1.34
13	6.65 ± 1.54	4.06 ± 0.29

[a] IC_{50} are the means ± standard deviations calculated from three independent assays.

The triterpenes **11–13** showed a potent activity, whereas triterpene **9** exhibited only a weak activity, suggesting that the presence of the acetoxy group in **9** could decrease its cytotoxic potential.

These results also suggest that the cis-configuration of the methyl groups in C-10 and C-11 of compounds **7** and **8** leads to a higher cytotoxic activity than the trans-configuration (compounds **1** and **6**).

3.5. Feature-Based Molecular Networking Analysis of the Crude Extract

A feature-based molecular networking (FBMN) [23] approach was performed in order to provide more information about the chemodiversity of the species and to detect additional cytotoxic metabolites by highlighting close analogues of the bioactive isolated compounds. For this purpose, leaf EtOAc extract was subjected to an UHPLC-HRESIMS/MS analysis and a molecular network (MN) was generated with the FBMN tool on the GNPS platform.

3.5.1. Chemodiversity of the Species

A molecular network (MN) comprising 520 features and 55 clusters (two features at least) was obtained (Figure 6). Squared orange nodes correspond to the isolated compounds **1**, **3**, **4**, **5**, **6**, **8** and **10**. Green nodes correspond to spectral matches on GNPS or ISDB databases. The edge thickness correlates with the cosine score (CS) value (0.7–1) between two nodes.

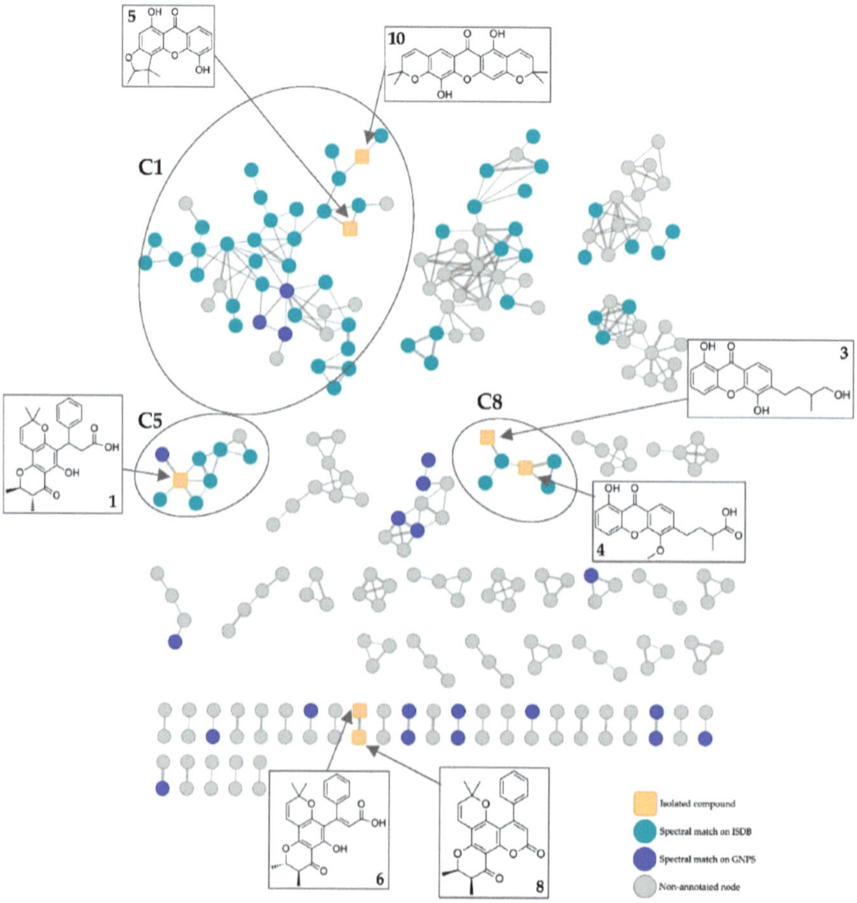

Figure 6. Molecular networking (MN) of the isolated compounds (orange squared nodes) and MN annotation based on GNPS and ISDB spectral matches (green and blue nodes).

Relatively few consistent spectral matches on GNPS or ISDB databases were obtained. Based on these matches, the largest cluster C1 (43 nodes) could correspond to xanthones. Two nodes correspond to the isolated xanthones **5** and **10**, and three nodes were putatively identified as xanthones previously reported in the genus *Calophyllum*: 6-deoxyisojacareubin, mammea B/BA and caloxanthone. Seven nodes could correspond to xanthones reported in close botanical families of Calophyllaceae: elliptoxanthone B (Hypericaceae), garcinexanthone C (Clusiaceae), celebixanthone (Hypericaceae), nigrolineaxanthone K (Clusiaceae), garcinone A (Clusiaceae), hypericumxanthone B (Hypericaceae) and garcimangosone C (Clusiaceae).

Cluster C8 is another cluster of xanthones, containing the isolated metabolites **3** and **4**, as well as one node putatively identified as caloxanthone H. The latter was previously reported in the genus *Calophyllum*.

The new compound **1** is located in cluster C5. In the latter, one node corresponds to a close analogue of **1** (m/z 423.1785, CS > 0.9). Based on ISDB matches, this close analogue was putatively identified as isochapelieric acid, a compound isolated from the species *Calophyllum calaba* [38].

These observations are consistent with the data in the literature, indicating that xanthones and chromanones are largely represented in the genus *Calophyllum*.

3.5.2. Detection of Additional Bioactive Metabolites

Two analogues of the cytotoxic isolated compound pyranojacareubin (**10**) have been detected in cluster C1 (Figure 7) at *m*/*z* 395.1475 and *m*/*z* 327.0854. Based on structure–activity relationship, these analogues could correspond to cytotoxic metabolites. They were putatively identified as muxiangrine I and elliptoxanthone B, according to ISDB matches. To the best of our knowledge, no cytotoxic properties have been reported in the literature for these compounds. As these identifications are highly hypothetical, it would be necessary to target the isolation of these two compounds, to identify them and assess their biological properties in a future work.

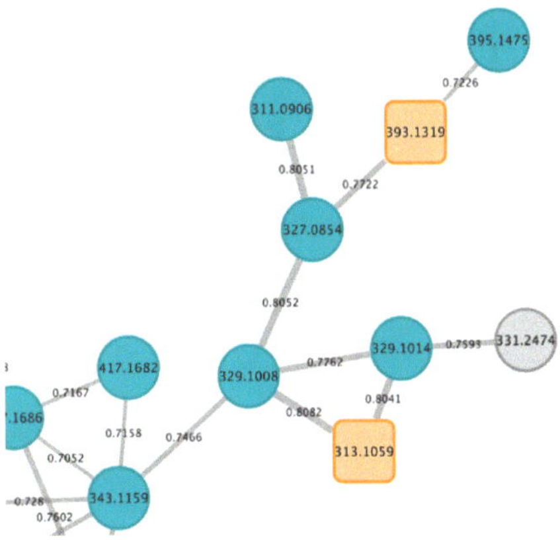

Figure 7. Part of cluster C1 containing analogues of compound **10** (orange squared node at *m*/*z* 393.1319). Ion parent mass is indicated in each node and cosine score value is indicated on each edge.

4. Conclusions

Fourteen metabolites (**1–14**) were isolated from the EtOAc leaf extract of *C. tacamahaca*. To the best of our knowledge, compound **1** was reported for the first time. Six compounds (**7**, **8**, **10**, **11**, **12** and **13**) showed a potent cytotoxicity against HepG2 and/or HT29 cell lines. The FBMN approach allowed the detection of a large amount of xanthones in the extract, including two close analogues of the cytotoxic compound **10**. Xanthones are well known for their cytotoxic properties [2], so the results of this study suggest that *C. tacamahaca* leaves are a significant source of cytotoxic metabolites. These compounds could be interesting candidates for future therapeutic applications. Nevertheless, further studies are needed to evaluate their in vivo anticancer activity, as well as their mechanism of action, and thus confirm their therapeutic potential.

Supplementary Materials: The following supporting information can be downloaded at: https://www.mdpi.com/article/10.3390/metabo13050582/s1, Figure S1: HRESIMS spectrum for isocaloteysmannic acid (**1**); Figure S2: ^{1}H NMR (600 MHz, CD_3OD) spectrum for isocaloteysmannic acid (**1**); Figure S3: ^{13}C NMR (150 MHz, CD_3OD) spectrum for isocaloteysmannic acid (**1**); Figure S4: ^{1}H-^{1}H COSY NMR (600 MHz, CD_3OD) spectrum for isocaloteysmannic acid (**1**); Figure S5: ^{1}H-^{13}C HSQC NMR (600 MHz, CD_3OD) spectrum for isocaloteysmannic acid (**1**); Figure S6: ^{1}H-^{13}C HMBC NMR (600 MHz, CD_3OD) spectrum for isocaloteysmannic acid (**1**); Figure S7: Plot of Boltzmann-weighted calculated NMR δ 13C of 4*R*,10*R*,11*R* and 4*S*,10*R*,11*R* isomers versus experimental NMR δ 13C of isocaloteysmanic acid (**1**); Figure S8: UV (left) and ECD (right) spectra calculated using SMD(CH_3OH)/CAM-B3LYP/6-31++G(d,p)//GD3BJ-B3LYP/6-311G(d,p) level for (4*R*,10*R*,11*R*) (red) and (4*S*,10*R*,11*R*) (blue); Figure S9: Conformations A_1 to A_6 selected to build the UV and ECD spectra of the diastereomer (4*R*,10*R*,11*R*). Geometries optimized to the level GD3BJ-B3LYP/6-311G(d,p); Figure S10: Conformations B_1 to B_6 selected to build the UV and ECD spectra of the diastereomer (4*S*,10*R*,11*R*). Geometries optimized to the level GD3BJ-B3LYP/6-311G(d,p); Figure S11: HRESIMS spectrum for 6-(4-Hydroxy-3-methylbutyl)-1,5-dihydroxyxanthone (**3**); Figure S12: ^{1}H

NMR (600 MHz, CD_3OD) spectrum for 6-(4-Hydroxy-3-methylbutyl)-1,5-dihydroxyxanthone (**3**); Figure S13: ^{1}H-^{1}H COSY NMR (600 MHz, CD_3OD) spectrum for 6-(4-Hydroxy-3-methylbutyl)-1,5-dihydroxyxanthone (**3**); Figure S14: ^{1}H-^{13}C HSQC NMR (600 MHz, CD_3OD) spectrum for 6-(4-Hydroxy-3-methylbutyl)-1,5-dihydroxyxanthone (**3**); Figure S15: ^{1}H-^{13}C HMBC NMR (600 MHz, CD_3OD) spectrum for 6-(4-Hydroxy-3-methylbutyl)-1,5-dihydroxyxanthone (**3**); Table S1: Boltzmann-weighted populations of conformations 1–36 used for DP4$^+$ analysis of isomer 4R,10R,11R; Table S2: Boltzmann-weighted populations of conformations 1–36 used for DP4$^+$ analysis of isomer 4*R*,10*R*,11*S*; Table S3: Boltzmann-weighted populations of conformations 1–36 used for DP4$^+$ analysis of isomer 4*S*,10*R*,11*R*; Table S4: Boltzmann-weighted populations of conformations 1–36 used for DP4$^+$ analysis of isomer 4*S*,10*R*,11*S*; Table S5: Coefficients of determination R^2 of the linear regressions made between experimental chemical shifts and theorical chemical shifts of each isomer; Table S6: Enthalpies and Boltzmann populations of conformations A_1–A_{14} of (4*R*,10*R*,11*R*) and conformations B_1–B_{14} of (4*S*,10*R*,11*R*), calculated using GD3BJ-B3LYP/6-311G(d,p) level.

Author Contributions: Conceptualization, E.G. (Elise Gerometta), A.G.-B. and I.G.; methodology, E.G. (Elise Gerometta), A.G.-B., I.G., E.G. (Elnur Garayev), G.H., P.-E.C., P.C., A.L. and A.M.; validation, A.G.-B., I.G., G.H., E.G. (Elnur Garayev), M.F., B.B. and P.-E.C.; formal analysis, E.G. (Elise Gerometta) and E.G. (Elnur Garayev); investigation, E.G. (Elise Gerometta), G.H., C.D.G., A.M. and J.-V.N.; resources, A.G.-B.; writing—original draft preparation, E.G. (Elise Gerometta); writing—review and editing, G.H., E.G. (Elnur Garayev), A.M., J.-V.N., C.D.G., P.-E.C., P.C., M.F., A.L., B.B., I.G. and A.G.-B.; visualization, E.G. (Elise Gerometta), I.G. and A.G.-B.; supervision, I.G. and A.G.-B.; project administration, A.G.-B.; funding acquisition, A.G.-B. All authors have read and agreed to the published version of the manuscript.

Funding: This work was supported by the European Regional Development Funds GURDTI 2018-1828-0002370 (FEDER PHAR, EU-Région Réunion-French State national counterpart). Elise Gerometta is a recipient of a fellowship from the Région Réunion.

Institutional Review Board Statement: Not applicable.

Informed Consent Statement: Not applicable.

Data Availability Statement: NMR raw data (^{1}H, ^{13}C, gCOSY, gHSQC, gHMBC) of compounds **1** and **3** are made freely available at https://doi.org/10.5281/zenodo.7728239. Raw MS/MS data (open format .mzXML) have been deposited on MassIVE (https://massive.ucsd.edu): MSV000089771.

Acknowledgments: We thank H. Thomas (Parc National de La Réunion) for locating and identifying the investigated plant species. We thank C. Simmler (IMBE), S. Greff (IMBE) and the Service of Chemical Ecology and Metabolomics (Aix-Marseille Univ) for the acquisition of the LC/MSMS data. This work was supported by the computing facilities of the CRCMM, 'Centre Régional de Compétences en Modélisation Moléculaire de Marseille'.

Conflicts of Interest: The authors declare no conflict of interest. The funders had no role in the design of the study; in the collection, analyses, or interpretation of data; in the writing of the manuscript; or in the decision to publish the results.

Appendix A

Based on NMR analysis, the UV and ECD spectra of the diastereomer of (**1**) were calculated for the two potential enantiomers of absolute configuration (4*R*,10*R*,11*R*) and (4*S*,10*R*,11*R*). The calculated enthalpies for the 14 most stable conformations discovered for each diastereomer and the corresponding Boltzmann populations are shown in Table S6. For each optimized conformation, a calculation of the vibrational frequencies established that they were local minima (no imaginary frequency).

The conformations A_1 to A_6 with Boltzmann populations greater than 5% (Figure S9) and the conformations B_1 to B_6 with Boltzmann populations greater than 5% (Figure S10) were selected to build the UV and ECD spectra of the diastereomer (4*R*,10*R*,11*R*) and (4*S*,10*R*,11*R*), respectively. The averaged UV and ECD spectra (Figure S8) of the diastereomers were built from the calculated spectra for the selected conformations weighted by their Boltzmann population evaluated from the calculated enthalpy. These averaged spectra were then compared to the measured spectra in order to establish the absolute configuration of (**1**).

References

1. Gupta, S.; Gupta, P. The Genus Calophyllum: Review of Ethnomedicinal Uses, Phytochemistry and Pharmacology. In *Bioactive Natural Products in Drug Discovery*; Singh, J., Meshram, V., Gupta, M., Eds.; Springer: Singapore, 2020; pp. 215–242. ISBN 9789811513930.
2. Zamakshshari, N.H.; Ee, G.C.L.; Ismail, I.S.; Ibrahim, Z.; Mah, S.H. Cytotoxic Xanthones Isolated from Calophyllum Depressinervosum and Calophyllum Buxifolium with Antioxidant and Cytotoxic Activities. *Food Chem. Toxicol.* **2019**, *133*, 110800. [CrossRef] [PubMed]
3. Leu, T.; Raharivelomanana, P.; Soulet, S.; Bianchini, J.P.; Herbette, G.; Faure, R. New Tricyclic and Tetracyclic Pyranocoumarins with an Unprecedented C-4 Substituent. Structure Elucidation of Tamanolide, Tamanolide D and Tamanolide P from *Calophyllum inophyllum* of French Polynesia. *Magn. Reson. Chem.* **2009**, *47*, 989–993. [CrossRef] [PubMed]
4. Zailan, A.A.D.; Karunakaran, T.; Bakar, M.H.A.; Mian, V.J.Y. The Malaysian Genus Calophyllum (Calophyllaceae): A Review on Its Phytochemistry and Pharmacological Activities. *Nat. Prod. Res.* 2022, *in press*. [CrossRef]
5. Leu, T. Contribution à la Connaissance de la Flore Polynésienne: ÉVALUATION de L'intérêt Pharmacologique de Quelques Plantes Médicinales et Étude Phytochimique du Tamanu (*Calophyllum inophyllum*, L.—Clusiaceae). Ph.D. Thesis, Université de la Polynésie Française, Tahiti, French Polynesia, 2009.
6. Gómez-Verjan, J.C.; Rodríguez-Hernández, K.D.; Reyes-Chilpa, R. Bioactive Coumarins and Xanthones From Calophyllum Genus and Analysis of Their Druglikeness and Toxicological Properties. In *Studies in Natural Products Chemistry*; Elsevier: Amsterdam, The Netherlands, 2017; Volume 53, pp. 277–307. ISBN 978-0-444-63930-1.
7. Dorla, E.; Grondin, I.; Hue, T.; Clerc, P.; Dumas, S.; Gauvin-Bialecki, A.; Laurent, P. Traditional Uses, Antimicrobial and Acaricidal Activities of 20 Plants Selected among Reunion Island's Flora. *S. Afr. J. Bot.* **2019**, *122*, 447–456. [CrossRef]
8. Adsersen, A.; Adsersen, H. Plants from Réunion Island with Alleged Antihypertensive and Diuretic Effects—An Experimental and Ethnobotanical Evaluation. *J. Ethnopharmacol.* **1997**, *58*, 189–206. [CrossRef]
9. Ledoux, A.; Cao, M.; Jansen, O.; Mamede, L.; Campos, P.-E.; Payet, B.; Clerc, P.; Grondin, I.; Girard-Valenciennes, E.; Hermann, T.; et al. Antiplasmodial, Anti-Chikungunya Virus and Antioxidant Activities of 64 Endemic Plants from the Mascarene Islands. *Int. J. Antimicrob. Agents* **2018**, *52*, 622–628. [CrossRef] [PubMed]
10. McKee, T.C.; Covington, C.D.; Fuller, R.W.; Bokesch, H.R.; Young, S.; Cardellina, J.H.; Kadushin, M.R.; Soejarto, D.D.; Stevens, P.F.; Cragg, G.M.; et al. Pyranocoumarins from Tropical Species of the Genus *Calophyllum*: A Chemotaxonomic Study of Extracts in the National Cancer Institute Collection. *J. Nat. Prod.* **1998**, *61*, 1252–1256. [CrossRef] [PubMed]
11. Bordignon, A.E. *Evaluation of Antioxidant and Anti-Inflammatory Effects of Medicinal Plants from Reunion Island against Obesity-Related Disorders*; Université de Liège: Liège, Belgium, 2014; p. 46.
12. Frisch, M.J.; Trucks, G.W.; Schlegel, H.B.; Scuseria, G.E.; Robb, M.A.; Cheeseman, J.R.; Scalmani, G.; Barone, V.; Mennucci, B.; Petersson, G.A.; et al. *Gaussian 09, Revision D.01*; Gaussian Inc.: Wallingford, CT, USA, 2013. Available online: http://www.gaussian.com (accessed on 1 June 2022).
13. Becke, A.D. Density-functional Thermochemistry. III. The Role of Exact Exchange. *J. Chem. Phys.* **1993**, *98*, 5648–5652. [CrossRef]
14. Stephens, P.J.; Devlin, F.J.; Chabalowski, C.F.; Frisch, M.J. Ab Initio Calculation of Vibrational Absorption and Circular Dichroism Spectra Using Density Functional Force Fields. *J. Phys. Chem.* **1994**, *98*, 11623–11627. [CrossRef]
15. AMPAC 11, 1992–2017 Semichem, Inc. 12456 W 62nd Terrace—Suite D, Shawnee, KS, USA, 66216. Available online: http://www.semichem.com (accessed on 6 June 2022).
16. Grimme, S.; Ehrlich, S.; Goerigk, L. Effect of the Damping Function in Dispersion Corrected Density Functional Theory. *J. Comput. Chem.* **2011**, *32*, 1456–1465. [CrossRef]
17. Marenich, A.V.; Cramer, C.J.; Truhlar, D.G. Universal Solvation Model Based on Solute Electron Density and on a Continuum Model of the Solvent Defined by the Bulk Dielectric Constant and Atomic Surface Tensions. *J. Phys. Chem. B* **2009**, *113*, 6378–6396. [CrossRef] [PubMed]
18. Frisch, M.J.; Trucks, G.W.; Schlegel, H.B.; Scuseria, G.E.; Robb, M.A.; Cheeseman, J.R.; Scalmani, G.; Barone, V.; Petersson, G.A.; Nakatsuji, H. *Gaussian 16, Revision A.03*; Gaussian Inc.: Wallingford, CT, USA, 2016.
19. Pluskal, T.; Castillo, S.; Villar-Briones, A.; Orešič, M. MZmine 2: Modular Framework for Processing, Visualizing, and Analyzing Mass Spectrometry-Based Molecular Profile Data. *BMC Bioinform.* **2010**, *11*, 395. [CrossRef] [PubMed]
20. Olivon, F.; Grelier, G.; Roussi, F.; Litaudon, M.; Touboul, D. MZmine 2 Data-Preprocessing To Enhance Molecular Networking Reliability. *Anal. Chem.* **2017**, *89*, 7836–7840. [CrossRef]
21. Smirnov, A.; Jia, W.; Walker, D.I.; Jones, D.P.; Du, X. ADAP-GC 3.2: Graphical Software Tool for Efficient Spectral Deconvolution of Gas Chromatography–High-Resolution Mass Spectrometry Metabolomics Data. *J. Proteome Res.* **2018**, *17*, 470–478. [CrossRef]
22. Wang, M.; Carver, J.J.; Phelan, V.V.; Sanchez, L.M.; Garg, N.; Peng, Y.; Nguyen, D.D.; Watrous, J.; Kapono, C.A.; Luzzatto-Knaan, T.; et al. Sharing and Community Curation of Mass Spectrometry Data with Global Natural Products Social Molecular Networking. *Nat. Biotechnol.* **2016**, *34*, 828–837. [CrossRef] [PubMed]

23. Nothias, L.F.; Petras, D.; Schmid, R.; Dührkop, K.; Rainer, J.; Sarvepalli, A.; Protsyuk, I.; Ernst, M.; Tsugawa, H.; Fleischauer, M.; et al. Feature-Based Molecular Networking in the GNPS Analysis Environment. *Nat. Methods* **2020**, *17*, 905–908. [CrossRef]
24. Allard, P.-M.; Péresse, T.; Bisson, J.; Gindro, K.; Marcourt, L.; Pham, V.C.; Roussi, F.; Litaudon, M.; Wolfender, J.-L. Integration of Molecular Networking and *In-Silico* MS/MS Fragmentation for Natural Products Dereplication. *Anal. Chem.* **2016**, *88*, 3317–3323. [CrossRef]
25. Yang, N.-Y.; Tao, W.-W.; Duan, J.-A. Antithrombotic Flavonoids from the Faeces of *Trogopterus xanthipes*. *Nat. Prod. Res.* **2010**, *24*, 1843–1849. [CrossRef]
26. Kijjoa, A.; Gonzalez, M.J.; Afonso, C.M.; Pinto, M.M.M.; Anantachoke, C.; Herz, W. Xanthones from *Calophyllum teysmannii* Var. Inophylloide. *Phytochemistry* **2000**, *53*, 1021–1024. [CrossRef]
27. Ito, C.; Miyamoto, Y.; Rao, K.S.; Furukawa, H. A Novel Dibenzofuran and Two New Xanthones from Calophyllum Panciflorum. *Chem. Pharm. Bull.* **1996**, *44*, 441–443. [CrossRef]
28. Patil, A.D.; Freyer, A.J.; Eggleston, D.S.; Haltiwanger, R.C.; Bean, M.F.; Taylor, P.B.; Caranfa, M.J.; Breen, A.L.; Bartus, H.R. The Inophyllums, Novel Inhibitors of HIV-1 Reverse Transcriptase Isolated from the Malaysian Tree, *Calophyllum inophyllum* Linn. *J. Med. Chem.* **1993**, *36*, 4131–4138. [CrossRef] [PubMed]
29. Kawazu, K.; Ohigashi, H.; Mitsui, T. The Psiscicidal Constituents of *Calophyllum inophyllum* Linn. *Tetrahedron Lett.* **1968**, *19*, 2383–2385. [CrossRef]
30. Laure, F.; Herbette, G.; Faure, R.; Bianchini, J.P.; Raharivelomanana, P.; Fogliani, B. Structures of New Secofriedelane and Friedelane Acids from *Calophyllum inophyllum* of French Polynesia. *Magn. Reson. Chem.* **2005**, *43*, 65–68. [CrossRef] [PubMed]
31. Cao, S.-G.; Lim, T.-B.; Sim, K.-Y.; Goh, S.H. A Highly Prenylated Xanthone from the Bark of *Calophyllum gracilipes* (Guttiferae). *Nat. Prod. Lett.* **1997**, *10*, 55–58. [CrossRef]
32. Ragasa, C.Y.; Ebajo, V., Jr.; Brkljača, R.; Urban, S. Triterpenes from *Calophyllum inophyllum* Linn. *Int. J. Pharmacogn. Phytochem. Res.* **2015**, *7*, 718–722.
33. Li, X.J.; Liu, Z.Z.; Kim, K.-W.; Wang, X.; Li, Z.; Kim, Y.-C.; Yook, C.S.; Liu, X.Q. Chemical Constituents from Leaves of *Pileostegia viburnoides* Hook.f.et Thoms. *Nat. Prod. Sci.* **2016**, *22*, 154–161. [CrossRef]
34. Jackson, B.; Locksley, H.D.; Scheinwans, F. Extractives from Guttiferae—VIII. The Isolation of 6-(3,3-Dimethylallyl)-1,5-Dihydroxyxanthone from Calophyllum Scriblitifolium Henderson and Wyatt-Smith. *Tetrahedron* **1967**, *24*, 3059–3068. [CrossRef]
35. Lim, C.K.; Subramaniam, H.; Say, Y.H.; Jong, V.Y.M.; Khaledi, H.; Chee, C.F. A New Chromanone Acid from the Stem Bark of *Calophyllum teysmannii*. *Nat. Prod. Res.* **2015**, *29*, 1970–1977. [CrossRef]
36. Huitric, A.C.; Carr, J.B.; Trager, W.F.; Nist, B.J. Configurational and Conformational Analysis. *Tetrahedron* **1963**, *19*, 2145–2151. [CrossRef]
37. Grimblat, N.; Zanardi, M.M.; Sarotti, A.M. Beyond DP4: An Improved Probability for the Stereochemical Assignment of Isomeric Compounds Using Quantum Chemical Calculations of NMR Shifts. *J. Org. Chem.* **2015**, *80*, 12526–12534. [CrossRef]
38. Gunatilaka, A.A.L.; De Silva, A.M.Y.J.; Sotheeswaran, S.; Balasubramaniam, S.; Wazeer, M.I.M. Terpenoid and Biflavonoid Constituents of Calophyllum Calaba and Garcinia Spicata from Sri Lanka. *Phytochemistry* **1984**, *23*, 323–328. [CrossRef]

metabolites

MDPI

Article

Co-Localization of Resistance and Metabolic Quantitative Trait Loci on Carrot Genome Reveals Fungitoxic Terpenes and Related Candidate Genes Associated with the Resistance to *Alternaria dauci*

Claude Emmanuel Koutouan [1], Valérie Le Clerc [1,*], Anita Suel [1], Latifa Hamama [1], Patricia Claudel [2], David Halter [2], Raymonde Baltenweck [2], Philippe Hugueney [2], Jean-François Chich [2], Sitti Anlati Moussa [1], Clémentine Champlain [1], Sébastien Huet [1], Linda Voisine [1], Sandra Pelletier [1], Sandrine Balzergue [1], Wilfried Chevalier [1], Emmanuel Geoffriau [1] and Mathilde Briard [1]

1 Institut Agro, Université d'Angers, INRAE, IRHS, SFR 4207 QUASAV, F-49000 Angers, France
2 Université de Strasbourg, INRAE, SVQV UMR-A 1131, F-68000 Colmar, France
* Correspondence: valerie.leclerc@institut-agro.fr; Tel.: +33-241-225-433

Abstract: Alternaria leaf blight, caused by the fungus *Alternaria dauci*, is the most damaging foliar disease of carrot. Some carrot genotypes exhibit partial resistance to this pathogen and resistance Quantitative Trait Loci (rQTL) have been identified. Co-localization of metabolic QTL and rQTL identified camphene, α-pinene, α-bisabolene, β-cubebene, caryophyllene, germacrene D and α-humulene as terpenes potentially involved in carrot resistance against ALB. By combining genomic and transcriptomic analyses, we identified, under the co-localization regions, terpene-related genes which are differentially expressed between a resistant and a susceptible carrot genotype. These genes include five terpene synthases and twenty transcription factors. In addition, significant mycelial growth inhibition was observed in the presence of α-humulene and caryophyllene.

Keywords: metabolomic; transcriptomic; antifungal activities; *Daucus carota*; leaf blight

Citation: Koutouan, C.E.; Le Clerc, V.; Suel, A.; Hamama, L.; Claudel, P.; Halter, D.; Baltenweck, R.; Hugueney, P.; Chich, J.-F.; Moussa, S.A.; et al. Co-Localization of Resistance and Metabolic Quantitative Trait Loci on Carrot Genome Reveals Fungitoxic Terpenes and Related Candidate Genes Associated with the Resistance to *Alternaria dauci*. *Metabolites* **2023**, *13*, 71. https://doi.org/10.3390/metabo13010071

Academic Editors: Ramona Paltinean and Irina Ielciu

Received: 8 December 2022
Revised: 15 December 2022
Accepted: 21 December 2022
Published: 2 January 2023

1. Introduction

Plants synthesize a large number of specialized metabolites (about 200,000) involved in many aspects of plant life [1,2]. Terpenoids, with an estimated 20,000 to 40,000 compounds, belong to an important family of these specialized metabolites [3,4]. Monoterpenes and sesquiterpenes are subfamilies of terpenoids and are synthetized through the methylerythritol phosphate (MEP) and the mevalonic acid (MVA) pathways, respectively. Geranyl diphosphate (GPP) and farnesyl diphosphate (FPP) are precursors to mono- and sesquiterpenes, respectively [4]. Monoterpenes and sesquiterpenes are involved in plant flavors, perfumes, thermotolerance, response to light stress, attraction of pollinators or predators of insect pests and defense against microbial pathogens [3,5–7]. Their involvement in defense response to fungal pathogens has been shown in different pathosystems [8–10]. Terpene synthases (TPS) are key enzymes in terpenoid biosynthesis, giving rise to a large diversity of terpene carbon skeletons. Plant TPS gene families may comprise several dozens of members, which have been classified into eight subfamilies, designated TPS-a to TPS-h, based on sequence and functions. In angiosperms, the TPS-a family contains mostly sesquiterpene and diterpene synthases, while monoterpene synthases belong mostly to the TPS-b and TPS-g clades [11]. In addition to terpene synthases, transcription factors (TF) can regulate gene expression involved in terpene biosynthesis. Since the initial characterization of the R2R3 MYB TF ODORANT1 as a regulator of fragrance biosynthesis in petunia flowers [12], a number of TF involved in the regulation of terpenoid biosynthesis have been characterized in a number of plant species [13].

Alternaria leaf blight is the most damaging foliar disease in carrot, causing burning symptoms on leaves, which make harvesting difficult or impossible in case of a major attack [14]. Symptoms of *A. dauci* appear as brown lesions surrounded or not by a chlorotic halo and then spread to the entire leaf under favorable conditions (average temperature 22–24 °C and high humidity in the range 96 to 100%). The fungus can also contaminate inflorescences, seeds and persist in infected soils [14]. Commercial varieties are only partially resistant and there is still a need for antifungal treatments. The objective of carrot breeders is to develop varieties with higher levels of resistance than those currently available. Deciphering the resistance mechanisms is therefore crucial for this purpose. In previous work, three QTL regions associated with resistance (rQTL) were identified within a segregating population obtained from a cross between a parent susceptible to ALB (S269) and a partially resistant one (R268) [15]. From two other connected populations, PC2 and PC3, obtained from the cross between a susceptible parent H1 and two partially resistant parents I2 and K3, respectively, Le Clerc et al. [16] highlighted new rQTL. Eleven rQTL were identified with four and five most favorable alleles coming from the two resistant parental lines, while two other favorable alleles came from the susceptible one. Global R^2 were high, i.e., 43% and 52% in 2014 and 2011, respectively. Nevertheless, the mechanisms underlying these rQTL remained unknown. In a previous work, we also showed significant differences in the accumulation of secondary metabolites belonging to the flavonoid and terpenes families in carrot genotypes with different resistance levels to ALB [17].

The objective of the present study is to evaluate the potential role of terpenes in carrot resistance based on a metabolite-QTL (mQTL)-rQTL co-localization approach. For this purpose (i) the terpene contents in plants of the PC3 segregating population were analyzed after *A. dauci* natural infestation in the field; (ii) these terpene contents were used for m-QTL mapping and the co-localization between mQTL and rQTL was analyzed; (iii) a microarray analysis was performed to identify candidate genes underlying co-localizing mQTL–rQTL; and (iv) finally, the antifungal activities of four terpene candidates were tested in vitro.

2. Materials and Methods

2.1. Plant Materials

A cross between H1, a French susceptible S3 line from the HM Clause company breeding program, and K3, a partially resistant Asiatic S2 line from the Institut Agro breeding program, was realized to develop a F2:3 segregating population named PC3. These two parental lines H1 and K3 are not only very different in terms of susceptibility to *Alternaria dauci*, but also on their metabolite and genetic profiles [16,17].

The development of the PC3 population was described in [16]. Briefly, in order to choose appropriate susceptible and resistant parental lines, different accessions were evaluated in different environments (France, Brazil, field and tunnel) from 1997 to 2006. Based on those trials and parental genetic distance, F1 hybrids were obtained from the cross between H1 and K3 and self-pollinated to provide F2 populations. About 180 individual plants from one of these best segregating F2 populations were self-pollinated to derive the F3 population called PC3 (i.e., 180 $F_{2:3}$ lines). In addition to the parental lines (H1 and K3) and the $F_{2:3}$ lines of PC3, Presto, a susceptible cultivar from Vilmorin, was also used for the trials to ensure pathogen attack. Boléro, another cultivar from Vilmorin, was used as a partially resistant reference.

2.2. Experimental Design and Crop Management

For metabolite-QTL detection, two field trials were performed, one in 2015 at Blagon (latitude 44.7835, longitude −0.9319; Gironde, France) and one in 2016 at Ychoux (latitude 44.3333, longitude −0.9667; Les Landes, France). During 2015 and 2016, 137 and 142 $F_{2:3}$ progenies from PC3 with enough seeds available were used, respectively. Two replicates per progeny and ten replicates of parental lines with about 180 seeds over a two linear meter row were sown in each field trial. Presto was sown all along a central row. Two blocks were designed on each side of this Presto row. Replicates of progenies and parental lines were randomly distributed within these two blocks. As described by [17], both trials were done in sandy soils during the optimal carrot growth period. The two fields are located in carrot production areas where *A. dauci* pressure is very high. Each replicate was harvested eight days after natural infestation in both experiments. During both field trials (2015, 2016), pathogen attacks were predicted using the Plant-Plus system® developed by Dacom (http://www.dacom.nl/ (accessed on 7 December 2022)), which considers weather conditions, plant development stage and pathogen concentration in this area of carrot production.

In 2017, the two parental lines (H1 and K3) were sown under tunnel at Angers (latitude 47.4711; longitude −0.5518; Maine et Loire, France) for microarray analysis. Four replicates of each line with 360 seeds were sown over a four linear meter row. The inoculum was prepared from four *Alternaria dauci* strains used by GEVES (Groupe d'Etude et de contrôle des Variétés et des Semences) for resistance testing of carrot varieties before their registration in the French official catalogue (GEVES, personal communication). The four replicates per line were harvested 48 h post inoculation (H48).

2.3. Sampling Design

Eight plants per replicate were harvested for mQTL and microarray analyses. The following steps were as described by [17]. Briefly, the plants were cold transported to laboratory, and then two intermediate leaves per plant were bulked from eight plants per replicate, ground in a mortar with liquid nitrogen to obtain fresh powder stored at −80 °C (called below "the roughly ground powder").

2.4. Headspace Solid-Phase Microextraction Followed by Gas Chromatography–Mass Spectrometry (HS-SPME-GC-MS) for Terpene Analyses

SPME vials (20 mL; Macherey–Nagel) containing 25 mg of fresh frozen roughly ground powder were filled with 2 mL of Na_2SO_3 solution (10 g/L). 3-octanol (50 µg) was added as an internal standard. Each sample was incubated for 15 min at 35 °C. The volatile compounds were extracted under agitation (250 rpm) with a divinylbenzene/Carboxen/polydimethylsiloxane fiber (1 cm, 23-gauge, 50/30 µm DVB/CAR/PDMS; Supelco, Bellefonte, PA, USA) at 35 °C for 15 min, fitted to a Gerstel MPS2 autosampler. The GC (Agilent 6890 Gas Chromatograph) was fitted with a DB-Wax column (i.d.: 30 m × 0.32 mm; film thickness: 0.5 µm). Helium was used as carrier gas with a column flow rate of 1.3 mL·min^{-1}. Volatiles were desorbed from the fiber in the GC inlet (220 °C) for 3 min and separated using the following temperature program: 40 °C for 5 min, increasing by 3 °C/min to 240 °C, and then held for 5 min. The MS (Agilent 5973N Mass Spectrometer) transfer line and ion source temperatures were set at 270 and 230 °C, respectively. The MS was operated in electron ionization mode and positive ions at 70 eV were recorded with a scan range from m/z 30 to m/z 300. ChemStation software (G1701DA, Rev D.03.00, Agilent, Santa Clara, CA, USA) was used for instrument control and data processing. The identity of the detected volatiles was determined by comparing their mass spectra with those of authentic standards and spectral libraries. The U.S. National Institute of Standards and Technology (NIST-05a, Gaithersburg, MA, USA), and the Wiley Registry 7th Edition mass spectral libraries were used for identification. Data are presented as normalized peak area per mg of fresh weight.

2.5. Correlation between Metabolite Accumulation and Disease Score

For disease evaluation, symptoms on leaves were scored for each replicate with 0–9 severity scale, 0 corresponds to no visible symptoms and 9 to leaves totally blight [18], which means the lower the disease score, the higher the resistance. In 2015, even if disease attack was confirmed by Dacom model and positive *A. dauci* isolation, it was not possible to evaluate symptoms due to too little disease development. Therefore, in order to rely on solid disease evaluation, the results of two previous exactly similar experiments were added to the 2016 results. Thus, three years of scoring were used for the rQTL analysis (2011, 2014 and 2016 in Les Landes), data were autoscaled, i.e., centered to mean, scaled to the standard deviation of the disease score [19] and averaged for each genotype. Terpene contents were expressed as the mean of autoscaled data (2015 and 2016). Spearman correlations were calculated between terpene contents and disease score obtained for each PC3 progeny. Significance of each correlation was estimated with p-value calculated from a Student's t-test. All analyses were performed using RStudio 1.0.136 software.

2.6. Resistance and Metabolite-QTL Detection

The genetic map was already described [20] with a few SSR added. The rQTL were already detected for 2011 and 2014 [20]. A new detection was performed to integrate the new markers and to identify the rQTL with the phenotypic evaluation performed in 2016. The QTL detection was performed by regression interval mapping using MCQTL-5.2.6-Linux.sh software [21]. Briefly, the QTL detections were performed on the two bi-parental populations (PC2 and PC3) in a connected way with the additive connected model for disease (rQTL) and with the PC3 population with the additive disconnected model for the detection of metabolite-QTLs. For each trait (disease score and secondary metabolites accumulation), a threshold value for QTL detection and co-factor selection was computed under an F test with 1000 permutations. Marker co-factors were selected in a forward method with 90% of the detection threshold value. Then, QTL was detected with the iterative QTL mapping procedure according to the detection threshold value. A QTL was indicated when the LOD (logarithm of odds) exceeded the threshold. A 1 and 1.5 LOD support interval (LOD SI) suitable for the 95% confidence interval were computed for all QTLs. The phenotypic variation (for the disease score) and the accumulation (for the metabolite) explained by each QTL and by all QTLs were calculated and referred as R^2 and global R^2, respectively. The mQTL detection was performed by regression interval mapping using MCQTL-5.2.6-Linux.sh software [21].

The co-localization zones were obtained using a two-step method: the mQTL and rQTL were mapped and then the genome zones common to their respective 1 LOD SI were defined by the SSR markers flanking these zones.

The mean of the autoscaled values (terpene contents), from the two-year experiments in field (2015 and 2016) were used for mQTL detection. For each trait, broad sense heritability (H^2) were estimated as $H^2 = \sigma G^2/\sigma P^2$, where σG^2 is the genotypic variance and σP^2 the phenotypic one. The phenotypic variance includes $\sigma P^2 = \sigma G^2 + \sigma GY^2/Y + \sigma\varepsilon^2/rY$, where σGY^2 is the genotype: environment variance, Y is the number of years, $\sigma\varepsilon^2$ is the residual variance and r is the number of replicates.

2.7. Transcriptomic Analysis

RNA extraction. Two randomly chosen H48 leaf samples for each of the two H1 and K3 lines harvested in the inoculated part of the 2017 tunnel experiment were used. About 1 g of each roughly ground powder was ground again with liquid nitrogen and iron beads using an MM2 Retsch mixer-mill to obtain very fine powder. RNA was extracted from 50 mg fine powder following the protocol of the NucleoSpin® RNA Plus kit (Macherey–Nagel, Hoerdt, France). RNA was quantified using a NanoDrop ND-1000 (NanoDrop Technologies, Wilmington, DE, USA), and RNA quality was assessed with an Agilent 2100 Bioanalyzer.

Microarray analysis. One hundred nanograms of RNA from each H48 inoculated replicate was amplified and labeled using a Low Input Quick Amp Labeling Kit (Agilent, Les Ulis, France, ref: 5190–2306) as follows. Briefly, mRNA was retro-transcribed into complementary-DNA (cDNA) and a cRNA was synthesized from the cDNA. The cRNA was amplified and labeled with two dyes, cyanine 3 or 5. Then, the cRNA was purified with Rneasy Mini Kit (Qiagen, Courtaboeuf, France, ref: 74106) and purified cRNA was hybridized onto the Agilent–*Daucus carota* v1 chip (Agilent ref: 084550-G4862A) and with Gene Expression Hybridization Kit (Agilent, Les Ulis, France, ref: 5188–5242). After hybridization, washing steps were performed with a Gene Expression Wash Buffer Kit

(Agilent, Les Ulis, France, ref: 5188–5327) and the slide was scanned using the InnoScan 710 (Innopsys, Carbone, France) scanner. Data were extracted using the Mapix® (Innopsys, Carbone, France) software.

The *Daucus carota* v1 chip was designed by the BIDefi-IRHS team using the contigs of transcriptome sequencing from [22]. It has been submitted to Gene Expression Omnibus under the accession number GPL25816. The microarray contains three sense and two complementary antisense sequences probes corresponding to 33,978 coding genes. All the probes were designed within the Coding DNA Sequence (CDS) of the corresponding genes and fixed on a 4 × 180 K microarray slide. For the comparison of H1 and K3 inoculated samples, two biological replicates with two technical repetitions per replicate were analyzed in dye-swap, as described by [23].

2.8. Statistical Analysis and Highlighting of Differentially Expressed Genes within the Co-Localization Area

Analyses were performed using the R package Limma from Bioconductor. First, data were normalized with the Loess method. Then, the Lmfit function and the eBayes function from the Limma package were used to highlight differentially expressed genes. Background noise (i.e., the average of the lowest intensities plus two times the standard deviation) was subtracted from normalized data [24]. Only transcripts with adjusted *p*-value (Benjamini–Hochberg) <0.01 were considered.

Molecular markers (SSR) flanking mQTL-rQTL co-localization areas were aligned to the carrot genome using the Geneious 10.2.3 software [22,25]. The locus number of each gene inside the co-localization area was extracted and associated with the corresponding differentially expressed gene already identified.

2.9. Fungal Growth Inhibition Assays

The *Alternaria dauci* P2 (FRA017) strain with medium aggressiveness [26] was cultivated on V8 agar medium in the dark at 22 °C for 10–15 days. After fungal growth, one infected agar plate was punched out with a 5 mm diameter punch, which was deposited on the side of a 5 cm diameter malt/agar Petri dish. A sterile antibiotic assay paper (Dominique Dutscher, Bernolsheim, France) was deposited at the opposite side of the Petri dish, as illustrated in Figure S1. Five microliters of pure terpenes were poured on the paper, the Petri dishes were closed with parafilm and conserved at 25 °C. Selected terpenes (α-pinene, camphene, caryophyllene and humulene) were purchased from Sigma-Aldrich (L'lsle-d'Abeau, France). Terpene concentrations were chosen so that they were of the same order of magnitude as the natural terpene concentrations in carrot leaves. Indeed, terpene concentrations used in vitro were around 1 mM. Individual terpene concentrations in leaves of the carrot genotypes used in this work reached 0.3 mM and estimated total terpene amounts exceeded 200 mg/kg (representing about 2 mM when calculated for monoterpenes). In addition, the sequestration of terpenes in glandular trichomes probably results in *Alternaria dauci* facing much higher local terpene concentrations when infecting carrot leaves.

Mycelial growth was monitored by using pictures taken with a Nikon D5600 camera with an AF-S Nikkor 50 mm 1:1.4, at 0, 3, 4, 5 and 6 days. *A. dauci* growth was evaluated by measuring mycelium surface using Fiji software [27]. Two independent experiments were performed, each with three replicates per treatment. *T*-tests were performed using Excel.

3. Results

3.1. Consistency of rQTL among Years

In 2016, we detected less rQTL (4 rQTL) than in 2011 (5 rQTL) and 2014 (7 rQTL), and the phenotypic variation explained by the rQTL (global R^2) was lower in 2016 (24%) than in 2011 (43%) and 2014 (58%). Nevertheless, the four rQTL found in 2016 overlapped with those identified during the two previous years with some differences in the confidence interval of each rQTL (Table S1). The most important and consistent QTL through the years was observed on chromosome 6. The heritability of disease score during the three years was very high, around 71%.

3.2. *mQTL-rQTL Co-Localization Analysis Reveals Candidate Terpenes for Resistance to ALB*

Using terpene profiling data determined by HS-SPME-GC-MS, we performed an mQTL detection for 30 terpenes (15 monoterpenes and 15 sesquiterpenes, Table S2). We found mQTLs for 25 terpenes (11 monoterpenes and 14 sesquiterpenes). Among these 25 terpenes, 22 had mQTLs which co-localized with r-QTLs on chromosome (chr) 1, 2, 4, 6 and 8 with two hotspots on chr 4 and one hotspot on chr 8 (Figure S2). Furthermore, we performed a correlation analysis between terpene accumulation levels and ALB disease scores. Seven terpenes, namely α-pinene, camphene, α-bisabolene, α-humulene, β-cubebene, caryophyllene and germacrene D, showed significant negative correlation with disease score, i.e., higher levels of these terpenes were associated with lower symptoms (Figure 1).

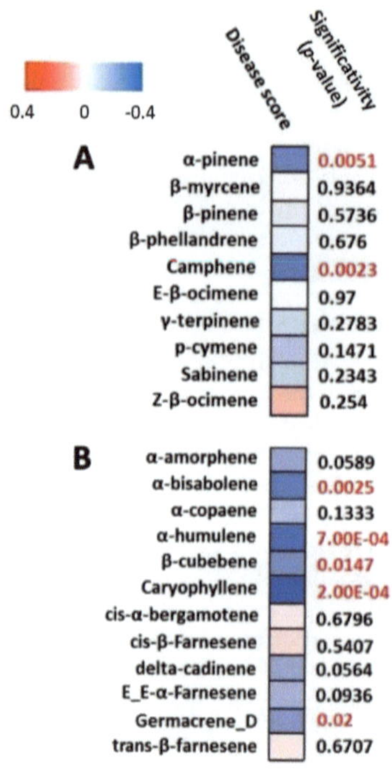

Figure 1. Spearman rank correlation between disease score and the contents of the 22 terpenes ((**A**) monoterpenes; (**B**) sesquiterpenes) that mQTL co-localized with rQTL. Red square indicates positive correlation and blue, negative correlation (i.e., the higher levels of these terpenes the lower the symptoms). Significance of correlation is given by the *p*-value. Red labels indicate *p*-values < 0.05.

mQTL associated with these seven terpenes co-localized with rQTL on chr 1, 4, 6 and 8 (Figure 2). For mQTL on chr 1, 4 and 6, favorable alleles for their accumulation and favorable alleles for resistance underlying the rQTL were brought by K3. For rQTLs on chr 8, H1 brought the favorable allele for resistance and the favorable alleles for terpene accumulation (α-humulene, caryophyllene and germacrene D). Metabolite accumulation explained by the mQTL (global R^2) ranged between 12% (for α-humulene) and 58% (for α-pinene and camphene) (Table 1). The heritability varied from 31% (β-cubebene) to 65 and 68% for camphene and α-pinene, respectively. Camphene and α-pinene exhibited the same max position for their respective two mQTL on chr4 with approximately the same R^2. These two metabolites shared the same max position with caryophyllene at the bottom of the chr 4. Germacrene D and β-cubebene also had the same max position on chr 1 and chr 4 (Figure 2 and Table 1).

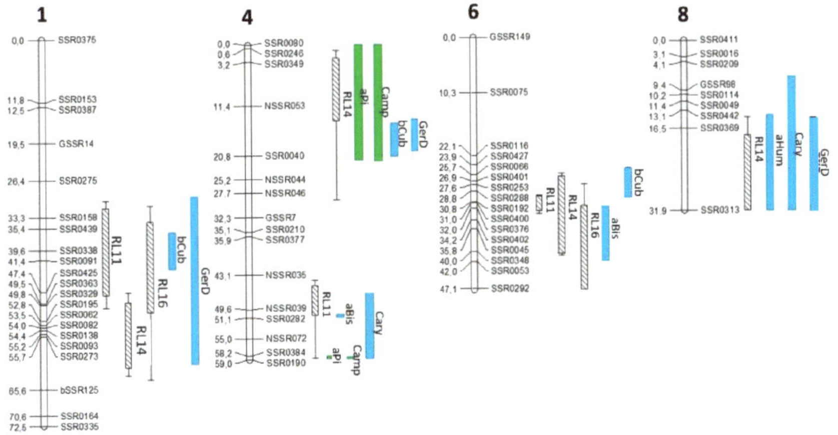

Figure 2. Co-localization between rQTL and mQTL of terpenes presenting significant negative correlation with ALB disease score. rQTL are named RL for resistance loci with the year of phenotyping. The confidence interval is bordered by two small vertical lines for the 1.5 LOD support interval (SI). A hatched area represents a 1 LOD SI. Metabolite-QTL is named with an abbreviation of each terpene: aPi = α-pinene, Camp = camphene, aBis = α-bisabolene, aHum = α-humulene, bCub = β-cubebene, Cary = caryophyllene, GerD = germacrene D. Monoterpene and sesquiterpene mQTL are represented with green and blue bars, respectively, with a 1 LOD SI.

Table 1. Characteristics of terpene mQTL co-localizing with rQTL and significantly correlated with disease score on carrot leaves after *A. dauci* attack. SI = support interval; Chr = chromosome; RL= resistance loci; R^2 = explained phenotypic variation; cM = centimorgan. aPi = α-pinene, Camp = camphene, aBis = α-bisabolene, aHum = α-humulene, bCub = β-cubebene, Cary = caryophyllene, GerD = germacrene D; ^{t} and ^{b} = top and bottom of Chr respectively.

Code of mQTL	Chr	1-LOD SI (cM)	Max Position (cM)	R^2 (%)	Global R^2 (%)	Heritability H^2 (%)	Additive Effect of Allele	
							H1	K3
aPi	4^t	0–21.6	20.4	11.4	58.3	68	−0.157	0.157
	4^b	58.3–58.7	58.7	42.9			−0.388	0.388
Camp	4^t	0–21.7	20.4	10.7	57.6	65	−0.153	0.153
	4^b	58.4–58.7	58.7	42.5			−0.389	0.389
aBis	4^b	50.6–51	50.8	51.7	53.1	34	−0.436	0.436
	6	31.8–41.9	40.2	9.7			−0.140	0.140
aHum	8	14.1–31.9	16.4	12.2	12.2	39	0.209	−0.209
bCub	1	36.3–43.2	39.6	20.5	41.7	31	−0.263	0.263
	4^t	14.7–20.8	16.4	26.9			−0.273	0.273
	6	24.6–30.1	28.9	12.6			−0.167	0.167
Cary	4^b	46.6–58.7	58.7	9.3	18.6	33.3	−0.161	0.161
	8	6.9–31.9	11.3	9.1			0.174	−0.174
GerD	1	29.7–61	39.6	10	46	36.2	−0.173	0.173
	4^t	13.9–19.8	16.4	26.4			−0.271	0.271
	8	14.5–31.9	21.4	18.7			0.235	−0.235

3.3. *Differential Expression Analysis of Genes Underlying the Co-Localization Regions*

We compared gene expression in H1 and K3 lines 48 h after *A. dauci* inoculation. We gave special attention to genes under mQTL-rQTL co-localization areas. Candidate genes potentially related to terpene accumulation were selected based on their belonging to either the terpene synthase (TPS) or the transcription factors (TF) gene families. Such genes were identified at all co-localization sites and some of them were significantly differentially expressed between the resistant K3 and the susceptible H1 line (Table 2).

Table 2. Number of all genes and genes differentially expressed (GDE) identified by microarray analysis under mQTL-rQTL co-localization with special emphasis to the GDE potentially related to terpenes. ^{t} and ^{b} = top and bottom of chromosome (Chr) respectively.

Chr	Located Genes	Genes Differentially Expressed between K3 and H1		
		Underexpressed	Overexpressed	Related to Terpenes
1	2181	207	206	5
4 ^{t}	1229	76	92	2
4 ^{b}	1718	154	196	10
6	577	63	45	3
8	1025	100	105	5

For chr 1, five genes coding for transcription factors (TF) were identified at the co-localization site, i.e., ethylene response factor (ERF), two basic leucine zipper (bZIP) and two WRKY. The genes coding for bZIPHY5 and the WRKY (WRKY33 and WRKY48) were overexpressed in K3 compared to H1, in contrast to those coding for ERF71 and bZIP53. However, no TPS gene was found at this site (Table 3).

Table 3. Terpene synthase and transcription factor genes underlying co-localization regions, which are differentially expressed between H1 (susceptible) and K3 (resistant) genotypes. TF = transcription factor; TPS = terpene synthase; ^{t} and ^{b} = top and bottom of chromosome (Chr), respectively. Dc refers to TPS naming system proposed by [28].

Chr	Gene Name (Locus Number)	Function	Relative Expression		*p*-Value
			H1	K3	
1	bZIPHY5 (108204232)	TF	3.94	4.48	0.004
	WRKY33 (108204915)	TF	2.65	4.64	$<2 \times 10^{-16}$
	WRKY48 (108204668)	TF	2.67	3.79	$<2 \times 10^{-16}$
	ERF71 (108206243)	TF	0.98	0.15	0.0004
	bZIP53 (108206338)	TF	3.28	2.23	$<2 \times 10^{-16}$
4 ^{t}	NAC29 (108218926)	TF	4.44	5.90	0.0004
	WRKY33 (108219317)	TF	4.00	6.21	$<2 \times 10^{-16}$
4 ^{b}	Terpene synthase 10-like (DcTPS55;108217470)	TPS	3.68	4.76	0.002
	Terpene synthase 10-like (DcTPS26;108217599)	TPS	4.19	5.22	$<2 \times 10^{-16}$
	Terpene synthase 10-like (DcTPS54;108217617)	TPS	5.80	6.92	$<2 \times 10^{-16}$
	β-bisabolene synthase-like (108216085)	TPS	6.67	5.67	0.0004
	WRKY7 (108215789)	TF	5.39	4.74	0.0008

Table 3. *Cont.*

Chr	Gene Name (Locus Number)	Function	Relative Expression		p-Value
			H1	K3	
	ERF054 (108216387)	TF	3.45	2.1	$<2 \times 10^{-16}$
	NAC2 (108215781)	TF	4.51	5.26	0.0004
	ERF4 (108217832)	TF	1.87	2.87	$<2 \times 10^{-16}$
	bZIP17 (108218833)	TF	3.98	4.76	0.0004
	bZIP27 (108217633)	TF	3.60	0.51	$<2 \times 10^{-16}$
6	bZIP61 (108225065)	TF	4.28	4.89	0.007
	ERF4 (108225207)	TF	0.82	3.29	0.002
	SPL1 (108224238)	TF	3.98	2.84	$<2 \times 10^{-16}$
8	α-farnesene synthase-like (DcTPS44; 108198720)	TPS	0.23	2.23	$<2 \times 10^{-16}$
	AP2/ERF (108198780)	TF	0.92	5.22	$<2 \times 10^{-16}$
	ERF1B-like (108198802)	TF	1.35	3.21	$<2 \times 10^{-16}$
	MYB (108197621)	TF	0.74	0.19	0.005
	ERF_like (108197006)	TF	4.38	2.66	$<2 \times 10^{-16}$

Under the rQTL region RL14 at the top of chromosome 4, only two genes coding for transcription factors NAC29 and WRKY33 were significantly overexpressed in K3. In the same region, we also identified a gene encoding a germacrene D synthase-like enzyme (108217562).

Among the ten genes potentially related to terpenes at the co-localization site at the bottom of the same chromosome 4, we identified four TPS and six transcription factor genes. Three TPS genes were annotated as coding for terpene synthase 10-like [22] and correspond to DcTPS55, DcTPS26 and DcTPS54 using the most recent naming system proposed by Keilwagen et al. (2017). These genes were functionally characterized by [29,30]. The last TPS gene was annotated as β-bisabolene synthase-like gene. DcTPS55, DcTPS26 and DcTPS54, NAC2, ERF4 and bZIP17 were significantly overexpressed in K3, while the genes coding for β-bisabolene synthase-like gene, WRKY7, ERF054 and bZIP27, were underexpressed in K3 compared to H1 (Tables 2 and 3). In addition, we identified a germacrene D synthase-like gene (108216912) and a TPS 10-like gene (108217598, DcTPS04), but they were not differentially expressed between resistant (K3) and susceptible (H1) genotypes.

In the co-localization region on chr 6 (Table 2), differentially expressed TF genes were bZIP61 and ERF4 with higher expression in K3 and SQUAMOSA-promoter binding protein 1 (SPL1) higher in H1 (Table 3). Finally, on chr 8, differentially expressed candidates were an α-farnesene synthase-like gene (Table 3) overexpressed in K3 and four genes coding for transcription factors with AP2/ERF and ERF1B-like higher in K3, while one MYB and ERF-like genes were underexpressed in K3 (Table 3).

3.4. *Further Analysis of TPS Genes Overexpressed in K3*

Among the genes overexpressed in K3 in comparison to H1 (Figure 3), we identified three genes encoding monoterpene synthases DcTPS55, DcTPS26 and DcTPS54 genes under rQTL on chr 4 between NSSR039 and SSR0282. The four terpenes whose mQTL co-localized with the rQTL in this region were α-bisabolene, α-pinene, camphene and caryophyllene (Figure 2 and Table 1). Under rQTL on chr 8, we detected an α-farnesene synthase-like DcTPS44 gene overexpressed for K3 (Figure 3). It was consistent with the mQTL detected for α-farnesene on this chr 8 (Figure S2) but in our study, α-farnesene level was not correlated with resistance (Figure 1).

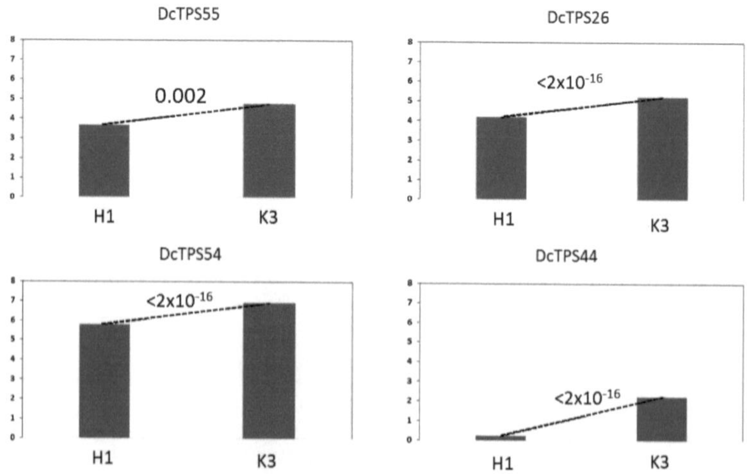

Figure 3. Relative expression of TPS genes located in the mQTL-rQTL co-localization regions on chromosomes 4 and 8 in H1 (susceptible) and K3 (resistant) genotypes. The number above the dotted line indicates significance of the difference (p-value).

3.5. *In Vitro Bioactivity of α-Pinene, Camphene, Caryophyllene and Humulene towards A. dauci*

The effect on fungal growth of four commercially available compounds among the seven candidate terpenes was assessed using an in vitro assay. By comparing mycelium growth in control conditions and in the presence of the selected terpenes, significant growth inhibition was observed with α-humulene and caryophyllene from the fourth day of the experiment. Growth inhibition persisted until the sixth day. Conversely, no significant growth inhibition was observed in these conditions with camphene and α-pinene (Figure 4).

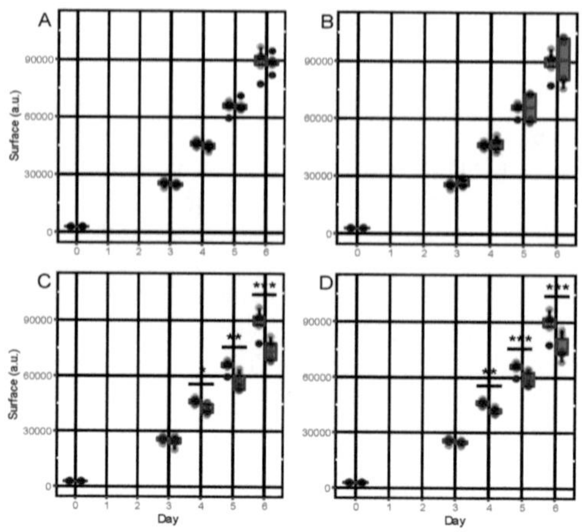

Figure 4. Inhibitory effect of terpenes on Alternaria dauci growth. Growth of Alternaria dauci was monitored as described in Section 2 (Materials and Methods). Two independent experiments with 3 replicates were performed. The surface of Alternaria mycelium was measured at days 0, 3 and 6. Controls are in gray and terpene assays in red. Tested terpenes were: (**A**) camphene, (**B**) α-pinene, (**C**) caryophyllene, (**D**) α-humulene. The symbols *, ** and *** indicate significantly different mycelium surfaces compared to control conditions using a T test, with p-value <0.05, <0.01 and <0.001, respectively.

4. Discussion

Quantitative trait loci detection is a good way to understand the genetic basis of variation associated with a quantitative trait [31]. Unfortunately, most of the time, the underlying mechanisms are poorly understood. In 2018, Koutouan et al. [17] identified fifteen terpenes which were more highly accumulated in the resistant genotype K3 than in the susceptible H1. The present study aimed at the identification of mQTL of terpenes co-localizing with rQTL in a segregating population obtained between H1 and K3. This mQTL-rQTL co-localization analysis was performed without a priori on all terpenes detected in carrot leaves, with no particular focus on the fifteen terpenes identified by [17].

This mQTL-rQTL co-localization approach identified camphene, α-pinene, α-bisabolene, α-humulene, β-cubebene, caryophyllene and germacrene D as candidate terpenes potentially associated with the resistance to *A. dauci*. Interestingly, these metabolites were previously identified as differentially accumulated between several resistant and susceptible genotypes by [17] except β-cubebene, and moreover they showed that camphene, α-pinene, α-bisabolene, α-humulene and caryophyllene were more accumulated in K3 compared to H1. For these five terpenes, we mapped mQTL but also β-cubebene and germacrene D, which were not differentially accumulated between parent H1 and K3. This last result was unexpected, as it seems more difficult to map QTL whatever the trait when there is no difference between parental lines for this trait. Keurentjes et al. [32] also identified mQTL for metabolites which were not differentially accumulated between parents. These detections could be explained by genome recombination between the two parental alleles. This result confirms that relying only on differential metabolite accumulation between parental lines is not sufficient to postulate the involvement of some metabolites in resistance, as the link between metabolite amounts and resistance may be fortuitous. For example, in the present study, only 7 of the 15 terpenes highlighted in the previous study by [17] were confirmed, while some new candidates have been identified, such as β-cubebene and germacrene D.

Through the mQTL-rQTL co-localization approach and a correlation test, we could associate seven metabolites with resistance. These seven terpenes were negatively correlated with disease score, which means that their accumulation is favorable for resistance. In addition, for a given chromosome, the parental origin of the favorable alleles (leading to metabolite accumulation) of mQTL was also the one bringing the favorable allele for resistance. Working on resistance of *Arabidopsis* to *Botrytis cinerea*, Rowe and Kliebenstein [33] used the same approach and highlighted that mQTL of camalexin co-localized with rQTL. They also indicated that both favorable alleles for camalexin accumulation and for resistance to *B. cinerea* were brought by the same genotype, which makes sense with the involvement of camalexin in resistance. On the same topic, Zhang et al. [34] showed co-localization between mQTLs of indole glucosinolates and rQTLs involved in *Arabidopsis* resistance to *B. cinerea*. They also showed that favorable alleles for resistance and indole glucosinolate accumulation came from the same resistant genotype.

In our analysis, mQTL for different terpenes often co-localized on the genome with the same maximal position and sometimes with the same R^2. This could be explained by the fact that a single terpene synthase may catalyze the synthesis of different products. For example, Gambliel and Croteau [35] and Huber et al. [36] showed that the activity of pinene synthase (EC:4.2.3.119) could lead to an equal amount of α-pinene, camphene or β-pinene. Using the carrot genome published by [22], we searched for genes potentially involved in terpene synthesis within the co-localizing loci. Among these genes, our microarray analysis revealed differential gene expression for four TPS genes under mQTL-rQTL at the bottom of chromosome 4; three of them (DcTPS55, DcTPS26 and DcTPS54) were the ones identified by [30]. It is well known that in many plant genomes, terpene synthase genes are organized in clusters [37,38]. Working on carrot terpenes, Keilwagen et al. [28] identified in the same area a cluster of TPS on chr 4 with five terpene synthase genes. In the present study, two other TPS genes were identified in this same area, but they were not differentially expressed between H1 and K3.

Recently, Reichardt et al. [30] investigating this region of carrot genome identified a QTL cluster for sabinene, α-terpinene, γ-terpinene and terpinen-4-ol in leaves while in vitro assay identified DcTPS54 as a single product sabinene synthase. Additionally, among the 19 TPS biochemically characterized by Muchlinski et al. [29], recombinant DcTPS55, DcTPS26 and DcTPS54 proteins produced in vitro similar volatiles profiles such as sabinene, limonene, β-myrcene, α-pinene and α-terpineol from geranyl diphosphate, while DcTPS26 also produced several sesquiterpenes including β-bisabolene. At this time, there is no evidence that one of these TPS could be responsible for the accumulation of the mapped metabolite identified in this chr 4 region, but it is very likely. On chr 8, we also identified a gene coding for an α-farnesene synthase-like that showed similarity with known terpene synthases genes coding for germacrene D, caryophyllene and α-humulene. However, additional characterization of these TPS genes is necessary to investigate its involvement in the accumulation of these three co-localizing metabolites.

In addition to the TPS 10-like and α-farnesene-like genes, we also identified transcription factor genes that can be involved in the regulation of terpene accumulation. The role of the WRKY transcription factor in regulation of specialized metabolite is well known [39]. Tholl [3] was the first to review the involvement of the WRKY transcription factor in TPS regulation and the presence of W-box, the binding site of WRKY over numerous TPS. For example, Spyropoulou et al. [40] showed that WRKY73 could activate three TPS promoters from *Solanum lycopersicum*. Herein, we identified three different WRKY genes, two on chr 1 (WRKY33 and WRKY48) where germacrene D and β-cubebene mQTL mapped, and one WRKY33 on the top part of chr 4 where α-pinene, camphene, germacrene D and β-cubebene mQTL mapped. These three WRKY genes were overexpressed in K3, which coincided with favorable alleles for accumulation on co-localizing mQTL (α-pinene, camphene, germacrene D and β-cubebene). As far as we know, no WRKY33 or 48 have been identified in regulation of TPS, but their role in resistance to pathogens has been highlighted, especially for WRKY33. Zheng et al. [41] indicated that WRKY33 was involved in *Arabidopsis* resistance to *Botrytis cinerea* and *Alternaria brassicicola* through activation of the jasmonate biosynthesis pathway, which itself regulated activation of defense genes. In a previous work, Rodriguez et al. [42] demonstrated that downregulation of limonene synthase in orange peel enhanced resistance to *Penicillium digitatum*. Afterward, they showed that orange line AS7 (antisense 7), which was downregulated for limonene synthase gene, overexpressed a putative gene encoding the WRKY33 [43]. In our case, the WRKY33 could be part of a signal cascade to induce resistance by triggering TPS synthesis.

In addition to WRKY, we also identified other transcription factor families. The bZIP TFs were found on almost all the co-localization sites except on chr 8. Among all, the HY5 overexpressed by K3 has been shown by Zhou et al. [44] to be a positive regulator of β-pinene synthase gene in *Artemisia annua* L. Two NAC domain transcription factors (TF), under chr 4, were overexpressed by K3. These NAC transcription factors were NAC2 and NAC29. Interestingly, the NAC2 was shown to bind to the promoter of a kiwi (*Actinidia arguta*) *Aa*TPS1 gene, which resulted in an overexpression of the gene, overaccumulation of the protein and a high level of terpinolene, myrcene, limonene, α-pinene and linalool [45]. Ethylene response factor (ERF) TF was found under all of our mQTL-rQTL co-localization areas except for the top of chr 4. Their overexpression was divergent, ERF4 on chr 4 and chr 6 (108217832 and 108225207) and ERF1B-like (108198802) were overexpressed in K3 at the opposite of the three others on chr 1, 4b and 8. One of these ERF, ERF71 has been identified by Li et al. [46] as upregulator (activator) of geraniol synthase, which catalyzes the synthesis of the monoterpene E-geraniol in *Citrus sinensis*. In our pathosystem, if ever ERF71 would also catalyze synthesis of monoterpens, it could result in fewer quantities of sesquiterpens. This could be consistent with the unfavorable alleles for accumulation of β-cubebene and germacrene D observed for H1 on chr 1. Clearly, further analyses are needed to decipher ERF impacts within carrot *A. dauci* interaction. Finally, the last family of TF identified was the SQUAMOSA binding protein 1-like (SPL1) found on chr 6 and overexpressed by H1. Yu et al. [47] indicated that an SPL TF was activator of a TPS21 that catalyzes the biosynthesis of caryophyllene in *Arabidopsis*. As previously suggested for ERF71, if ever this SPL TF would also play a role in the accumulation of the α-bisabolene and β-cubebene in carrot, which mQTL co-localized with this TF gene, we hypothesized SPL TF would be a downregulator of this accumulation because favorable allele for accumulation of mQTL of α-bisabolene and β-cubebene came from K3.

Based on the co-localization between metabolic and resistance QTL, we proposed that some terpenes could be part of the resistance mechanism against the fungus *A. dauci*. In order to substantiate this hypothesis, we tested the potential fungitoxic properties of four of these terpenes, which were commercially available. Indeed, significant mycelial growth inhibition was observed in the presence of α-humulene and caryophyllene. Conversely, camphene and α-pinene did not show any significant effect on fungal growth. However, relative quantification of these two last compounds in leaf samples showed three times higher concentration in the resistant K3 line than in the susceptible H1 [17] suggesting together with mQTL/rQTL co-localization that they could play a role in the resistance of carrot to *A. dauci*. These apparently contradictory results suggest that depending on the metabolite, different stages of the life cycle of the fungus could be affected, which makes sense when considering the mQTL/rQTL co-localization with the four terpenes: while the two monoterpene α-pinene and camphene mQTL co-localized with rQTL on chromosome 4, the two sesquiterpene α-humulene and caryophyllene mQTL were mainly co-localizing with rQTL on chromosome 8. We previously concluded that these different rQTLs could act on different components of quantitative resistance, some of them being able to reduce the efficiency of fungal penetration while others delay the invasion and growth of the pathogen in the leaf [16]. In this hypothesis, α-humulene and caryophyllene may impact fungal growth, as determined by our assay, while α-pinene and camphene may act on a different, but not yet clearly identified fungal target. Preliminary results suggested that it could be around conidia germination, as illustrated in Figures S3 and S4.

Literature already reported the involvement of some of these terpenes in resistance to some pathogens using in vitro assay. Hammer et al. [48] showed fungistatic and fungicidal activity of α-pinene against *Aspergillus niger, A. flavus* and *A. fumigatus*. Sati et al. [49] demonstrated that essential oil of *Artemisia nilagirica*, in which germacrene D, caryophyllene and camphene are main constituents, had significant antifungal effect against the plant pathogenes *Rhizoctonia solani*, *Sclerotium rolfsii* and *Macrophomina phaseolina*. Caryophyllene alone inhibited mycelial development of *Aspergillus niger*, which may be a pathogen of both plants and humans [50]. García-Rellán et al. [51] showed that essential oil from *Satureja cuneifolia*, containing camphene and camphor, exhibited strong antifungal activity against *Phytophthora palmivora* and *Phaemoniella chlamydospora*. Interestingly, the reported antifungal activity of terpenes was mostly shown through essential oils with different components, suggesting that the seven terpenes identified here could act in a synergistic way for carrot resistance to *A. dauci*.

The roles of metabolite in plant resistance to pathogens have been suggested across an abundance of literature; however, few studies have combined genetic and metabolomic approaches to investigate metabolite-associated defenses [52]. Herein, through an mQTL-rQTL co-localization approach, we highlighted seven terpenes that were strongly associated with carrot resistance to *A. dauci*. Using transcriptomics and genome analysis, we identified potential enzymes and regulators involved in the biosynthesis of these terpenes. Finally, functional analyses revealed that two of these terpene candidates, α-humulene and caryophyllene, exhibited fungitoxic properties, consistent with a direct role of these compounds in disease resistance. Future work will be aimed at evaluating the role of other candidate metabolites or combinations of them on carrot resistance to *A. dauci*.

Supplementary Materials: The following supporting information can be downloaded at: https://www.mdpi.com/article/10.3390/metabo13010071/s1, Figure S1: Illustration of the fungal growth inhibition assay; Figure S2: Co-localization of 22 terpene mQTL with rQTL; Figure S3: Evaluation of the inhibitory effect of α-pinene and camphene (7.34 mM) on germination of *Alternaria dauci* P2 strain after 4 and 70 h of incubation; Figure S4: Evaluation of the inhibitory effect of α-pinene and camphene on growth of *Alternaria dauci* P2 strain after 8 days of incubation; Table S1: List of quantitative resistance loci (rQTL) detected by connected analysis; Table S2: List of all terpenes per family.

Author Contributions: C.E.K., V.L.C., M.B. and A.S. initiated the work and created the workplan. A.S. performed the inoculation in the tunnel trial. C.E.K., S.A.M., A.S., V.L.C., L.H., S.H. and M.B. performed the phenotyping experiments. C.E.K., S.A.M., P.C., L.V. and R.B. performed the metabolic analyses. C.E.K. analyzed all data with the participation of D.H. for metabolic analyses. S.P. and S.B. designed the microarray. C.C. analyzed the microarray. J.-F.C. and A.S. performed the in vitro assay. C.E.K., V.L.C., M.B. and P.H. wrote the manuscript. W.C. and E.G. reviewed the manuscript and provided comments. All authors have read and agreed to the published version of the manuscript.

Funding: This research was funded by a grant from Ministry for Higher Education and Research (France) and conducted in the framework of the regional program "Objectif Végétal", Research, Education and Innovation in Pays de la Loire, supported by the French Region Pays de la Loire, Angers Loire Métropole and the European Regional Development Fund.

Institutional Review Board Statement: Not applicable.

Informed Consent Statement: Not applicable.

Data Availability Statement: The data presented in this study are available in the article and Supplementary Materials.

Acknowledgments: Mathilde Piquet, Marie-Hélène Bouvet Merlet, Kerigan Tahari, Safae Ouchetto and Mylène Giteau are warmly acknowledged for their technical assistance during the laboratory experiments. Invenio and Imorphen are also acknowledged for their help during the field and tunnel trials. We are thankful to Anan for their support in the design of the microarray and Imac for their help with the microscopic analyses. We thank the RFI Objectif Vegetal for financing the Metabocar project.

Conflicts of Interest: The authors declare no conflict of interest.

References

1. Fiehn, O. Metabolomics—The link between genotypes and phenotypes. *Plant Mol. Biol.* **2002**, *48*, 155–171. [CrossRef] [PubMed]
2. Goodacre, R.; Vaidyanathan, S.; Dunn, W.B.; Harrigan, G.G.; Kell, D.B. Metabolomics by numbers: Acquiring and understanding global metabolite data. *Trends Biotechnol.* **2004**, *22*, 245–252. [CrossRef] [PubMed]
3. Tholl, D. Terpene synthases and the regulation, diversity and biological roles of terpene metabolism. *Curr. Opin. Plant Biol.* **2006**, *9*, 297–304. [CrossRef] [PubMed]
4. Tholl, D. Biosynthesis and biological functions of terpenoids in plants. *Biotechnol. Isoprenoids* **2015**, *148*, 63–106. [CrossRef]
5. Sharkey, T.D.; Yeh, S. Isoprene emission from plants. *Annu. Rev. Plant Physiol. Plant Mol. Biol.* **2001**, *52*, 407–436. [CrossRef]
6. Pichersky, E.; Gershenzon, J. The formation and function of plant volatiles: Perfumes for pollinator attraction and defense. *Curr. Opin. Plant Biol.* **2002**, *5*, 237–243. [CrossRef]
7. Gil, M.; Pontin, M.; Berli, F.; Bottini, R.; Piccoli, P. Metabolism of terpenes in the response of grape (*Vitis vinifera* L.) leaf tissues to UV-B radiation. *Phytochemistry* **2012**, *77*, 89–98. [CrossRef]
8. Rodríguez, A.; Andrés, V.S.; Cervera, M.; Redondo, A.; Alquézar, B.; Shimada, T.; Gadea, J.; Rodrigo, M.; Zacarías, L.; Palou, L.; et al. The monoterpene limonene in orange peels attracts pests and microorganisms. *Plant Signal. Behav.* **2011**, *6*, 1820–1823. [CrossRef]
9. Pontin, M.; Bottini, R.; Burba, J.L.; Piccoli, P. Allium sativum produces terpenes with fungistatic properties in response to infection with *Sclerotium cepivorum*. *Phytochemistry* **2015**, *115*, 152–160. [CrossRef]
10. Abbas, F.; Ke, Y.; Yu, R.; Yue, Y.; Amanullah, S.; Jahangir, M.M.; Fan, Y. Volatile terpenoids: Multiple functions, biosynthesis, modulation and manipulation by genetic engineering. *Planta* **2017**, *246*, 803–816. [CrossRef]
11. Chen, F.; Tholl, D.; Bohlmann, J.; Pichersky, E. The family of terpene synthases in plants: A mid-size family of genes for specialized metabolism that is highly diversified throughout the kingdom. *Plant J.* **2011**, *66*, 212–229. [CrossRef]
12. Verdonk, J.C.; Haring, M.A.; Van Tunen, A.J.; Schuurink, R.C. ODORANT1 regulates fragrance biosynthesis in petunia flowers. *Plant Cell* **2005**, *17*, 1612–1624. [CrossRef]

13. Picazo-Aragonés, J.; Terrab, A.; Balao, F. Plant volatile organic compounds evolution: Transcriptional regulation, epigenetics and polyploidy. *Int. J. Mol. Sci.* **2020**, *21*, 8956. [CrossRef]
14. Farrar, J.J.; Pryor, B.M.; Davis, R.M. Alternaria diseases of carrot. *Plant Dis.* **2004**, *88*, 776–784. [CrossRef]
15. Le Clerc, V.; Pawelec, A.; Birolleau-Touchard, C.; Suel, A.; Briard, M. Genetic architecture of factors underlying partial resistance to Alternaria leaf blight in carrot. *Theor. Appl. Genet.* **2009**, *118*, 1251–1259. [CrossRef]
16. Le Clerc, V.; Marques, S.; Suel, A.; Huet, S.; Hamama, L.; Voisine, L.; Auperpin, E.; Jourdan, M.; Barrot, L.; Prieur, R.; et al. QTL mapping of carrot resistance to leaf blight with connected populations: Stability across years and consequences for breeding. *Theor. Appl. Genet.* **2015**, *128*, 2177–2187. [CrossRef]
17. Koutouan, C.E.; Le Clerc, V.; Baltenweck, R.; Claudel, P.; Halter, D.; Hugueney, P.; Hamama, L.; Suel, A.; Huet, S.; Bouvet Merlet, M.H.; et al. Link between carrot leaf secondary metabolites and resistance to *Alternaria dauci*. *Sci. Rep.* **2018**, *8*, 13746. [CrossRef]
18. Pawelec, A.; Dubourg, C.; Briard, M. Evaluation of carrot resistance to Alternaria leaf blight in controlled environments. *Plant Pathol.* **2006**, *55*, 68–72. [CrossRef]
19. van den Berg, R.A.; Hoefsloot, H.C.; Westerhuis, J.A.; Smilde, A.K.; van der Werf, M.J. Centering, scaling, and transformations: Improving the biological information content of metabolomics data. *BMC Genom.* **2006**, *7*, 1–15. Available online: http://www.biomedcentral.com/1471-2164/7/142 (accessed on 7 December 2022). [CrossRef]
20. Le Clerc, V.; Aubert, C.; Cottet, V.; Yovanopoulos, C.; Piquet, M.; Suel, A.; Huet, S.; Koutouan, C.E.; Hamama, L.; Chalot, G.; et al. Breeding for carrot resistance to *Alternaria dauci* without compromising taste. *Mol. Breed.* **2019**, *39*, 59. [CrossRef]
21. Jourjon, M.F.; Jasson, S.; Marcel, J.; Ngom, B.; Mangin, B. MCQTL: Multi-allelic QTL mapping in multi-cross design. *Bioinformatics* **2005**, *21*, 128–130. [CrossRef] [PubMed]
22. Iorizzo, M.; Ellison, S.; Senalik, D.; Zeng, P.; Satapoomin, P.; Huang, J.; Bowman, M.; Lovene, M.; Sanseverino, W.; Cavagnaro, P.; et al. A high-quality carrot genome assembly provides new insights into carotenoid accumulation and asterid genome evolution. *Nat. Genet.* **2016**, *48*, 657–666. [CrossRef] [PubMed]
23. Gagnot, S.; Tamby, J.P.; Martin-Magniette, M.L.; Bitton, F.; Taconnat, L.; Balzergue, S.; Aubourg, S.; Renou, J.P.; Lecharny, A.; Brunaud, V. CATdb: A public access to Arabidopsis transcriptome data from the URGV-CATMA platform. *Nucleic Acids Res.* **2007**, *36*, 986–990. [CrossRef] [PubMed]
24. Celton, J.M.; Gaillard, S.; Bruneau, M.; Pelletier, S.; Aubourg, S.; Martin-Magniette, M.L.; Navarro, L.; Laurens, F.; Renou, J.P. Widespread anti-sense transcription in apple is correlated with si RNA production and indicates a large potential for transcriptional and/or post-transcriptional control. *New Phytol.* **2014**, *203*, 287–299. [CrossRef]
25. Kearse, M.; Moir, R.; Wilson, A.; Stones-Havas, S.; Cheung, M.; Sturrock, S.; Buxton, S.; Copper, A.; Markowitz, S.; Duran, C.; et al. Geneious Basic: An integrated and extendable desktop software platform for the organization and analysis of sequence data. *Bioinformatics* **2012**, *28*, 1647–1649. [CrossRef]
26. Boedo, C.; Benichou, S.; Berruyer, R.; Bersihand, S.; Dongo, A.; Simoneau, P.; Lecomte, M.; Briard, M.; Le Clerc, V.; Poupard, P. Evaluating aggressiveness and host range of *Alternaria dauci* in a controlled environment. *Plant Pathology* **2012**, *61*, 63–75. [CrossRef]
27. Schindelin, J.; Arganda-Carreras, I.; Frise, E.; Kaynig, V.; Longair, M.; Pietzsch, T.; Preibisch, S.; Rueden, C.; Saalfeld, S.; Schmid, B.; et al. Fiji: An open-source platform for biological-image analysis. *Nat. Methods* **2012**, *9*, 676–682. [CrossRef]
28. Keilwagen, J.; Lehnert, H.; Berner, T.; Budahn, H.; Nothnagel, T.; Ulrich, D.; Dunemann, F. The terpene synthase gene family of carrot (*Daucus carota* L.): Identification of QTLs and candidate genes associated with terpenoid volatile compounds. *Front. Plant Sci.* **2017**, *8*, 1930. [CrossRef]
29. Muchlinski, A.; Ibdah, M.; Ellison, S.; Yahyaa, M.; Nawade, B.; Laliberte, S.; Senalik, D.; Simon, P.; Whitehead, S.R.; Tholl, D. Diversity and function of terpene synthases in the production of carrot aroma and flavor compounds. *Sci. Rep.* **2020**, *10*, 9989. [CrossRef]
30. Reichardt, S.; Budahn, H.; Lamprecht, D.; Riewe, D.; Ulrich, D.; Dunemann, F.; Kopertekh, L. The carrot monoterpene synthase gene cluster on chromosome 4 harbours genes encoding flavour-associated sabinene synthases. *Hortic. Res.* **2020**, *7*, 190. [CrossRef]
31. Kearsey, M.J. The principles of QTL analysis (a minimal mathematics approach). *J. Exp. Bot.* **1998**, *49*, 1619–1623. [CrossRef]
32. Keurentjes, J.J.; Fu, J.; De Vos, C.H.; Lommen, A.; Hall, R.D.; Bino, R.J.; van der Plas, L.H.W.; Jansen, R.C.; Vreugdenhil, D.; Koornneef, M. The genetics of plant metabolism. *Nat. Genet.* **2006**, *38*, 842–849. [CrossRef]
33. Rowe, H.C.; Kliebenstein, D.J. Complex genetics control natural variation in *Arabidopsis thaliana* resistance to *Botrytis cinerea*. *Genetics* **2008**, *180*, 2237–2250. [CrossRef]
34. Zhang, W.; Kwon, S.T.; Chen, F.; Kliebenstein, D.J. Isolate dependency of *Brassica rapa* resistance QTLs to *Botrytis cinerea*. *Front. Plant Sci.* **2016**, *7*, 161. [CrossRef]
35. Gambliel, H.; Croteau, R. Pinene cyclases I and II. Two enzymes from sage (*Salvia officinalis*) which catalyze stereospecific cyclizations of geranyl pyrophosphate to monoterpene olefins of opposite configuration. *J. Biol. Chem.* **1984**, *259*, 740–748. [CrossRef]
36. Huber, D.P.; Philippe, R.N.; Godard, K.A.; Sturrock, R.N.; Bohlmann, J. Characterization of four terpene synthase cDNAs from methyl jasmonate-induced Douglas-fir, *Pseudotsuga menziesii*. *Phytochemistry* **2005**, *66*, 1427–1439. [CrossRef]

37. Matsuba, Y.; Nguyen, T.T.; Wiegert, K.; Falara, V.; Gonzales-Vigil, E.; Leong, B.; Schäfer, P.; Kudrna, D.; Wing, R.A.; Bolger, A.M.; et al. Evolution of a complex locus for terpene biosynthesis in Solanum. *Plant Cell* **2013**, *25*, 2022–2036. [CrossRef]
38. Nützmann, H.W.; Osbourn, A. Gene clustering in plant specialized metabolism. *Curr. Opin. Biotechnol.* **2014**, *26*, 91–99. [CrossRef]
39. Schluttenhofer, C.; Yuan, L. Regulation of Specialized Metabolism by WRKY Transcription Factors. *Plant Physiol.* **2015**, *167*, 295–306. [CrossRef]
40. Spyropoulou, E.A.; Haring, M.A.; Schuurink, R.C. RNA sequencing on *Solanum lycopersicum* trichomes identifies transcription factors that activate terpene synthase promoters. *BMC Genom.* **2014**, *15*, 402. Available online: http://www.biomedcentral.com/1471-2164/15/402 (accessed on 7 December 2022). [CrossRef]
41. Zheng, Z.; Qamar, S.A.; Chen, Z.; Mengiste, T. Arabidopsis WRKY33 transcription factor is required for resistance to necrotrophic fungal pathogens. *Plant J.* **2006**, *48*, 592–605. [CrossRef] [PubMed]
42. Rodríguez, A.; San Andrés, V.; Cervera, M.; Redondo, A.; Alquézar, B.; Shimada, T.; Gadea, J.; Rodrigo, M.J.; Zacarias, L.; Palou, L.; et al. Terpene down-regulation in orange reveals the role of fruit aromas in mediating interactions with insect herbivores and pathogens. *Plant Physiol.* **2011**, *156*, 793–802. [CrossRef] [PubMed]
43. Rodríguez, A.; Shimada, T.; Cervera, M.; Alquézar, B.; Gadea, J.; Gómez-Cadenas, A.; José De Ollas, C.; Rodrigo, M.J.; Zacarias, L.; Peña, L. Terpene down-regulation triggers defense responses in transgenic orange leading to resistance against fungal pathogens. *Plant Physiol.* **2014**, *164*, 321–339. [CrossRef] [PubMed]
44. Zhou, F.; Sun, T.H.; Zhao, L.; Pan, X.W.; Lu, S. The bZIP transcription factor HY5 interacts with the promoter of the monoterpene synthase gene QH6 in modulating its rhythmic expression. *Front. Plant Sci.* **2015**, *6*, 304. [CrossRef] [PubMed]
45. Nieuwenhuizen, N.J.; Chen, X.; Wang, M.Y.; Matich, A.J.; Perez, R.L.; Allan, A.C.; Green, S.A.; Atkinson, R.G. Natural variation in monoterpene synthesis in kiwifruit: Transcriptional regulation of terpene synthases by NAC and ETHYLENE-INSENSITIVE3-like transcription. *Plant Physiol.* **2015**, *167*, 1243–1258. [CrossRef]
46. Li, X.; Xu, Y.; Shen, S.; Yin, X.; Klee, H.; Zhang, B.; Chen, K.; Hancock, R. Transcription factor CitERF71 activates the terpene synthase gene CitTPS16 involved in the synthesis of E-geraniol in sweet orange fruit. *J. Exp. Bot.* **2017**, *68*, 4929–4938. [CrossRef]
47. Yu, Z.X.; Wang, L.J.; Zhao, B.; Shan, C.M.; Zhang, Y.H.; Chen, D.F.; Chen, X.Y. Progressive regulation of sesquiterpene biosynthesis in Arabidopsis and Patchouli (*Pogostemon cablin*) by the miR156-targeted SPL transcription factors. *Mol. Plant* **2015**, *8*, 98–110. [CrossRef]
48. Hammer, K.A.; Carson, C.F.; Riley, T.V. Antifungal activity of the components of *Melaleuca alternifolia* (tea tree) oil. *J. Appl. Microbiol.* **2003**, *95*, 853–860. [CrossRef]
49. Sati, S.C.; Sati, N.; Ahluwalia, V.; Walia, S.; Sati, O.P. Chemical composition and antifungal activity of Artemisia nilagirica essential oil growing in northern hilly areas of India. *Nat. Prod. Res.* **2013**, *27*, 45–48. [CrossRef]
50. Dahham, S.S.; Tabana, Y.M.; Iqbal, M.A.; Ahamed, M.B.; Ezzat, M.O.; Majid, A.S.; Majid, A.M. The anticancer, antioxidant and antimicrobial properties of the sesquiterpene β-caryophyllene from the essential oil of *Aquilaria crassna*. *Molecules* **2015**, *20*, 11808–11829. [CrossRef]
51. Garcia-Rellán, D.; Verdeguer, M.; Salamone, A.; Blázquez, M.A.; Boira, H. Chemical composition, herbicidal and antifungal activity of *Satureja cuneifolia* essential oils from Spain. *Nat. Prod. Commun.* **2016**, *11*, 1934578X1601100636. [CrossRef]
52. Feiner, A.; Pitra, N.; Matthews, P.; Pillen, K.; Wessjohann, L.A.; Riewe, D. Downy mildew resistance is genetically mediated by prophylactic production of phenylpropanoids in hop. *Plant Cell Environ.* **2021**, *44*, 323–338. [CrossRef]

metabolites

Article

LC-MS/MS and GC/MS Profiling of *Petroselinum sativum* Hoffm. and Its Topical Application on Burn Wound Healing and Related Analgesic Potential in Rats

Meryem Slighoua [1,*], Ismail Mahdi [2], Fatima Zahrae Moussaid [1], Omkulthom Al Kamaly [3,*], Fatima Ez-zahra Amrati [1], Raffaele Conte [4], Aziz Drioiche [5], Asmaa Saleh [3], Abdelilah Iraqi Housseini [1], Amina Bari [1] and Dalila Bousta [1]

1 Laboratory of Biotechnology, Environment, Agro-Food, and Health (LBEAS), Faculty of Sciences, University Sidi-Mohamed-Ben-Abdellah (USMBA), Fez 30050, Morocco
2 AgroBioSciences Research Program, Mohammed VI Polytechnic University, Lot 660-Hay Moulay Rachid, 11, Ben-Guerir 43150, Morocco
3 Department of Pharmaceutical Sciences, College of Pharmacy, Princess Nourah bint Abdulrahman University, P.O. Box 84428, Riyadh 11671, Saudi Arabia
4 Research Institute on Terrestrial Ecosystems (IRET)—CNR, Via Pietro Castellino 111, 80131 Naples, Italy
5 Laboratory of Innovative Materials and Biotechnology of Natural Resources, Faculty of Sciences, Moulay 19 Ismail University, Meknes 50000, Morocco
* Correspondence: slighoua.meryem@gmail.com (M.S.); omalkmali@pnu.edu.sa (O.A.K.)

Abstract: Parsley (*Petroselinum sativum* Hoffm.) is renowned for its ethnomedicinal uses including managing pain, wound, and dermal diseases. We previously highlighted the estrogenic and anti-inflammatory properties of parsley and profiled the phytochemistry of its polyphenolic fraction using HPLC-DAD. To extend our investigation, we here characterized the phytochemical composition of the hydro-ethanolic extract using LC-MS/MS and GC-MS upon silylation, and evaluated the antioxidant, analgesic, antimicrobial, and wound healing activities of its hydro-ethanolic and polyphenolic fraction. The antioxidant property was assessed using FRAP, DPPH, and TAC assays. The antimicrobial activity was tested against four wound infectious microbes (*Staphylococcus aureus*, *Pseudomonas aeruginosa* and *Candida albicans*). The analgesic effect was studied using acetic acid (counting the number of writhes) and formalin (recording the licking and biting times) injections while the wound healing activity was evaluated using burn model in vivo. The LC-MS/MS showed that the hydro-ethanolic contains four polyphenols (oleuropein, arbutin, myricetin, and naringin) while GC-MS revealed that it contains 20 compounds including malic acid, D-glucose, and galactofuranoside. The hydro-ethanolic (1000 mg/kg) decreased abdominal writhes (38.96%) and licking time (37.34%). It also elicited a strong antioxidant activity using DPPH method (IC_{50} = 19.38 ± 0.15 µg/mL). Polyphenols exhibited a good antimicrobial effect (MIC = 3.125–12.5 mg/mL). Moreover, both extracts showed high wound contraction by 97.17% and 94.98%, respectively. This study provides evidence that *P. sativum* could serve as a source of bio-compounds exhibiting analgesic effect and their promising application in mitigating ROS-related disorders, impeding wound infections, and enhancing burn healing.

Keywords: *Petroselinum sativum* Hoffm.; LC-MS/MS; GC-MS; analgesic activity; burn wound healing

check for updates

Citation: Slighoua, M.; Mahdi, I.; Moussaid, F.Z.; Kamaly, O.A.; Amrati, F.E.-z.; Conte, R.; Drioiche, A.; Saleh, A.; Housseini, A.I.; Bari, A.; et al. LC-MS/MS and GC/MS Profiling of *Petroselinum sativum* Hoffm. and Its Topical Application on Burn Wound Healing and Related Analgesic Potential in Rats. *Metabolites* **2023**, *13*, 260. https://doi.org/10.3390/metabo13020260

Academic Editors: Ramona Paltinean and Irina Ielciu

Received: 30 January 2023
Revised: 2 February 2023
Accepted: 6 February 2023
Published: 11 February 2023

1. Introduction

Pain and inflammation are strongly involved in the healing process of the wound. Therefore, conventional medicines are often used to mitigate the intensity of pain and inflammation [1]. After an injury, the skin regenerates through the process of wound healing. The interaction between several cellular elements (fibroblasts, keratinocytes, endothelial cells, and macrophages/monocytes), and the constituents of the extracellular environment (collagen and fibronectin) during the wound healing phases, promote the wound contraction, and restore tissue integrity [2]. The wound healing provides the ideal micro-environment at the injured surface to achieve the maximum skin repair [3]. However, several disorders may affect the ability of healing such as mechanical stresses, toxic agents, or infections. In most situations, the molecular events that accelerate the nociception response are similar whether the pain is of extrinsic or intrinsic origin [2].

By reacting with biological components such as proteins and nucleic acids, excessive levels of reactive oxygen species (ROS) disrupt intrinsic tissue disability and lead to loss of function [4,5]. For instance, high concentrations of H_2O_2 can cause oxidative damage and thus delay healing, while low concentrations can act as a signaling molecule and promotes healing. However, all the advancements achieved against oxidative stress, the production of ROS during injuries, and their impact on the healing process still constitute a major health challenge [6].

Parsley, whose scientific name is *Petroselinum sativum* Hoffm., is a plant belonging to the Apiaceae family. Currently, it is cultivated all over the world and has been used as food, cosmetic ingredient, and for pharmaceutical purposes [7]. Parsley has several biological and pharmacological activities, mainly spasmolytic, antioxidant, immunomodulating, gastrointestinal, and antidiabetic attributes [8]. These various virtues are due to its bioactive phytoconstituents such as carotenoids, flavonoids, coumarin, and vitamins [9].

Several previous investigations have stated that herbal and plant-based ingredients activate the wound and cutaneous healing process. These include many medicinal and aromatic plants (MAP) such as turmeric (*Curcuma longa*), centella (*Centella asiatica*), tree peony (*Paeonia suffruticosa*), and aloe vera (*Aloe barbadensis*) [10,11]. Additionally, in traditional medicine, *Petroselinum* species including *P. sativum* were reported to be used in Anatolia, Turkey for wound healing purposes [12] as well as in treating some dermal diseases [13,14]. However, very limited studies were devoted to providing scientifically sound data to test this claim. Very recently, Thangavelu et al. (2022) showed that the leaf methanolic extracts of *P. crispum*, whose synonym is *P. sativum*, exhibited potent wound healing and anti-inflammatory activities on the human lung cancer cell lines by enhancing cell migration [15]. Interestingly, in another study, parsley was used as a maintenance diet for cutaneous closure of wounds in rabbits as a postoperative care after the surgery [16]. Elsewhere, *P. crispum* was also investigated for immunomodulatory and wound healing activities [17]. Moreover, the analgesic use of parsley in folklore medicine was also reported and demonstrated in vivo using seeds hydroalcoholic extract [14,18]. Other studies corroborated the same findings [19,20].

The growing antimicrobial resistance of microbes responsible for skin infections blew up the research on the potential of MAP preparations in antimicrobial therapeutics. In fact, many phytochemicals are shown to be effective against microbial infections [21,22]. In wounds, the skin barrier is breached and becomes susceptible to microbial infections by bacteria, fungi, and/or viruses, that delay the healing process [23]. These pathogens include Gram-positive bacteria such as *Staphylococcus aureus* and *Streptococcus pyogenes*, and Gram-negative bacteria including *Escherichia coli*, *Pseudomonas aeruginosa*, and *Klebsiella species*, and fungi—mainly *Candida* and *Aspergillus* [24]. Hence, minimizing the predisposition of wounds to infections is required in wound surveillance to reduce the infection rate [25].

We previously explored the estrogenic as well as the anti-inflammatory activities of *P. sativum* Hoffm. in vivo and characterized the chemical composition of its polyphenolic fraction [26]. Recognizing the promising therapeutic potential of this plant species and to follow up on our previous findings, we conducted this investigation that aimed to (1) annotate the phytoconstituents of the hydro-ethanolic extract of *P. sativum* Hoffm. using both LC-MS/MS and GC-MS to identify the maximum number of phytocompounds and (2) monitor the analgesic, antioxidant, antimicrobial, and burn healing effects of both the hydro-ethanolic extract and polyphenolic fraction of the plant. This study is the first to demonstrate the pharmacological relevance of parsley as a source of wound healing and analgesic biochemicals.

2. Materials and Methods

2.1. Plant Material

Aerial parts of *P. sativum* Hoffm. were collected before sunrise in the Taounate region (north of Morocco). The botanical name was checked by Pr. Bari Amina, at the laboratory of Biotechnology, Environment, Agro-Food, and Health, at Sidi Mohamed Ben Abdellah University. A sample was deposited under a voucher number (18TA5001) at the herbarium of the Faculty of Sciences-Fez.

2.2. Animal Material

In this study, we used Swiss albino mice rearing in an animal house that has a relative humidity of 50 to 55%, with an average temperature of 22 ± 2 °C, and a day/night cycle of about 12/12 h. Animals were given free access to water and food. The average weight of mice used was 30 ± 4 g. The handling and manipulation of animals were up to the standards of the directive EEC/86/EEC of the European community [26,27]. The experiment was approved by the institutional ethical committee of the Faculty of Sciences Dhar El Mehrez, Sidi Mohamed Ben Abdallah Fez University, Morocco (#04/2019/LBEAS).

2.3. Preparation of the Hydro-Ethanolic Extract

Aerial parts were left to the laboratory to dry for one week, then grinded using a blender. The grind (30 g) was macerated in 210 mL of ethanol 70% for three days [26]. The macerate was filtered, evaporated, and dried (rotary evaporator, 37 °C) [26,28].

2.4. Preparation of the Polyphenolic Fraction

One hundred grams of well ground *P. sativum* powder was subjected to three extractions in methanol (300 mL × 3) at 50 °C for 3 h. Subsequently, solvent was evaporated, and the obtained extract was dissolved in distilled water (500 mL) and extracted in hexane (200 mL × 3) and then in chloroform (200 mL × 3) to remove caffeine and chlorophyll residues. Next, the aqueous phase was extracted in ethyl acetate (200 mL × 3) which was evaporated later. Using 300 mL of distilled water, the residue was dissolved and then lyophilized [29,30].

2.5. Phytochemical Analysis by LC-MS/MS

The chemical profiling of the hydro-ethanolic extract of *P. sativum* was performed using ultra-high performance liquid chromatography (Shimadzu, Nexera XR LC 40) coupled with mass spectrometry (LCMS 8060, Shimadzu Italy, Milan, Italy). The heating and nebulization gas flow was set to 10 and 2.9 L/min, respectively. The drying gas flow was at 10 L/min, the DL temperature was at 250 °C, the heating block temperature was 400 °C, and the interface temperature was 300 °C. Separation of compounds was carried out using C18 column, 3 × 100 mm, 2.6 µm (Phenomenex, Torrance, CA, USA). The mobile phase consisted of acetonitrile (A) and water containing 0.01% formic acid (B). The extract was added to acetonitrile and water (1:1) and then diluted (1/50) in acetonitrile and injected [31]. The ion currents' acquisition was carried out in single ion monitoring (MRM) mode in negative ESI ionization. The analyzed molecular adducts were, respectively, 579, 539, 317, and 271.2 for Naringin, Oleuropein, Myricetin, and Arbutin.

2.6. Phytochemical Analysis by GC-MS

Phytochemical identification of the hydro-ethanolic extract of *P. sativum* was carried out using GC-MS after silylation. This latter is based on dissolving 1 mg of the grind in 100 mL of HMDS-TMCS-Pyridine 3:1:9 (*v*/*v*/*v*) reagent. After 30 min incubation [32], the extract was injected into the GC-MS apparatus (Agilent Technologies MASS Selective Detector, 5973 Network) with a capillary column Agilent 19091S-433 model, 30 m in nominal length—0.25 mm in diameter and 0.25 µm thick. Helium served as the carrier gas, and the total flow of 31.4 mL/min and a split ratio of 30:1 was used. The temperature program was between 60 and 300 °C and maintained for 20 min of the run time. The detector temperature was set to 260 °C. Splitless injection was used [33].

2.7. Determination of Total Phenol and Flavonoid Contents

The quantification of total phenol content (TPC) and total flavonoid content (TFC) was carried out calorimetrically using the methods described by Slinkard and Singleton (1977) [34]. TPC content was expressed in milligrams (mg) of gallic acid equivalents per gram (g) of dry weight of extract (mg GAE/g DW) while the values of TFC were expressed in mg of quercetin equivalent per g of dry extract (mg EQ/g DW).

2.8. Assessment of Antioxidant Activity

2.8.1. Scavenging of the Free Radical (DPPH)

In this study, the DPPH test performed by Brand-Williams in 1995 [35] was followed. Methanol (100 µL) was mixed with 750 µL of DPPH solution, incubated for 30 min, and then absorbance was measured at OD_{517nm}. BHT (butylated hydroxytoluene) was used as the standard antioxidant. To calculate the percentage of inhibition (IP) of DPPH, the following formula was used:

$$\mathrm{IP}\ (\%) = \left(\mathrm{A0} - \frac{\mathrm{A}}{\mathrm{A0}}\right) \times 100$$

IP: Inhibition Percentage.
A0: OD of DPPH solution in the absence of the extract (negative control).
A: OD of DPPH solution containing the extract.

2.8.2. Ferric Reducing Antioxidant Power (FRAP)

The FRAP assay was performed according to the method described by Oyazu (1986) [36]. Briefly, 200 µL of each extract was mixed with the buffer solution (0.5 mL) (0.2 M, pH = 6.6) and potassium ferricyanide $[K_3Fe\ (CN)_6]$ at 1% (0.5 mL). The solution was kept at 50 °C in a water bath for 20 min. Next, to acidify to solution, 500 µL of trichloracetic acid at 10% was added to the solution and centrifuged for 10 min at 3000 rpm. Five hundred microliter of the top layer of the solution was mixed with distilled water (500 µL) and $FeCl_3$ 0.1% (100 µL). Ascorbic acid was used as standard, and the absorbance was read at OD_{700nm}. The values were expressed as EC_{50} (mg/mL). The EC_{50} was calculated using the standard curve.

2.8.3. Total Antioxidant Capacity Test (TAC)

The TAC was evaluated by mixing 25 µL of every studied extract with 1 mL of liquid reactive solution (28 mM Na_3PO_4, 0.6 M H_2SO_4, and 4 mM $(NH_4)_2MoO_4$). Following incubation (90 min, 95 °C), the absorbance values were read at OD_{695nm}. The antioxidant potential was determined in mg of equivalent of ascorbic acid per gram of extracts (mg EAA/g of extracts) [37,38].

2.9. Antimicrobial Activity

The extracts were evaluated for their potential antimicrobial effect against three human pathogenic strains including one Gram-negative bacterium (*Pseudomonas aeruginosa* CECT118), one Gram-positive bacterium (*Staphylococcus aureus* CECT976), and one fungal species (*Candida albicans* ATCC 10231). The antimicrobial susceptibility of the three strains was evaluated using agar well diffusion and broth dilution methods.

2.9.1. Agar Well Diffusion Assay

The method of diffusion in agar wells was used to carry out the qualitative test, a widely-known assay to check the antimicrobial activity of herbal extracts [39,40]. From fresh overnight cultures, microbial suspensions were prepared and adjusted to 0.5 McFarland corresponding to 10^6 CFU/mL [41,42]. Afterwards, 5 mL of soft agar (agar 4 g/L) inoculated with 100 µL of the microbial suspension (10^6 CFU/mL) of each strain were poured over the surface of each plate. After solidification, wells of 8 mm diameter were aseptically punched using glass Pasteur pipets. Then, 50 µL of each extract was dissolved in the appropriate solvent (40 mg/mL) and introduced into the appropriate wells [43]. Under the same conditions, the controls were established using the solvent only. The Petri dishes were then incubated at 37 ± 2 °C for 48 h. The extracts diffuse through the agar medium and the formation of inhibition zones surrounding the wells indicate positive antimicrobial activities. Streptomycin (1 mg/mL) and fluconazole (5 mg/mL) served as positive controls for bacteria and fungi, respectively.

2.9.2. Determination of the Minimum Inhibitory Concentration (MIC)

The MIC of the hydro-ethanolic and polyphenolic extract was checked using microdilution assays according to the standards of the NCCLS [44] in 96-well microtiter microplates. First, 50 µL of the culture broth was introduced into each well of the microplates. Next, seven concentrations of the hydro-ethanolic extract and the polyphenolic fraction (0.78–50 mg/mL), streptomycin and fluconazole (0.078–5 mg/mL) were prepared in both LB and YPG in sterile haemolysis tubes. Each microplate well was inoculated by 100 µL of LB liquid culture medium for bacteria and YPG liquid culture medium for yeast strains, 50 µL of each extract, and then concentrations were carried out by successive two-fold dilutions. Afterwards, 50 µL of the microbial suspensions, whose turbidity was checked in the same way as described above, were inoculated into the microplate's wells. The microplates were incubated at 37 °C for the bacterial strains and at 30 °C for *Candida albicans* ATCC 10231 under 150 agitation rpm for 24 h. By the end of the incubation, we added 20 µL of 2,3,5-triphenyltetrazolium chloride into each plate's well and incubated the plate for 2 h. The formation of a pinkish coloration indicates that the growth is due to the activity of the dehydrogenases. The MIC corresponds to the lowest concentration that does not produce a red colour [45].

2.10. Analgesic Activity In Vivo

2.10.1. Abdominal Writhes

Five groups containing five mice each (25 mice) were prepared. The control mice were given 10 mL/kg of NaCl 0.9% and Tramadol (10 mg/kg) was used as a standard (reference drug) [46]. The other groups received two doses of the hydro-ethanolic extract (500–1000 mg/kg) and one dose from the polyphenolic fraction (200 mg/kg). One hour later, 1% acetic acid was injected by intraperitoneal route at 10 mL/kg rate. Ten minutes later, the number of writhes were determined over a duration of 20 min [47,48].

2.10.2. Formalin Induced Pain

First, we injected 10% formalin (20 mL) into the right posterior paw of each animal. With the help of a stopwatch, we recorded the licking and biting times. The initial nociceptive response is the sum of the seconds passed in licking and biting from 0 to 5 min after the injection of formalin while the second phase was from 15 to 30 min [49,50]. Half an hour beforehand, the animals were subjected to oral prior treatment by the test extract, NaCl or Tramadol [51].

2.11. Wound Healing Activity In Vivo

2.11.1. Ointments Preparation

The preparation of the ointment was carried out at 10% (*w*/*w*) by adding 1 g of the extract to 9 g of Vaseline, and melted using a bain-marie at a temperature of 50 °C. After homogenization, the preparations were kept at 4 °C in sealed containers [4].

2.11.2. Induction of Burn Injuries

In this experiment, 4 groups containing 5 rats each were prepared. The first group represents the negative control (Vaseline), the second group represents the positive control

(Madecassol (1%)), and the third and fourth groups represent the groups treated with the hydro-ethanolic and polyphenolic extracts of *P. sativum* Hoffm., respectively. According to the protocol of Heidari et al., (2019), after anesthesia of rats with pentobarbital (50 mg/kg) and shaving the dorsal part with an electric clipper [52], the induction of the burn was carried out on the shaved part using an aluminum rod (1.7 cm) heated to 110 °C for 10 s. After 24 h, the treatment was started by applying the ointments to the burned zones for 25 days while photographing the healing progress using a digital camera and the ruler used as a scale. *ImageJ*® software was used to analyze the images and measure the rate of wound contraction using the following formula [53]:

$$WC\ (\%) = \left[\frac{(WS0 - WSSD)}{WS0}\right] \times 100$$

WC (%) = Rate of wound contraction.
WS0 = Size of the wound at the first day.
WSSD = Size of the wound at each specific day.

2.12. Statistical Analysis

Results obtained from each experiment were treated by using a one-way ANOVA followed by the post-hoc analysis with Tukey's test in GraphPad Prism 6 software. Values were expressed as mean ± SD and the significance level was set at "$p < 0.05$". The significant differences between treatments were shown using different superscript letters (a, b, c, etc.).

3. Results

3.1. Phytochemical Analysis by LC-MS/MS

The determination of parsley phytoconstituents was performed according to the molecular weight of the fragments generated. The analysis of the hydro-ethanolic extract revealed the presence of four molecules namely oleuropein, arbutin, myricetin, and naringin, all classified as polyphenols (Figure 1 and Table 1). For instance, negative ion ESI-MS/MS spectra of oleuropein resulted in the formation of *m/z* 539, a pseudomolecular ion, as the sole base peak of the ESI-MS spectra, while the MS/MS products were abundant (e.g., 307 and 275). Similarly, myricetin (*m/z* 317) was fragmented to four main products (150.8, 178.8, 270.9, and 286.9). The MS/MS spectra of identified compounds, annotations, and their characteristic fragmentation patterns are presented in Table 1. XIC chromatograms of identified compounds are provided in the Supplementary Material S1.

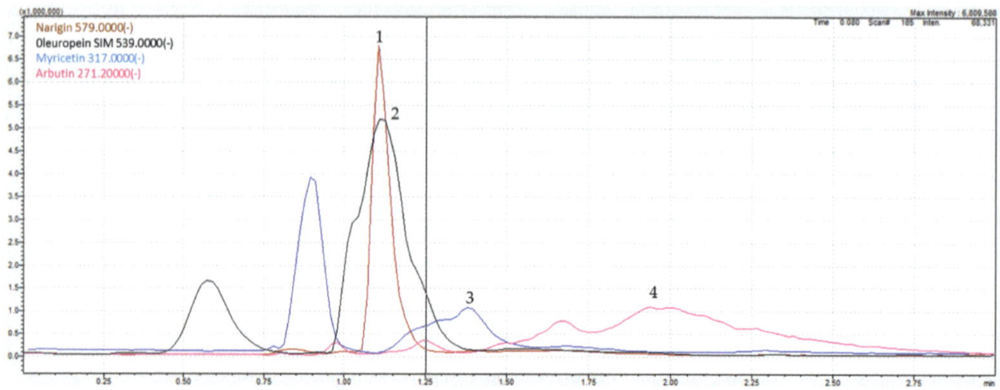

Figure 1. LC-MS/MS chromatogram of the hydro-ethanolic extract of *P. sativum*. (**1**) Naringin, (**2**) Oleuropein, (**3**) Myricetin, (**4**) Arbutin.

Table 1. LC-MS/MS identification composition of *P. sativum* Hoffm. hydro-ethanolic extract.

Peak	Molecules	Formula	R_t (min)	[M-H]- (*m/z*)	SI	Typical MS/MS	Area under Curve
1	Naringin	$C_{27}H_{32}O_{14}$	1.108	579.00	579.2 → 124.89	271	27,397,987
2	Oleuropein	$C_{25}H_{32}O_{13}$	1.115	539.00	539 → 179	307	56,757,543
3	Myricetin	$C_{15}H_{10}O_8$	1.384	317.00	317 →150.8	287	3,417,929
4	Arbutin	$C_{12}H_{16}O_7$	1.939	271.20	271.2 →144.9	162	28,442,471

3.2. Phytochemical Analysis by GC-MS

Silylation is the appropriate method to check for thermolabile and non-volatile phytochemicals by GC-MS. It consists of replacing the active hydrogen in =NH, $-NH_2$, –OH, –COOH, or –SH by a trimethylsilyl group. This showed that the hydro-ethanolic of *P. sativum* contained twenty compounds with a total of 99.9%. According to the area percentage, the most dominant compounds were malic acid (13.52%), D-glucose (13.529%), D-mannitol (10.957%), and talose (9.758%) (Table 2).

Table 2. Identified compounds by GC-MS in the hydro-ethanolic extract of *P. sativum* Hoffm.

Peak	Name	Formula	R_t (min)	Area %
1	l-Alanine, *N*-(trimethylsilyl)	$C_9H_{23}NO_2Si_2$	7.430	0.51
2	Cyclotetrasiloxane, octamethyl	$C_8H_{24}O_4Si_4$	8.148	1.60
3	Propanedioic acid, bis(trimethylsilyl)	$C_9H_{20}O_4Si_2$	8.417	1.08
4	L-Valine, *N*-trimethylsilyl	$C_{11}H_{27}NO_2Si_2$	8.558	0.72
5	Propanephosphonic acid, bis(trimethylsilyl)	$C_9H_{25}O_3PSi_2$	9.088	12.57
6	L-Isoleucine, *N*-(trimethylsilyl)	$C_{12}H_{29}NO_2Si_2$	9.248	0.62
7	Butanedioic acid, bis(trimethylsilyl)	$C_{10}H_{22}O_4Si_2$	9.317	0.75
8	L-Proline, 1-(trimethylsilyl)	$C_{11}H_{25}NO_2Si_2$	9.361	0.98
9	Benzonitrile	$C_{18}H_{18}N_2$	9.594	0.62
10	Malic acid, tris(trimethylsilyl) ester	$C_{13}H_{30}O_5Si_3$	10.730	13.52
11	2,3,4-Trihydroxybutyric acid tetraTMS	$C_{16}H_{40}O_5Si_4$	11.260	0.56
12	D-Ribofuranose, 1,2,3,5-tetrakis-*O*-(trimethylsilyl)	$C_{17}H_{42}O_5Si_4$	12.918	5.62
13	β-D-Galactofuranoside, ethyl 2,3,5,6-tetrakis-*O*-(trimethylsilyl)	$C_{20}H_{48}O_6Si_4$	13.019	13.29
14	9*H*-Carbazole, 9-phenyl- alcaloide	$C_{18}H_{13}N$	13.220	3.04
15	Talose, 2,3,4,5,6-pentakis-*O*-(trimethylsilyl)	$C_{21}H_{52}O_6Si_5$	13.400	9.75
16	D-Mannitol, 1,2,3,4,5,6-hexakis-*O*-(trimethylsilyl)	$C_{24}H_{62}O_6Si_6$	13.653	10.95
17	D-Glucose, 2,3,4,5,6-pentakis-*O*-(trimethylsilyl)	$C_{21}H_{52}O_6Si_5$	13.782	13.52
18	D-gluconic acid 6TMS	$C_{24}H_{60}O_7Si_6$	14.071	1.31
19	Myo-Inositol, 1,2,3,4,5,6-hexakis-*O*-(trimethylsilyl)	$C_{24}H_{60}O_6Si_6$	14.549	1.19
20	Mannoonic acid, 2,3,5,6-tetrakis-*O*-(trimethylsilyl)	$C_{18}H_{42}O_6Si_4$	18.050	7.70
	Total			99.9

3.3. Estimation of Total Phenol and Flavonoid Contents

The total phenols and flavonoids contents contained in the hydro-ethanolic extract of *P. sativum* were 34.55 ± 3.74 mg GAE/g of extract and 16.46 ± 0.06 mg QE/g of extract, respectively.

3.4. Antioxidant Activity

3.4.1. DPPH and FRAP Assays

The antioxidant activity of our extracts was tested by DPPH and FRAP tests. The inhibition of the free radical DPPH by the hydro-ethanolic was greater than that of the polyphenols (Figure 2a). The 50% inhibition concentration (IC_{50}) of the hydro-ethanolic extract was seen at 19.38 ± 0.15 μg/mL, and that of the polyphenols was obtained at 40.36 ± 1.47 μg/mL (Table 3). However, these results are significantly lower than those obtained using butylated hydroxytoluene (BHT) (IC_{50} = 1.97 ± 0.1 μg/mL) (Figure 2a and Table 3). The ferric reducing power of our extracts showed that hydro-ethanolic extract was more effective than polyphenols, but the potential of both extracts was slightly lower than that of the standard antioxidant ascorbic acid (Figure 2b & Table 3).

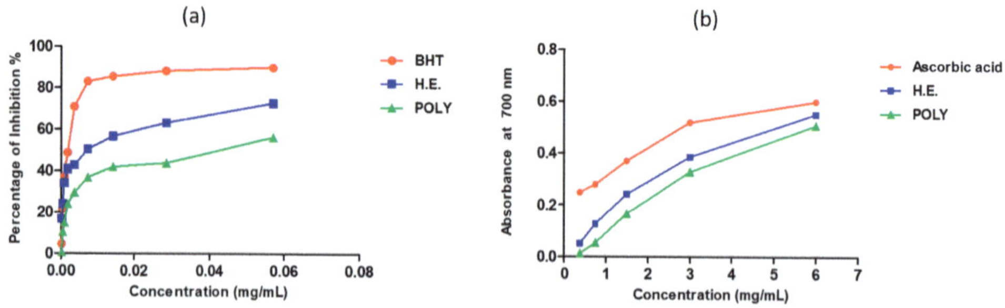

Figure 2. Antioxidant activity of the hydro-ethanolic and polyphenols extracts of *P. sativum* Hoffm. using DPPH (**a**) and FRAP (**b**) assays.

Table 3. Antioxidant activity of hydro-ethanolic and polyphenols of *P. sativum* Hoffm. using DPPH and FRAP methods.

Extract	DPPH	FRAP
	IC_{50} (µg/mL)	EC_{50} (mg/mL)
Hydro-ethanolic extract	19.38 ± 0.15	5.34 ± 0.40
Polyphenols	40.36 ± 1.47	4.91 ± 0.40
BHT	01.97 ± 0.10	–
Ascorbic acid	–	1.43 ± 0.02

3.4.2. Total Antioxidant Capacity (TAC)

TAC assay showed that the hydro-ethanolic has a greater total antioxidant capacity (175.2 ± 6.360 mg EAA/g) comparing to polyphenols (148.2 ± 13.86 mg EAA/g).

3.5. Antimicrobial Activity

The antimicrobial assay on the plate showed that the hydro-ethanolic extract elicited an antibacterial and anti-fungal activity against *P. aeruginosa* and *C. albicans*, respectively, while *S. aureus* was showed to be resistant. In contrast, the polyphenolic fraction exhibited a higher inhibition zone diameter against the three tested pathogens with a noticeable inhibitory effect toward *P. aeruginosa* (Table 4). Nevertheless, positive controls, streptomycin (1 mg/mL) and fluconazole (5 mg/mL), were relatively more potent comparatively to the tested extracts.

Table 4. Inhibition zone diameters of *P. sativum* Hoffm. extracts tested against bacterial and fungal species.

Fractions	Inhibition Zone Diameter in mm		
	Gram-Negative Bacteria	Gram-Positive Bacteria	Fungi
	P. aeruginosa	*S. aureus*	*C. albicans*
Hydro-ethanolic extract	12.33 ± 0.33 ^{a}	0.00 ± 0.00 ^{b}	9.33 ± 0.33 ^{ab}
Polyphenols	14.00 ± 0.33 ^{a}	9 ± 0.16 ^{a}	13 ± 0.57 ^{a}
Streptomycin (1 mg/mL)	14.67 ± 0.17 ^{a}	16 ± 0.57 ^{a}	–
Fluconazole (5 mg/mL)	–	–	21 ± 1.2 ^{a}

The different letters in superscript (a, b) indicate the significant difference between treatments at $p < 0.05$.

The antimicrobial effects against the three species were noticed at MIC values ranging from 3.125 to 12.5 mg/mL. The polyphenols were the most active as they inhibited the three species with a prominent effect towards *S. aureus* (MIC = 3.125 mg/mL). Worth noting is that the broth dilution assay corroborated the non-toxic effect of the hydro-ethanolic extract against *S. aureus* even at the highest concentration tested, 50 mg/mL (Table 5).

Table 5. MIC of *P. sativum* Hoffm. fractions tested against bacterial and fungal species.

Fractions	Minimum Inhibitory Concentration (MIC) in mg/mL		
	Gram-Negative Bacteria	Gram-Positive Bacteria	Fungi
	P. aeruginosa	*S. aureus*	*C. albicans*
Hydro-ethanolic extract	12.5	Resistant	6.25
Polyphenols	6.25	3.125	6.25
Streptomycin	0.625	0.15	—
Fluconazole	—	—	0.31

3.6. Analgesic Activity

3.6.1. Abdominal Writhes

Analgesic activity was evaluated using acetic acid method. Although less efficient than the standard drug (Tramadol), the hydro-ethanolic (1000 mg/kg) and polyphenols (200 mg/kg) induced a significant decrease in the number of writes by 38.96 and 29.23%, respectively, followed by the hydro-ethanolic extract at 500 mg/kg by 23.07% as compared to the negative control animals which we have taken as reference to calculate the percent of inhibition (Table 6 and Figure 3).

Table 6. Effect of *P. sativum* Hoffm. on acetic acid-induced writhing in mice (n = 5).

Treatment	Dose (mg/kg)	Number of Writes
Control		65.00 ± 2.88^{c}
Tramadol	10	17.33 ± 1.45^{a}
Hydro-ethanolic extract	500	50.00 ± 0.57^{bc}
	1000	39.67 ± 0.88^{b}
Polyphenols	200	46.00 ± 3.05^{b}

The different letters in superscript (a, b, c) indicate the significant difference between treatments at $p < 0.05$.

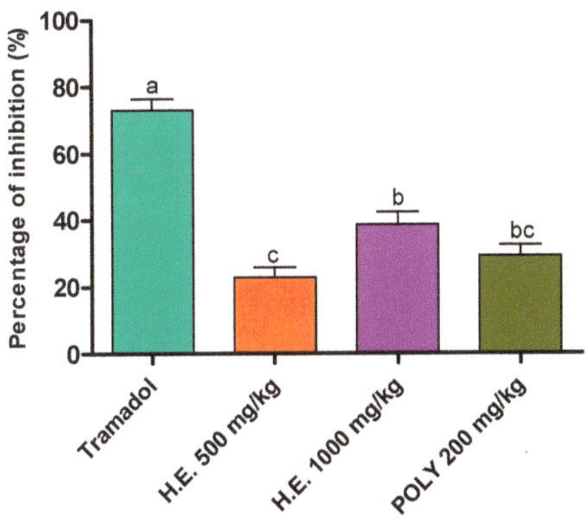

Figure 3. Inhibitory effect of the hydro-ethanolic (H.E.) and polyphenolic (Ploy) extracts of *P. sativum* Hoffm. and Tramadol (positive control) on contortions in mice. The different letters in superscript (a, b, c) indicate the significant difference between treatments at $p < 0.05$.

3.6.2. Formalin Induced Pain

In the formalin test, the hydro-ethanolic extract at the dose 500 and 1000 mg/kg of *P. sativum* Hoffm. elicited a significant reduction in response to nociception during the first phases (0–5 min) and the second phases at the doses 500 and 1000 mg/kg of the hydro-ethanolic extract and polyphenols comparatively to the control mice. Worth noting is that Tramadol was the most potent in both phases by up to 85.7% reduction in response time. During the first phase, hydro-ethanolic extract reduced the response time by 33.32% and 25.86% using 1000 and 500 mg/kg, respectively. In the second phase, both hydro-ethanolic and polyphenols extracts showed a significant reduction in response time by 28.55% using hydro-ethanolic at 500 mg/kg, 37.35% using hydro-ethanolic at 1000 mg/kg, and 30.76% using polyphenols at 200 mg/kg (Table 7 and Figure 4).

Table 7. Effect of the hydro-ethanolic and polyphenolic extracts of *P. sativum* Hoffm. on the response of mice upon the formalin-induced pain.

Treatment	Dose (mg/kg)	Licking Time (s)	
		First Phase (0–5 min)	Second Phase (15–30 min)
Control		58.00 ± 0.5 [c]	30.33 ± 2.5 [c]
Tramadol	10	10.33 ± 0.8 [a]	4.33 ± 1.7 [a]
Hydro-ethanolic extract	500	49.67 ± 1.4 [b]	21.67 ± 2.1 [b]
	1000	38.67 ± 1.7 [b]	19.00 ± 1.1 [b]
Polyphenols	200	43.00 ± 1.1 [bc]	21.00 ± 0.5 [b]

The different letters in superscript (a, b, c) indicate the significant difference between treatments at $p < 0.05$.

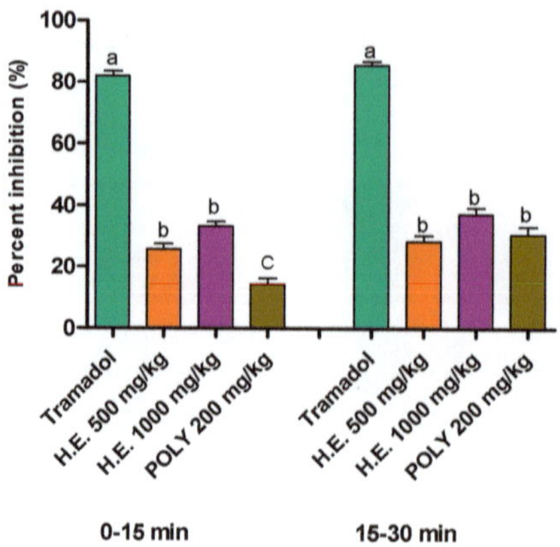

Figure 4. Percent of inhibition of formalin-induced pain in mice following the application of hydro-ethanolic (H.E.) and polyphenolic (Ploy) extracts of *P. sativum* Hoffm. and Tramadol (positive control). Inhibition rates were determined compared to the non-treated animals. The different letters in superscript (a, b, c) indicate the significant difference between treatments at $p < 0.05$.

3.7. Wound Healing Activity

The effects of the ointments prepared from hydro-ethanolic and polyphenols extracts are presented in Table 8. We showed that both applied extracts induced a significant healing activity as compared the untreated animals. Figure 5 illustrates the healing progression using the extracts and the controls from the 1st to the 25th day. The negative control group (Vaseline®), positive control group (Madecassol®), and polyphenols did not induce a complete wound closure. In contrast, the hydro-ethanolic extract induced a complete cicatrization of the wound at the 20th day.

Table 8. Wound size (cm^2) of the hydro-ethanolic and polyphenols extracts of *P. sativum* Hoffm. from day 1 till day 25.

	Wound Size in cm^2					
Treatments	Day 1	Day 5	Day 10	Day 15	Day 20	Day 25
Negative Control	1.91 ± 0.15 [a]	1.70 ± 0.18 [a]	1.26 ± 0.03 [a]	0.62 ± 0.05 [a]	0.46 ± 0.03 [a]	0.37 ± 0.06 [a]
Madecassol® (1%)	2.27 ± 0.12 [a]	1.59 ± 0.09 [a]	1.14 ± 0.11 [a]	0.46 ± 0.10 [a]	0.37 ± 0.04 [a]	0.13 ± 0.03 [b]
Hydro-ethanolic extract (10%)	2.83 ± 0.03 [a]	1.52 ± 0.06 [a]	0.49 ± 0.10 [b]	0.36 ± 0.05 [a]	0.23 ± 0.05 [a]	0.08 ± 0.02 [b]
Polyphenolic extract	2.79 ± 0.46 [a]	1.71 ± 0.23 [a]	0.83 ± 0.15 [ab]	0.58 ± 0.19 [a]	0.36 ± 0.13 [a]	0.14 ± 0.03 [b]

The different letters in superscript (a and b) indicate the significant difference between treatments at $p < 0.05$.

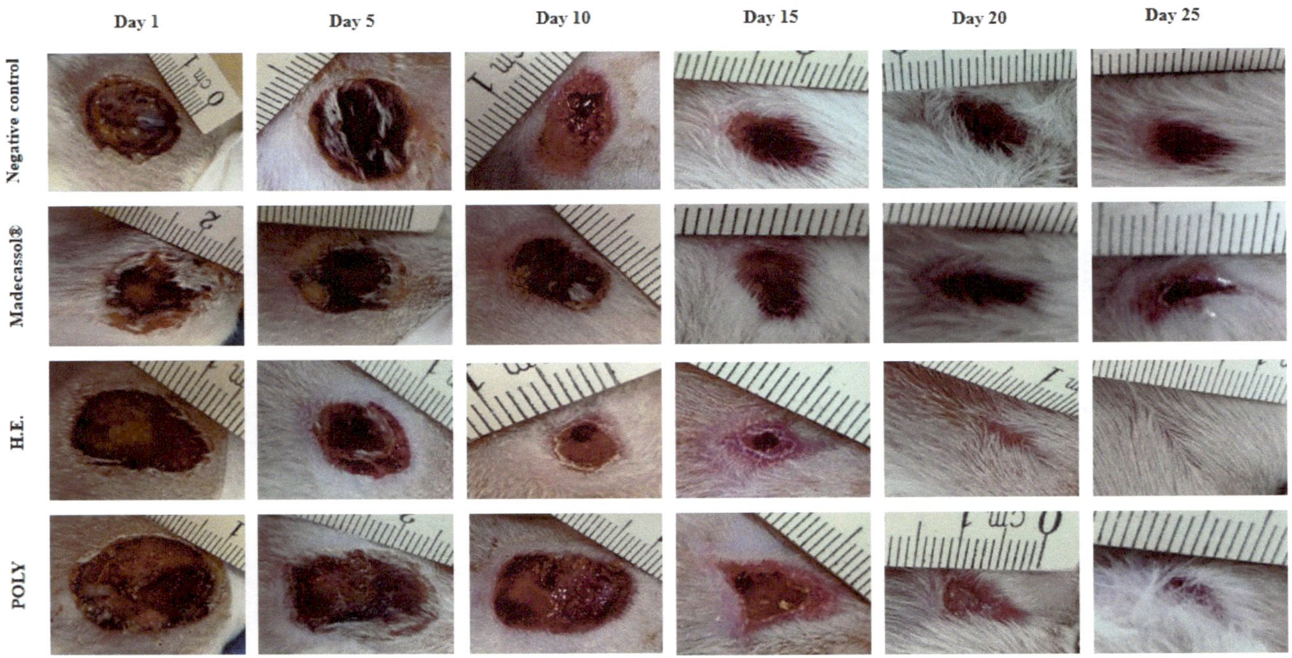

Figure 5. Morphological aspects of cutaneous burn healing process upon application of the hydro-ethanolic (H.E) and Poly (Polyphenols) extracts of *P. sativum* Hoffm. and the control groups during 25 days of treatment. The rate of the healing activities induced by each application were compared to each other. As shown in Figure 6, the hydro-ethanolic and polyphenols extracts elicited a high wound contraction at the 5th day by 46.45% and 38.7%, respectively. At the 25th day, the hydro-ethanolic extract elicited the highest healing (up to 97.17%), followed by the polyphenols and Madecassol ointment with almost the same percentage of inhibition (94.98% and 94.27%, respectively) while the negative control group was improved by only 80.62%.

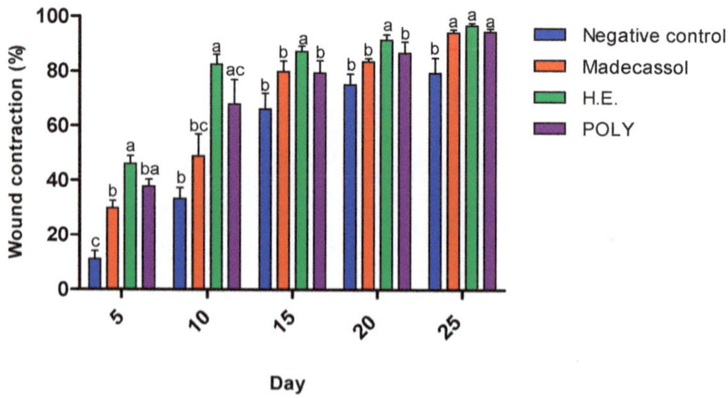

Figure 6. Burn healing activity of the hydro-ethanolic (H.E.) and polyphenolic (POLY) extracts of *P. sativum* Hoffm. and the control groups. Superscript letters (a, b, c) indicate the significant difference between treatments at $p < 0.05$.

4. Discussion

Parsley is a medicinal plant largely used a garnish and food-flavoring agent but also in traditional pharmacopoeia to treat several diseases such as inflammation, diabetes, cancer, digestive disorders, and kidney stones [14,54]. Nevertheless, many of these traditional applications remain to be discovered and proven. Here, we annotated the phytoconstituents of its hydro-ethanolic extract and studied the analgesic, antioxidant, and antimicrobial of the hydro-ethanolic and polyphenol extracts of parsley (*P. sativum* Hoffm.). Most importantly, we believe that this is the first study to assess the wound healing properties of *P. sativum* Hoffm. extracts using animal models.

The LC-MS/MS analysis of *P. sativum* corroborated the presence of some polyphenols known for their biological effects such as oleuropein (antioxidant, anti-atherogenic, antimicrobial, anti-cancer, antiviral activity, anti-inflammatory, hypoglycemic, and hypolipidemic activities) [55], arbutin (wound healing, anti-inflammatory, antioxidant, analgesic, anticancer, antiparkinsonic, and hypoglycemic effects) [56], myricetin (antimutagen, anti-ulcer, anticarcinogen, antioxidant, antibacterial, anti-diabetic, cardioprotective, anti-amyloidogenic, anti-inflammatory, and antiviral activities) [57,58], and naringin (anti-atherosclerosis, anti-diabetic, neuroprotective, cardioprotective, rheumatologic, and osteoporosis disorders). Moreover, the GC-MS analysis of the hydro-ethanolic extract of *P. sativum* Hoffm. showed the presence of some compounds renowned for their pharmacological activities like malic acid (anti-thrombotic) [59], D-glucose (anti-cancer, anti-diabetic) [60,61], carbazoles (antifungal) [62], and myo-inositol (against both male and female infertilities) [54]. The chemical compounds identified here are likely to be behind the biological and pharmacological activities demonstrated in this study. In line with other works, our results show the presence of phytochemicals in parsley such as alkaloids, polyphenols, and sugars that contribute to the structure of the flavonoid glycosides and amino acids [9,14,62].

The analgesic activity was tested by two protocols: abdominal writhes and formalin induced pain as described by Ganguly et al., 2016 [63]. Our findings suggest that the analgesic activity of *P. sativum* Hoffm. could be related to its effect on prostaglandin biosynthesis [64]. In fact, acetic acid induces the secretion of prostaglandins (PGE2 and PGE2α), partly involving peritoneal receptors and inflammatory discomfort [65]. Other studies showed the analgesic effect of the ethanolic extract of parsley prepared at 100, 150, and 200 mg/kg acts by decreasing both phases of pain using the formalin test [19]. Using this assay, there is two distinct biphasic nociceptive responses known as early and late phases [66]. Molecules that target precisely the central nervous system (CNS) can inhibit both phases by a similar mechanism. However, drugs acting on the peripheric nervous system (PNS) inhibit the late phases only [67,68]. The early phase could be triggered by the nociceptors induction in the paw reflecting the centrally mediated pain, while the late phase is activated by the release of pro-inflammatory agents such as serotonin, bradykinin, histamine, and prostaglandins [68,69]. However, this could be also due to the central nociceptive neuron's activation [69,70].

Balance between the release of anti and pro-inflammatory cytokines and analgesic mediators induces the chronicity of pain [71,72]. In a recent study, we demonstrated the anti-inflammatory effect of the hydro-ethanolic and polyphenolic extracts from *P. sativum* Hoffm. [26]. As the healing process is strongly associated with the inflammatory response by the intervention phase of monocytes and neutrophils, the proliferation of epithelial cells and fibroblasts, the synthesis of collagen, and the action of keratinocytes fibroblasts [73]. We evaluated the potential healing activity of our extracts in rats and showed that application of ointments prepared from the hydro-ethanolic extract and polyphenolics induced a significant cicatrizing effect as compared the untreated animals. This is most likely due to the presence of bioactive compounds that support inflammation to repair lesions and accelerate cell regeneration in damaged tissues. In fact, several phytochemicals such as polysaccharides, alkaloids, and saponins have been demonstrated to have wound healing properties [74]. For instance, triterpenes isolated from *Centella asiatica* stimulated glycosaminoglycans synthesis and ameliorated collagen remodeling [75]. In addition, madecassoside from this plant administered orally improved both collagen synthesis and angiogenesis. Worth noting is that *Arctium lappa* L. was able to monitor adhesion of dermal fibroblasts and regulate their gene expression by targeting the Wnt/β-catenin signaling pathway which is well documented as a major wound regulator [76]. Other molecules including apigenin are known for controlling the wound healing process [77].

As nutrients are important factors in wound healing, many studies showed that nutrient shortage is likely linked to the delayed healing of wounds [78]. For instance, vitamin K is mandatory during the first phase (hemostasis). Its deficiency alters wound repair, hemorrhage, and infection [79]. Interestingly, parsley is known as one of the leafy green vegetables to be rich in vitamin K [78]. In addition, as malic acid is one of the most abundant molecules identified in this study, its action as a wound healing agent could corroborate a previous study that showed that purified fractions from the leaves of *Sempervivum tectorum* L. harboring high contents of malic acid promote cellular proliferation and migration [80,81]. Other studies have investigated the role of malic acid derived polymers on muscle regeneration and bone repair [82].

Flavonoid contents of *P. sativum* Hoffm. were different from those obtained by Pereira et al., (2014) [83] who showed a low TFC in the hydro-ethanolic extract of parsley. This can be explained by the difference in the extraction method used, geographical regions of growth, seasonal variations, harvesting time, and postharvest treatment [5]. Previous studies have shown similar results regarding the TFC of parsley (27.2 mg QE/g) [84,85]. Other works revealed many factors that can influence the content of TPC such as genetic and extrinsic factors namely climatic and geographic ones [86]. There is also the duration of storage, chemotype, and the degree of maturation of the plant which have a strong influence on the polyphenols contents [87].

Here, the antioxidant activity was monitored using DPPH, FRAP and TAC tests. The difference seen between the hydro-ethanolic extract and the polyphenolics could be related to the presence of other chemicals having antioxidant activity other than the polyphenols. As previously reported, this antioxidant ability can be due to the presence of malic acid in *P. sativum* Hoffm. [88,89]. Other studies have shown the significant antioxidant capacity of the aerial part of parsley due to the presence of flavonoid [90,91]. For instance, Marin et al. (2016) demonstrated that the water extract of parsley exhibited a low oxidation inhibitory effect using FRAP test, with an EC_{50} of 0.93 mmol/L [92,93]. Moreover, it has been proposed that a high TAC may be closely related to the presence of a high content of polyphenolics [94].

The antimicrobial activity of *P. sativum* Hoffm. polyphenolic fraction towards *S. aureus*, *P. aeruginosa*, and *C. albicans* highlights the potential use of this plant's extract in treating bacterial and fungal wound infections. In fact, these pathogenic species along with others such as *S. epidermidis*, *Escherichia coli*, *Klebsiella pneumonia*, and *Proteus* are commonly isolated from infected wounds. Therefore, our findings provide useful information on the wound healing properties of plant-based chemicals through controlling infectious agents [95].

To explain the mechanisms of action of plant-derived antimicrobial compounds, several studies have attempted to correlate their antibacterial effect with their phytochemical composition [96]. Some scientists have suggested that the antibacterial action is associated with high concentrations of phenols, monoterpenes, aldehydes, and ketones that perturb the integrity of microbial membranes [97]. This may be due to the hydrophobic nature of some phytocompounds, allowing their accumulation on the cell membranes and disrupt their structure and function. This also weakens the microbial enzyme machinery, allows intracellular components to leak, and leads to apoptosis [98,99]. Other investigations have reported that these products can coagulate the microbial cytoplasm and bring down lipids and proteins [100]. Mostafa et al., (2020) recently reported that the stem bark extract from *Salix tetrasperma* impaired the virulence of *P. aeruginosa* by hindering its swimming and swarming on plates and by inhibiting its hemolytic and proteolytic activities [101]. Similarly, Ben bakrim et al., (2022) showed that the leaf extract from *Ximenia americana* var. *caffra* has the ability to inhibit the biofilm formation by the skin pathogen *P. aeruginosa* and reduced its mobilities in a dose-dependent manner [13]. Overall, the profile of a plant chemical composition may influence its targets, mechanisms of action, and consequently, its antibacterial activity.

5. Conclusions

This study profiled the phytochemical composition of the hydro-ethanolic extract of *P. sativum* Hoffm. and highlighted some biological and pharmacological activities of its hydro-ethanolic extract and polyphenolic fraction. Our findings corroborate many previous investigations on the role of parsley-based phytochemicals such as polyphenols as analgesic, antioxidant, and antimicrobial agents. However, as far as we know, this is the first study to deliver proof that parsley could serve as a source of wound healing bioactive principles. Nevertheless, more evidence is needed to prove their direct influence using fractionation and guided bioassays and individual compounds isolation. In addition, further in-depth assessments should address the underpinning molecular and physiological mechanisms of observed analgesic and healing activities in vivo. Lastly, our study shed light on the potential and promising role of *P. sativum* Hoffm. as a source of analgesic, antimicrobial, and wound healing plant-based agents.

Supplementary Materials: The following supporting information can be downloaded at: https://www.mdpi.com/article/10.3390/metabo13020260/s1, Figure S1: XIC chromatograms of identified compounds.

Author Contributions: Conceptualization: D.B. and M.S.; methodology: A.D. and F.Z.M.; software: A.I.H.; validation: A.B.; formal analysis: R.C.; investigation: O.A.K.; data curation: F.E.-z.A. and A.S.; writing—original draft preparation: M.S.; writing—review and editing: M.S. and I.M.; supervision: D.B. All authors have read and agreed to the published version of the manuscript.

Funding: This study was funded by Princess Nourah bint Abdulrahman University Researchers Supporting Project number (PNURSP2023R141). Princess Nourah bint Abdulrahman University, Riyadh, Saudi Arabia.

Institutional Review Board Statement: The institutional ethical committee of care and use of the laboratory animals at the Faculty of Sciences Dhar El Mehraz, Sidi Mohamed Ben Abdallah Fez University, Morocco, reviewed and approved the present study #04/2019/LBEAS.

Informed Consent Statement: Not applicable.

Data Availability Statement: The data presented in this study are available in the main article and the supplementary materials.

Acknowledgments: The authors extend their appreciation to Princess Nourah bint Abdulrahman University Researchers Supporting Project number (PNURSP2023R141), Princess Nourah bint Abdulrahman University, Riyadh, Saudi Arabia.

Conflicts of Interest: The authors have not declared any conflict of interests.

References

1. Zhang, Y.; Li, X.; Zou, D.; Liu, W.; Yang, J.; Zhu, N.; Huo, L.; Wang, M.; Hong, J.; Wu, P.; et al. Treatment of Type 2 Diabetes and Dyslipidemia with the Natural Plant Alkaloid Berberine. *J. Clin. Endocrinol. Metab.* **2008**, *93*, 2559–2565. [CrossRef]
2. Zeilhofer, H.U. Prostanoids in Nociception and Pain. *Biochem. Pharmacol.* **2007**, *73*, 165–174. [CrossRef]
3. Guo, S.; DiPietro, L.A. Factors Affecting Wound Healing. *J. Dent. Res.* **2010**, *89*, 219–229. [CrossRef]
4. Apak, R.; Güçlü, K.; Özyürek, M.; Karademir, S.E. Novel Total Antioxidant Capacity Index for Dietary Polyphenols and Vitamins C and E, Using Their Cupric Ion Reducing Capability in the Presence of Neocuproine: CUPRAC Method. *J. Agric. Food Chem.* **2004**, *52*, 7970–7981. [CrossRef] [PubMed]
5. Mahdi, I.; Bakrim, W.B.; Bitchagno, G.T.M.; Annaz, H.; Mahmoud, M.F.; Sobeh, M. Unraveling the Phytochemistry, Traditional Uses, and Biological and Pharmacological Activities of Thymus Algeriensis Boiss. & Reut. *Oxid. Med. Cell. Longev.* **2022**, *2022*, 6487430.
6. Rojkind, M.; Dominguez-Rosales, J.-A.; Nieto, N.; Greenwel, P. Role of Hydrogen Peroxide and Oxidative Stress in Healing Responses. *Cell. Mol. Life Sci. CMLS* **2002**, *59*, 1872–1891. [CrossRef] [PubMed]
7. López, M.G.; Sánchez-Mendoza, I.R.; Ochoa-Alejo, N. Compartive Study of Volatile Components and Fatty Acids of Plants and in Vitro Cultures of Parsley (Petroselinum Crispum (Mill) Nym Ex Hill). *J. Agric. Food Chem.* **1999**, *47*, 3292–3296. [CrossRef] [PubMed]
8. Aziz, N.; Kim, M.-Y.; Cho, J.Y. Anti-Inflammatory Effects of Luteolin: A Review of in Vitro, in Vivo, and in Silico Studies. *J. Ethnopharmacol.* **2018**, *225*, 342–358. [CrossRef]
9. Chaves, D.S.A.; Frattani, F.S.; Assafim, M.; de Almeida, A.P.; Zingali, R.B.; Costa, S.S. Composition Chimique Phénolique de l' Extrait de Petroselinum Crispum et Son Effet Sur l'hémostase. *Nat. Prod. Commun.* **2011**, *6*, 1934578X1100600709. [CrossRef]
10. Haque, A.; Pant, A.B. Mitigating Covid-19 in the Face of Emerging Virus Variants, Breakthrough Infections and Vaccine Hesitancy. *J. Autoimmun.* **2022**, 102792. [CrossRef]
11. Slighoua, M.; Mahdi, I.; Amrati, F.E.; Boukhira, S.; Youbi, A.E.H.E.; Bari, A.; Bousta, D. Ethnopharmacological Survey of Medicinal Plants Used in the Traditional Treatment of Female Infertility in Fez Region, Morocco. *Phytothérapie* **2020**, *18*, 321–339. [CrossRef]
12. Tümen, G.; Malyer, H.; Başer, K.H.C.; Öz Aydın, S. Plants Used in Anatolia for Wound Healing. In Proceedings of the IVth International Congress of Ethnobotany (ICEB 2005), Istanbul, Turkey, 21–26 August 2005; Volume 217, p. 221.
13. Bakrim, W.B.; Nurcahyanti, A.D.R.; Dmirieh, M.; Mahdi, I.; Elgamal, A.M.; El Raey, M.A.; Wink, M.; Sobeh, M. Phytochemical Profiling of the Leaf Extract of Ximenia Americana Var. Caffra and Its Antioxidant, Antibacterial, and Antiaging Activities In Vitro and in Caenorhabditis Elegans: A Cosmeceutical and Dermatological Approach. *Oxid. Med. Cell. Longev.* **2022**, *2022*, e3486257. [CrossRef]
14. Farzaei, M.H.; Abbasabadi, Z.; Ardekani, M.R.S.; Rahimi, R.; Farzaei, F. Parsley: A Review of Ethnopharmacology, Phytochemistry and Biological Activities. *J. Tradit. Chin. Med.* **2013**, *33*, 815–826. [CrossRef] [PubMed]
15. Thangavelu, S.; Balasubramanian, B.; Palanisamy, S.; Shanmugam, V.; Natchiappan, S.; Kalibulla, S.I.; Rathinasamy, B.; Arumugam, V.A. Characterization and Phytoconstituents of Petroselinum Crispum (Mill) and Coriandrum Sativum (Linn) and Their Impacts on Inflammation—An in Vitro Analysis against Human Adenocarcinoma Cells with Molecular Docking. *S. Afr. J. Bot.* **2022**, *146*, 776–788. [CrossRef]
16. AbdelKhalek, A.S.; Youssef, H.A.; Ali, M.F.; Ali, M.M.; Abdel-Hakiem, M.A.H.; Mahmoud, H.F.F. An Assessment of Clinical, Biometric, Cosmetic and Microscopic Outcomes of Four Suture Techniques for Cutaneous Closure of Laparotomy Wounds: An Experimental Study in Rabbits. *J. Dairy Vet. Anim. Res.* **2019**, *8*, 42–53. [CrossRef]
17. Esther, T.L.H. Investigation of the Antioxidant, Anticancer, Wound Healing, Immunomodulatory and Dna Protective Activities of Coriandrum Sativum and Petroselinum Crispum/Esther Tang Lai Har. Ph.D. Thesis, University of Malaya, Kuala Lumpur, Malaysia, 2014.
18. Behtash, N.; Kargarzadeh, F.; Shafaroudi, H. Analgesic Effects of Seed Extract from Petroselinum Crispum (Tagetes Minuta) in Animal Models. *Toxicol. Lett.* **2008**, *180*, S127–S128. [CrossRef]
19. Eidi, A.; Eidi, M.; Badiei, L. Antinociceptive Effects of Ethanolic Extract of Parsley (Petroselinum Crispum L.) Leaves in Mice. *Med. Sci. J. Islam. Azad Univesity Tehran Med. Branch* **2009**, *19*, 181–186.
20. Moazedi, A.A.; Mirzaie, D.N.; Seyyednejad, S.M.; Zadkarami, M.R.; Amirzargar, A. Spasmolytic Effect of Petroselinum Crispum (Parsley) on Rat's Ileum at Different Calcium Chloride Concentrations. *Pak. J. Biol. Sci. PJBS* **2007**, *10*, 4036–4042. [CrossRef] [PubMed]

21. Annaz, H.; Sane, Y.; Bitchagno, G.T.M.; Ben Bakrim, W.; Drissi, B.; Mahdi, I.; El Bouhssini, M.; Sobeh, M. Caper (*Capparis Spinosa* L.): An Updated Review on Its Phytochemistry, Nutritional Value, Traditional Uses, and Therapeutic Potential. *Front. Pharmacol.* **2022**, *13*. [CrossRef]
22. Ngaffo, C.M.; Tankeo, S.B.; Guefack, M.-G.F.; Nayim, P.; Wamba, B.E.; Kuete, V.; Mbaveng, A.T. Phytochemical Analysis and Antibiotic-Modulating Activity of Cocos Nucifera, Glycine Max and Musa Sapientum Methanol Extracts against Multidrug Resistant Gram-Negative Bacteria. *Investig. Med. Chem. Pharmacol.* **2021**, *4*, 53.
23. Terekhov, R.P.; Selivanova, I.A.; Anurova, M.N.; Zhevlakova, A.K.; Nikitin, I.D.; Cong, Z.; Ma, S.; Yang, F.; Dong, Z.; Liao, Y. Comparative Study of Wound-Healing Activity of Dihydroquercetin Pseudopolymorphic Modifications. *Bull. Exp. Biol. Med.* **2021**, *170*, 444–447.
24. Dahm, H. Silver Nanoparticles in Wound Infections: Present Status and Future Prospects. In *Nanotechnology in Skin, Soft Tissue, and Bone Infections*; Rai, M., Ed.; Springer International Publishing: Cham, Switzerland, 2020; pp. 151–168; ISBN 978-3-030-35147-2.
25. Sawyer, R.G.; Pruett, T.L. Wound Infections. *Surg. Clin. N. Am.* **1994**, *74*, 519–536. [CrossRef]
26. Slighoua, M.; Mahdi, I.; Amrati, F.; Di Cristo, F.; Amaghnouje, A.; Grafov, A.; Boucetta, N.; Bari, A.; Bousta, D. Assessment of in Vivo Estrogenic and Anti-Inflammatory Activities of the Hydro-Ethanolic Extract and Polyphenolic Fraction of Parsley (*Petroselinum Sativum* Hoffm.). *J. Ethnopharmacol.* **2021**, *265*, 113290. [CrossRef] [PubMed]
27. Lilienblum, W.; Dekant, W.; Foth, H.; Gebel, T.; Hengstler, J.G.; Kahl, R.; Kramer, P.-J.; Schweinfurth, H.; Wollin, K.-M. Alternative Methods to Safety Studies in Experimental Animals: Role in the Risk Assessment of Chemicals under the New European Chemicals Legislation (REACH). *Arch. Toxicol.* **2008**, *82*, 211–236. [CrossRef] [PubMed]
28. Wafa, G.; Amadou, D.; Larbi, K.M.; Héla, E.F.O. Larvicidal Activity, Phytochemical Composition, and Antioxidant Properties of Different Parts of Five Populations of Ricinus Communis L. *Ind. Crops Prod.* **2014**, *56*, 43–51. [CrossRef]
29. Amrati, F.E.-Z.; Bourhia, M.; Saghrouchni, H.; Slighoua, M.; Grafov, A.; Ullah, R.; Ezzeldin, E.; Mostafa, G.A.; Bari, A.; Ibenmoussa, S.; et al. Caralluma Europaea (Guss.) N.E.Br.: Anti-Inflammatory, Antifungal, and Antibacterial Activities against Nosocomial Antibiotic-Resistant Microbes of Chemically Characterized Fractions. *Molecules* **2021**, *26*, 636. [CrossRef] [PubMed]
30. Slighoua, M.; Mahdi, I.; ez-zahra Amrati, F.; Boucetta, N.; Cristo, F.D.; Boukhira, S.; El youbi el Hamsas, A.; Tattou, M.I.; Grafov, A.; Bari, A.; et al. Pharmacological Effects of Lavandula Officinalis Chaix and Its Polyphenols: Focus on Their in Vivo Estrogenic and Anti-Inflammatory Properties. *South Afr. J. Bot.* **2022**, *146*, 354–364. [CrossRef]
31. Amaghnouje, A.; Mechchate, H.; Es-safi, I.; Boukhira, S.; Aliqahtani, A.S.; Noman, O.M.; Nasr, F.A.; Conte, R.; Calarco, A.; Bousta, D. Subacute Assessment of the Toxicity and Antidepressant-Like Effects of *Origanum Majorana* L. Polyphenols in Swiss Albino Mice. *Molecules* **2020**, *25*, 5653. [CrossRef]
32. Griffiths, W.J.; Wang, Y. Analysis of Neurosterols by GC–MS and LC–MS/MS. *J. Chromatogr. B* **2009**, *877*, 2778–2805. [CrossRef]
33. Kabran, G.R.; Mamyrbekova-Bekro, J.A.; Pirat, J.-L.; Bekro, Y.-A.; Sommerer, N.; Verbaere, A.; Meudec, E. Identification de composés phénoliques extraits de deux plantes de la pharmacopée ivoirienne */Identification of phenolic compounds from two plants of ivorian pharmacopeia *. *J. Société Ouest-Afr. Chim.* **2014**, *38*, 57–63.
34. Slinkard, K.; Singleton, V.L. Total Phenol Analysis: Automation and Comparison with Manual Methods. *Am. J. Enol. Vitic.* **1977**, *28*, 49–55. [CrossRef]
35. Brand-Williams, W.; Cuvelier, M.E.; Berset, C. Use of a Free Radical Method to Evaluate Antioxidant Activity. *LWT Food Sci. Technol.* **1995**, *28*, 25–30. [CrossRef]
36. Oyaizu, M. Studies on Products of Browning Reaction Antioxidative Activities of Products of Browning Reaction Prepared from Glucosamine. *Jpn. J. Nutr. Diet.* **1986**, *44*, 307–315. [CrossRef]
37. Baali, F.; Boumerfeg, S.; Napoli, E.; Boudjelal, A.; Righi, N.; Deghima, A.; Baghiani, A.; Ruberto, G. Chemical Composition and Biological Activities of Essential Oils from Two Wild Algerian Medicinal Plants: *Mentha Pulegium* L. and *Lavandula Stoechas* L. *J. Essent. Oil Bear. Plants* **2019**, *22*, 821–837. [CrossRef]
38. Mašković, P.Z.; Manojlović, N.T.; Mandić, A.I.; Mišan, A.Č.; Milovanović, I.L.; Radojković, M.M.; Cvijović, M.S.; Solujić, S.R. Phytochemical Screening and Biological Activity of Extracts of Plant Species Halacsya Sendtneri (Boiss.) Dörfl. *Hem. Ind.* **2012**, *66*, 43–51. [CrossRef]
39. Balouiri, M.; Sadiki, M.; Ibnsouda, S.K. Methods for in Vitro Evaluating Antimicrobial Activity: A Review. *J. Pharm. Anal.* **2016**, *6*, 71–79. [CrossRef]
40. Mabadahanye, K. *Isolation and Analysing Chemical Profiles of Bioactive Compounds from South Africa Medicinal Plants with Activity against Pathogenic Organisms*; University of Johannesburg: Johannesburg, South Africa, 2020; ISBN 9798544213154.
41. Athanassiadis, B.; Abbott, P.V.; George, N.; Walsh, L.J. An in Vitro Study of the Antimicrobial Activity of Some Endodontic Medicaments and Their Bases Using an Agar Well Diffusion Assay. *Aust. Dent. J.* **2009**, *54*, 141–146. [CrossRef]
42. Nalawade, T.M.; Bhat, K.G.; Sogi, S. Antimicrobial Activity of Endodontic Medicaments and Vehicles Using Agar Well Diffusion Method on Facultative and Obligate Anaerobes. *Int. J. Clin. Pediatr. Dent.* **2016**, *9*, 335.
43. Hayet, E.; Maha, M.; Samia, A.; Mata, M.; Gros, P.; Raida, H.; Ali, M.M.; Mohamed, A.S.; Gutmann, L.; Mighri, Z. Antimicrobial, Antioxidant, and Antiviral Activities of Retama Raetam (Forssk.) Webb Flowers Growing in Tunisia. *World J. Microbiol. Biotechnol.* **2008**, *24*, 2933–2940. [CrossRef]
44. Wayne, P.A. Clinical and Laboratory Standards Institute: Performance Standards for Antimicrobial Susceptibility Testing: 20th Informational Supplement. *CLSI Doc. M100-S20* **2010**.

45. Adeoyo, O.R.; Pletschke, B.I.; Dames, J.F. Molecular Identification and Antibacterial Properties of an Ericoid Associated Mycorrhizal Fungus. *BMC Microbiol.* **2019**, *19*, 1–8. [CrossRef] [PubMed]
46. Scott, L.J.; Perry, C.M. Tramadol. *Drugs* **2000**, *60*, 139–176. [CrossRef]
47. França, D.S.; Souza, A.L.; Almeida, K.R.; Dolabella, S.S.; Martinelli, C.; Coelho, M.M. B Vitamins Induce an Antinociceptive Effect in the Acetic Acid and Formaldehyde Models of Nociception in Mice. *Eur. J. Pharmacol.* **2001**, *421*, 157–164. [CrossRef] [PubMed]
48. Hajhashemi, V.; Ghannadi, A.; Sharif, B. Anti-Inflammatory and Analgesic Properties of the Leaf Extracts and Essential Oil of Lavandula Angustifolia Mill. *J. Ethnopharmacol.* **2003**, *89*, 67–71. [CrossRef] [PubMed]
49. Manglik, A.; Lin, H.; Aryal, D.K.; McCorvy, J.D.; Dengler, D.; Corder, G.; Levit, A.; Kling, R.C.; Bernat, V.; Hübner, H. Structure-Based Discovery of Opioid Analgesics with Reduced Side Effects. *Nature* **2016**, *537*, 185–190. [CrossRef]
50. Wang, Y.; Hu, X.; Huang, H.; Jin, Z.; Gao, J.; Guo, Y.; Zhong, Y.; Li, Z.; Zong, X.; Wang, K. Optimization of 4-Arylthiophene-3-Carboxylic Acid Derivatives as Inhibitors of ANO1: Lead Optimization Studies toward Their Analgesic Efficacy for Inflammatory Pain. *Eur. J. Med. Chem.* **2022**, *237*, 114413. [CrossRef]
51. De Miranda, F.G.G.; Vilar, J.C.; Alves, I.A.N.; de Holanda Cavalcanti, S.C.; Antoniolli, Â.R. Antinociceptive and Antiedematogenic Properties and Acute Toxicity of Tabebuia Avellanedae Lor. Ex Griseb. Inner Bark Aqueous Extract. *BMC Pharmacol.* **2001**, *1*, 1–5. [CrossRef]
52. Heidari, M.; Bahramsoltani, R.; Abdolghaffari, A.H.; Rahimi, R.; Esfandyari, M.; Baeeri, M.; Hassanzadeh, G.; Abdollahi, M.; Farzaei, M.H. Efficacy of Topical Application of Standardized Extract of Tragopogon Graminifolius in the Healing Process of Experimental Burn Wounds. *J. Tradit. Complement. Med.* **2019**, *9*, 54–59. [CrossRef]
53. Slighoua, M.; Chebaibi, M.; Mahdi, I.; Amrati, F.E.; Conte, R.; Cordero, M.A.W.; Alotaibi, A.; Saghrouchni, H.; Agour, A.; Zair, T.; et al. The LC-MS/MS Identification and Analgesic and Wound Healing Activities of Lavandula Officinalis Chaix: In Vivo and In Silico Approaches. *Plants* **2022**, *11*, 3222. [CrossRef]
54. Charles, D.J. *Antioxidant Properties of Spices, Herbs and Other Sources [Electronic Resource]*; Springer: Berlin/Heidelberg, Germany, 2013.
55. Omar, S.H. Oleuropein in Olive and Its Pharmacological Effects. *Sci. Pharm.* **2010**, *78*, 133–154. [CrossRef]
56. Bhalla, M. Pharmacological Aspects of a Bioactive Compound Arbutin: A Comprehensive Review. *Biointerface Res. Appl. Chem.* **2022**, *13*, 119. [CrossRef]
57. Gupta, G.; Siddiqui, M.A.; Khan, M.M.; Ajmal, M.; Ahsan, R.; Rahaman, M.A.; Ahmad, M.A.; Arshad, M.; Khushtar, M. Current Pharmacological Trends on Myricetin. *Drug Res.* **2020**, *70*, 448–454. [CrossRef]
58. Ong, K.C.; Khoo, H.-E. Biological Effects of Myricetin. *Gen. Pharmacol. Vasc. Syst.* **1997**, *29*, 121–126. [CrossRef]
59. Zhang, Q.-C.; Zhao, Y.; Bian, H.-M. Anti-Thrombotic Effect of a Novel Formula from Corni Fructus with Malic Acid, Succinic Acid and Citric Acid. *Phytother. Res.* **2014**, *28*, 722–727. [CrossRef]
60. Wokoun, U.; Hellriegel, M.; Emons, G.; Gründker, C. Co-Treatment of Breast Cancer Cells with Pharmacologic Doses of 2-Deoxy-D-Glucose and Metformin: Starving Tumors. *Oncol. Rep.* **2017**, *37*, 2418–2424. [CrossRef]
61. Zhang, J.; Li, L.; Kim, S.-H.; Hagerman, A.E.; Lü, J. Anti-Cancer, Anti-Diabetic and Other Pharmacologic and Biological Activities of Penta-Galloyl-Glucose. *Pharm. Res.* **2009**, *26*, 2066–2080. [CrossRef] [PubMed]
62. Thevissen, K.; Marchand, A.; Chaltin, P.; Meert, E.M.K.; Cammue, B.P.A. Antifungal Carbazoles. *Curr. Med. Chem.* **2009**, *16*, 2205–2211. [CrossRef] [PubMed]
63. Ganguly, A.; Al Mahmud, Z.; Kumar Saha, S.; Abdur Rahman, S.M. Evaluation of Antinociceptive and Antidiarrhoeal Properties of Manilkara Zapota Leaves in Swiss Albino Mice. *Pharm. Biol.* **2016**, *54*, 1413–1419. [CrossRef] [PubMed]
64. de Fátima Arrigoni-Blank, M.; Dmitrieva, E.G.; Franzotti, E.M.; Antoniolli, A.R.; Andrade, M.R.; Marchioro, M. Anti-Inflammatory and Analgesic Activity of *Peperomia Pellucida* (L.) HBK (Piperaceae). *J. Ethnopharmacol.* **2004**, *91*, 215–218. [CrossRef] [PubMed]
65. Silva, V.G.; Silva, R.O.; Damasceno, S.R.; Carvalho, N.S.; Prudêncio, R.S.; Aragão, K.S.; Guimarães, M.A.; Campos, S.A.; Véras, L.M.; Godejohann, M. Anti-Inflammatory and Antinociceptive Activity of Epiisopiloturine, an Imidazole Alkaloid Isolated from Pilocarpus Microphyllus. *J. Nat. Prod.* **2013**, *76*, 1071–1077. [CrossRef]
66. Fischer, L.G.; Leitão, R.; Etcheverry, S.R.; de Campos-Buzzi, F.; Vãzquez, A.A.; Heinzen, H.A.; Filho, V.C. Analgesic Properties of Extracts and Fractions from Erythrina Crista-Galli (Fabaceae) Leaves. *Nat. Prod. Res.* **2007**, *21*, 759–766. [CrossRef]
67. Ahmadiani, A.; Hosseiny, J.; Semnanian, S.; Javan, M.; Saeedi, F.; Kamalinejad, M.; Saremi, S. Antinociceptive and Anti-Inflammatory Effects of Elaeagnus Angustifolia Fruit Extract. *J. Ethnopharmacol.* **2000**, *72*, 287–292. [CrossRef] [PubMed]
68. Daud, A.; Habib, N.; Riera, A.S. Anti-Inflammatory, Anti-Nociceptive and Antipyretic Effects of Extracts of Phrygilanthus Acutifolius Flowers. *J. Ethnopharmacol.* **2006**, *108*, 198–203. [CrossRef] [PubMed]
69. McNamara, C.R.; Mandel-Brehm, J.; Bautista, D.M.; Siemens, J.; Deranian, K.L.; Zhao, M.; Hayward, N.J.; Chong, J.A.; Julius, D.; Moran, M.M. TRPA1 Mediates Formalin-Induced Pain. *Proc. Natl. Acad. Sci. USA* **2007**, *104*, 13525–13530. [CrossRef]
70. Parvizpur, A.; Ahmadiani, A.; Kamalinejad, M. Probable Role of Spinal Purinoceptors in the Analgesic Effect of Trigonella Foenum (TFG) Leaves Extract. *J. Ethnopharmacol.* **2006**, *104*, 108–112. [CrossRef]
71. Crovetti, G.; Martinelli, G.; Issi, M.; Barone, M.; Guizzardi, M.; Campanati, B.; Moroni, M.; Carabelli, A. Platelet Gel for Healing Cutaneous Chronic Wounds. *Transfus. Apher. Sci.* **2004**, *30*, 145–151. [CrossRef] [PubMed]
72. Jung, M.J.; Heo, S.-I.; Wang, M.-H. Free Radical Scavenging and Total Phenolic Contents from Methanolic Extracts of Ulmus Davidiana. *Food Chem.* **2008**, *108*, 482–487. [CrossRef] [PubMed]
73. Haney, E.F.; Mansour, S.C.; Hancock, R.E. Antimicrobial Peptides: An Introduction. *Antimicrob. Pept.* **2017**, 3–22.

74. Shedoeva, A.; Leavesley, D.; Upton, Z.; Fan, C. Wound Healing and the Use of Medicinal Plants. *Evid. Based Complement. Alternat. Med.* **2019**, *2019*. [CrossRef]
75. Babu, M.K.; Prasad, O.S.; Murthy, T.E. Comparison of the Dermal Wound Healing of Centella Asiatica Extract Impregnated Collagen and Cross Linked Collagen Scaffolds. *J. Chem. Pharm. Res.* **2011**, *3*, 353–362.
76. Ramnath, V.; Sekar, S.; Sankar, S.; Sastry, T.P.; Mandal, A.B. In Vivo Evaluation of Composite Wound Dressing Material Containing Soya Protein and Sago Starch. *Int. J. Pharm. Sci.* **2012**, *4*, 414–419.
77. Motealleh, B.; Zahedi, P.; Rezaeian, I.; Moghimi, M.; Abdolghaffari, A.H.; Zarandi, M.A. Morphology, Drug Release, Antibacterial, Cell Proliferation, and Histology Studies of Chamomile-loaded Wound Dressing Mats Based on Electrospun Nanofibrous Poly (ε-caprolactone)/Polystyrene Blends. *J. Biomed. Mater. Res. B Appl. Biomater.* **2014**, *102*, 977–987. [CrossRef]
78. Brown, K.L.; Phillips, T.J. Nutrition and Wound Healing. *Clin. Dermatol.* **2010**, *28*, 432–439. [CrossRef] [PubMed]
79. Jenkins, M.E.; Gottschlich, M.M.; Kopcha, R.; Khoury, J.; Warden, G.D. A Prospective Analysis of Serum Vitamin K in Severely Burned Pediatric Patients. *J. Burn Care Rehabil.* **1998**, *19*, 75–81. [CrossRef] [PubMed]
80. Caruelle, J.-P.; Barritault, D.; Jeanbat-Mimaud, V.; Cammas-Marion, S.; Langlois, V.; Guerin, P.; Barbaud, C. Bioactive Functionalized Polymer of Malic Acid for Bone Repair and Muscle Regeneration. *J. Biomater. Sci. Polym. Ed.* **2000**, *11*, 979–991. [CrossRef]
81. Jeanbat-Mimaud, V.; Barbaud, C.; Caruelle, J.-P.; Barritault, D.; Cammas-Marion, S.; Guérin, P. Functionalized Polymers of Malic Acid Stimulate Tissue Repair Presumably by Regulating Heparin Growth Factors Bioavailability. In *Biomedical Polymers and Polymer Therapeutics*; Springer: Berlin/Heidelberg, Germany, 2002; pp. 243–251.
82. Cattaneo, F.; De Marino, S.; Parisi, M.; Festa, C.; Castaldo, M.; Finamore, C.; Duraturo, F.; Zollo, C.; Ammendola, R.; Zollo, F. Wound Healing Activity and Phytochemical Screening of Purified Fractions of *Sempervivum Tectorum* L. Leaves on HCT 116. *Phytochem. Anal.* **2019**, *30*, 524–534. [CrossRef]
83. Pereira, M.P.; Tavano, O.L. Use of Different Spices as Potential Natural Antioxidant Additives on Cooked Beans (*Phaseolus Vulgaris*). Increase of DPPH Radical Scavenging Activity and Total Phenolic Content. *Plant Foods Hum. Nutr.* **2014**, *69*, 337–343. [CrossRef]
84. Marinova, D.; Ribarova, F.; Atanassova, M. Total phenolics and total flavonoids in bulgarian fruits and vegetables. *JU Chem. Metal* **2005**, *40*, 255–260.
85. Papuc, C.; Predescu, C.; Nicorescu, V.; Stefan, G.; Nicorescu, I. Antioxidant Properties of a Parsley (*Petroselinum crispum*) Juice Rich in Polyphenols and Nitrites. *Curr. Res. Nutr. Food Sci. J.* **2016**, *4*, 114–118. [CrossRef]
86. Lafraxo, H.; Bakour, M.; Laaroussi, H.; El Ghouizi, A.; Ousaaid, D.; Aboulghazi, A.; Lyoussi, B. The Synergistic Beneficial Effect of Thyme Honey and Olive Oil against Diabetes and Its Complications Induced by Alloxan in Wistar Rats. *Evid. Based Complement. Alternat. Med.* **2021**, *2021*. [CrossRef]
87. Lee, C.E.; Petersen, C.H. Effects of Developmental Acclimation on Adult Salinity Tolerance in the Freshwater-Invading Copepod Eurytemora Affinis. *Physiol. Biochem. Zool.* **2003**, *76*, 296–301. [CrossRef] [PubMed]
88. Hanachi, P.; Golkho, S.H. Using HPLC to Determination the Composition and Antioxidant Activity of Berberis Vulgaris. *Eur. J. Sci. Res.* **2009**, *29*, 47–54.
89. Kazemi, M.; Hadavi, E.; Hekmati, J. Effect of Salicylic Acid, Malic Acid, Citric Acid and Sucrose on Antioxidant Activity, Membrane Stability and ACC-Oxidase Activity in Relation to Vase Life of Carnation Cut Flowers. *J. Plant Sci.* **2012**, *7*, 78–84. [CrossRef]
90. Al-Juhaimi, F.; Ghafoor, K. Total Phenols and Antioxidant Activities of Leaf and Stem Extracts from Coriander, Mint and Parsley Grown in Saudi Arabia. *Pak J Bot* **2011**, *43*, 2235–2237.
91. Wong, P.Y.Y.; Kitts, D.D. Studies on the Dual Antioxidant and Antibacterial Properties of Parsley (*Petroselinum Crispum*) and Cilantro (*Coriandrum Sativum*) Extracts. *Food Chem.* **2006**, *97*, 505–515. [CrossRef]
92. Agyare, C.; Appiah, T.; Boakye, Y.D.; Apenteng, J.A. Chapter 25 - *Petroselinum Crispum*: A Review. In *Medicinal Spices and Vegetables from Africa*; Kuete, V., Ed.; Academic Press: Cambridge, MA, USA, 2017; pp. 527–547. ISBN 978-0-12-809286-6.
93. Marín, I.; Sayas-Barberá, E.; Viuda-Martos, M.; Navarro, C.; Sendra, E. Chemical Composition, Antioxidant and Antimicrobial Activity of Essential Oils from Organic Fennel, Parsley, and Lavender from Spain. *Foods* **2016**, *5*, 18. [CrossRef] [PubMed]
94. Herken, E.N.; Guzel, S. Total Antioxidant Capacity and Total Phenol Contents of Selected Commercial Fruit Juices in Turkey. *Int. J. Food Prop.* **2010**, *13*, 1373–1379. [CrossRef]
95. Atef, N.M.; Shanab, S.M.; Negm, S.I.; Abbas, Y.A. Evaluation of Antimicrobial Activity of Some Plant Extracts against Antibiotic Susceptible and Resistant Bacterial Strains Causing Wound Infection. *Bull. Natl. Res. Cent.* **2019**, *43*, 144. [CrossRef]
96. Keita, K.; Darkoh, C.; Okafor, F. Secondary Plant Metabolites as Potent Drug Candidates against Antimicrobial-Resistant Pathogens. *SN Appl. Sci.* **2022**, *4*, 1–10. [CrossRef]
97. Chaillot, J.; Tebbji, F.; Remmal, A.; Boone, C.; Brown, G.W.; Bellaoui, M.; Sellam, A. The Monoterpene Carvacrol Generates Endoplasmic Reticulum Stress in the Pathogenic Fungus Candida Albicans. *Antimicrob. Agents Chemother.* **2015**, *59*, 4584–4592. [CrossRef]
98. Lv, F.; Liang, H.; Yuan, Q.; Li, C. In Vitro Antimicrobial Effects and Mechanism of Action of Selected Plant Essential Oil Combinations against Four Food-Related Microorganisms. *Food Res. Int.* **2011**, *44*, 3057–3064. [CrossRef]
99. Moussii, I.M.; Nayme, K.; Timinouni, M.; Jamaleddine, J.; Filali, H.; Hakkou, F. Synergistic Antibacterial Effects of Moroccan Artemisia Herba Alba, Lavandula Angustifolia and Rosmarinus Officinalis Essential Oils. *Synergy* **2020**, *10*, 100057. [CrossRef]

100. Viuda-Martos, M.; Mohamady, M.A.; Fernández-López, J.; Abd ElRazik, K.A.; Omer, E.A.; Pérez-Alvarez, J.A.; Sendra, E. In Vitro Antioxidant and Antibacterial Activities of Essentials Oils Obtained from Egyptian Aromatic Plants. *Food Control* **2011**, *22*, 1715–1722. [CrossRef]
101. Mostafa, I.; Abbas, H.A.; Ashour, M.L.; Yasri, A.; El-Shazly, A.M.; Wink, M.; Sobeh, M. Polyphenols from Salix Tetrasperma Impair Virulence and Inhibit Quorum Sensing of Pseudomonas Aeruginosa. *Molecules* **2020**, *25*, 1341. [CrossRef] [PubMed]

metabolites

MDPI

Article

Chemical Composition of *Ducrosia flabellifolia* L. Methanolic Extract and Volatile Oil: ADME Properties, In Vitro and In Silico Screening of Antimicrobial, Antioxidant and Anticancer Activities

Mejdi Snoussi [1,2,*], Ramzi Hadj Lajimi [3,4], Riadh Badraoui [1,5], Mousa Al-Reshidi [1,6], Mohammad A. Abdulhakeem [1], Mitesh Patel [7], Arif Jamal Siddiqui [1], Mohd Adnan [1], Karim Hosni [8], Vincenzo De Feo [9,*], Flavio Polito [9], Adel Kadri [10,11] and Emira Noumi [1,2]

1 Department of Biology, College of Science, University of Hail, P.O. Box 2440, Hail 2440, Saudi Arabia
2 Laboratory of Genetics, Biodiversity and Valorization of Bio-Resources (LR11ES41), Higher Institute of Biotechnology of Monastir, University of Monastir, Avenue Tahar Haddad, BP74, Monastir 5000, Tunisia
3 Department of Chemistry, College of Science, University of Hail, P.O. Box 2440, Hail 2440, Saudi Arabia
4 Laboratory of Water, Membranes and Environmental Biotechnologies, Center of Research and Water Technologies, P. B 273, Soliman 8020, Tunisia
5 Section of Histology Cytology, Medicine Faculty of Tunis, University of Tunis El Manar, La Rabta 1007, Road Djebal Lakhdhar, Tunis 1007, Tunisia
6 Molecular Diagnostics and Personalized Therapeutics Unit, University of Hail, P.O. Box 2440, Hail 2440, Saudi Arabia
7 Department of Biotechnology, Parul Institute of Applied Sciences, Centre of Research for Development, Parul University, Vadodara 391760, India
8 Laboratoire des Substances Naturelles, Institut National de Recherche et d'Analyse Physico-Chimique, Biotechpôle de Sidi Thabet 2020, Tunisia
9 Department of Pharmacy, University of Salerno, Via Giovanni Paolo II, 132, Fisciano, 84084 Salerno, Italy
10 Faculty of Science and Arts in Baljurashi, Albaha University, P.O. Box 1988, Albaha 65527, Saudi Arabia
11 Faculty of Science of Sfax, Department of Chemistry, University of Sfax, B.P. 1171, Sfax 3000, Tunisia
* Correspondence: m.snoussi@uoh.edu.sa (M.S.); defeo@unisa.it (V.D.F.)

Citation: Snoussi, M.; Lajimi, R.H.; Badraoui, R.; Al-Reshidi, M.; Abdulhakeem, M.A.; Patel, M.; Siddiqui, A.J.; Adnan, M.; Hosni, K.; De Feo, V.; et al. Chemical Composition of *Ducrosia flabellifolia* L. Methanolic Extract and Volatile Oil: ADME Properties, In Vitro and In Silico Screening of Antimicrobial, Antioxidant and Anticancer Activities. *Metabolites* **2023**, *13*, 64. https://doi.org/10.3390/metabo13010064

Academic Editors: Ramona Paltinean and Irina Ielciu

Received: 29 November 2022
Revised: 24 December 2022
Accepted: 27 December 2022
Published: 31 December 2022

Abstract: In the present study, the chemical composition of the volatile oil and methanolic extract from *Ducrosia flabellifolia* Boiss. was investigated. The antimicrobial, antioxidant, and anticancer activities of the methanolic extract from *D. flabellifolia* aerial parts were screened using experimental and computational approaches. Results have reported the identification of decanal (28.31%) and dodecanal (16.93%) as major compounds in the essential oil obtained through hydrodistillation. Farnesyl pyrophosphate, Methyl 7-desoxypurpurogallin-7-carboxylate trimethyl ether, Dihydro-Obliquin, Gummiferol, 2-Phenylaminoadenosine, and 2,4,6,8,10-dodecapentaenal, on the other hand, were the dominant compounds in the methanolic extract. Moreover, the tested extract was active against a large collection of bacteria and yeast strains with diameter of growth inhibition ranging from 6.67 ± 0.57 mm to 17.00 ± 1.73 mm, with bacteriostatic and fungicidal activities against almost all tested microorganisms. In addition, *D. flabellifolia* methanolic extract was dominated by phenolic compounds (33.85 ± 1.63 mg of gallic acid equivalent per gram of extract) and was able to trap DPPH• and ABTS•+ radicals with IC_{50} about 0.05 ± 0 mg/mL and 0.105 ± 0 mg/mL, respectively. The highest percentages of anticancer activity were recorded at 500 μg/mL for all cancer cell lines with IC_{50} about 240. 56 μg/mL (A-549), 202.94 μg/mL (HCT-116), and 154.44 μg/mL (MCF-7). The *in-silico* approach showed that *D. flabellifolia* identified compounds bound 1HD2, 2XCT, 2QZW, and 3LN1 with high affinities, which together with molecular interactions and the bond network satisfactorily explain the experimental results using antimicrobial, antioxidant, and anticancer assays. The obtained results highlighted the ethnopharmacological properties of the rare desertic *D. flabellifolia* plant species growing wild in Hail region (Saudi Arabia).

Keywords: *Ducrosia flabellifolia*; essential oil; methanolic extract; antimicrobial; antioxidant; anticancer; ADME; *in-silico*; molecular interactions

1. Introduction

The *Ducrosia flabellifolia* Boiss. (*D. flabellifolia*) plant species belongs to Apiaceae family and is popularly known in Saudi Arabia as Haza [1]. The aerial parts of this species are smoked in form of cigarettes and have been used in traditional medicine by the local people as a sedative agent, and for the treatment of dental pain [2]. Volatile oil of *D. flabellifolia* has been reported for its potent antimicrobial potential towards *C. albicans* and *S. aureus* and moderate activity against *E. coli* and *P. aeruginosa*. The essential oil of *D. flabellifolia* has been documented for its moderate-to-weak anti-proliferative activity against three human cancer cell lines, MCF-7, K562 and LS180 [3]. In addition, the *D. flabellifolia* ethanol extract has been reported for its apoptotic effect toward breast cancer [4]. More recently, it has been demonstrated that D. *flabellifolia* hydroalcoholic extract collected from Hail region (Saudi Arabia) is a rich source of chlorogenic acid, ferulic acid, caffeic acid, and sinapic acid with good antimicrobial and antioxidant activities [5].

Cancer as a non-communicable disease was classified as the most common cause of death after cardiovascular diseases [6]. According to the data from the international agency for research on cancer (IARC), in Saudi Arabia, the incidence of different types of cancers has increased over the past decade and the number of new cancer cases was estimated to be 27,885, including 13,069 deaths with colorectum, breast, thyroid, non-Hodgkin and lymphoma remaining the most common type of cancers [6]. The drastic spread of cancer disease with high mortality rate is due to its ability to metastasize and its migrating effect among multiple organs [7]. Despite the current treatments, it is still considered the second most devastating cause of death worldwide [8,9]. On the other hand, the emergence of multidrug resistance (MDR) causing infections has promoted the development of novel antibacterial agents [10,11]. Infections and cancerous diseases are enhanced and aggravated by oxidative stress, which is a key factor due to its dramatic and fulgurant impact on the generation of free radicals which overcome suffering and stimulate various diseases [12]. Moreover, the increase in the number of infections caused by pathogenic microorganisms in cancer patients has encouraged scientists to search for novel therapeutic agents [13]. Conventional cancer chemotherapeutic drugs are based on the administration of drugs with the power to hinder the proliferation of tumor cells by inducing their apoptosis. They are mostly non-selective to cancer cells with detrimental side effects that cause serious health diseases, and multidrug-resistant microorganisms that contribute to the development of drug resistance in cancer cells [14,15].

For the above-mentioned reasons, and due to a lack of new anti-cancer and anti-infective agents with high therapeutic efficacy, no drug resistance and low side effects, the exploration of a bioassay-guided approach for the discovery of anti-cancer and antimicrobial natural products has been intensified and become a focus of interest for scientists and pharmaceutical research. Herbal treatment is the greatest gift that humans can use to improve their health [16–18]. Moreover, aromatic and medicinal herbs still remain the ultimate choice and a source of promising molecules in all areas of health, both in the treatment and prevention of certain pathologies [19].

Based on traditional claims regarding the use of *D. flabellifolia*, the purpose of the present study was to assess for the first time the phytochemical composition, the in vitro antioxidant, and antimicrobial activities of *D. flabellifolia* methanolic extract and its anti-proliferative effectiveness against colon, lung, and breast cancer cell lines, supported with the ADME and molecular docking studies of their major constituents. The latter focused on targeting 1HD2, 2XCT, 2QZW, AND 3LN1 macromolecules, which correspond to Human peroxiredoxin 5, *S. aureus* Gyrase complex, Secreted aspartic proteinase (Sap) 1 from *C. albicans*, and cyclooxygenase-2 (COX-2) to assess the antioxidant, antibacterial and anticancer/anti-inflammatory potentials, respectively.

2. Results

2.1. Chemical Composition of D. flabellifolia EO

The yield of extraction of the essential oil was about 0.24 ± 0 mL/100 g of plant material. In addition, out of 98.46% of bioactive compounds identified with GC-MS technique, miscellaneous compounds are the dominant group (64.10%), followed by monoterpene hydrocarbons (16.60%), oxygenated monoterpenes (10.89%), and oxygenated monoterpenes (6.87%). The main most volatile constituents were decanal (28.31%) and dodecanal (16.93%), and 1-heptadecne (8.30%) were the dominant compounds followed by β-eudesmol (6.87%), α-pinene (5.83%), β-phellandrene (5.76%), and 1-decanol (4.34%) (Table 1).

Table 1. Chemical composition of *D. flabellifolia* aerial parts identified by GC-MS technique.

N°	Compound	RI *	*D. flabellifolia* EO	Molecular Weight	Chemical Formula
1	Nonane	900	0.68	128.259	C_9H_{20}
2	α-pinene	940	5.83	136.238	$C_{10}H_{16}$
3	Sabinene	969	0.85	136.23	$C_{10}H_{16}$
4	o-cymene	976	0.51	134.22	$C_{10}H_{14}$
5	β-pinene	983	0.38	136.278	$C_{10}H_{16}$
6	β-myrcene	992	2.87	136.238	$C_{10}H_{16}$
7	α-phellandrene	1005	0.40	136.23	$C_{10}H_{16}$
8	β-phellandrene	1029	5.76	136.23	$C_{10}H_{16}$
9	Fenchone	1093	1.81	152.23	$C_{10}H_{16}O$
10	4-undecene	1076	0.72	154.292	$C_{11}H_{22}$
11	Undecane	1100	0.33	156.313	$C_{11}H_{24}$
12	Citronellal	1152	1.81	154.25	$C_{10}H_{18}O$
13	Decanal	1202	28.31	153.26	$C_{10}H_{20}O$
14	Citronellol	1236	1.95	156.269	$C_{10}H_{20}O$
15	Verbenyl acetate	1269	2.84	194.270	$C_{12}H_{18}O_2$
16	1-decanol	1274	4.34	158.28	$C_{10}H_{21}OH$
17	Thymol	1298	1.84	150.22	$C_{10}H_{14}O$
18	Undecanal	1307	0.26	170.296	$C_{11}H_{22}O$
19	Geranyl acetate	1382	0.64	196.29	$C_{10}H_{20}O_2$
20	Dodecanal	1412	16.93	184.32	$C_{12}H_{24}O$
21	2-dodecenal	1476	0.80	182.3	$C_{12}H_{22}O$
22	1-hexadecene	1593	1.59	224.42	$C_{16}H_{32}$
23	Tetradecanal	1614	1.14	212.37	$C_{14}H_{28}O$
24	β-eudesmol	1654	6.87	222.37	$C_{15}H_{26}O$
25	1-heptadecne	1697	8.30	238.5	$C_{17}H_{34}$
26	2-Hydroxycyclopentadecanone	1852	0.70	240.38	$C_{15}H_{28}O_2$

* RI: Retention index relative to (C_8-C_{24}) *n*-alkanes on HP-5MS column.

2.2. Chemical Composition of D. flabellifolia Methanolic Extract

The obtained methanolic extract was oily with a black color. The yield of extraction was about 20.06 ± 0.19 g of dry extract/100 g of plant material. Seventeen tripeptides with molecular weights ranging from (288.1552) to (432.2024) g/mol were tentatively identified through comparison of spectrum data of the extract with that of known compounds. Details of identified peptides are given in Table S1.

It is important to note that all compounds were first reported in this study for *D. flabellifolia* aerial parts methanolic extract analyzed with HR-LC/MS. The complete list of identified chemical bioactive compounds is summarized in Table 2.

Table 2. Phytochemical compounds identified by the HR-LCMS technique in *D. flabellifolia* methanolic extract.

N°	Compound Name	Chemical Class	RT (mn)	MW (g/mol)	Chemical Formula	[*m*/*z*]-	[*m*/*z*]+
1	10-Hydroxyloganin	Terpenoids	0.963	406.1437	$C_{17}H_{26}O_{11}$	387.1277	-
2	2,4,6,8,10-dodecapentaenal	Fatty Acyls	1.060	174.105	$C_{12}H_{14}O$	191.0638	-
3	2-Phenylaminoadenosine	Glycosides	3.827	358.1398	$C_{16}H_{18}N_6O_4$	357.1324	-
4	Atractyloside	Glycosides	4.260	726.2192	$C_{30}H_{46}O_{16}S_2$	707.201	-
5	Cortisol 21-sulfate	Sterols lipids	5.549	442.1627	$C_{21}H_{30}O_8S$	459.1217	-
6	5,8,11-heptadecatriynoic acid	Fatty Acyls	6.468	258.156	$C_{17}H_{22}O_2$	239.1382	-
7	Galactan	Polysaccharides	8.436	680.2045	$C_{24}H_{40}O_{22}$	679.199	-
8	Harderoporphyrin	Pigment	9.807	608.2509	$C_{35}H_{36}N_4O_6$	643.2205	-
9	Ergoline-1,8-dimethanol, 10-methoxy-6-methyl-, (8b)-	Alkaloid	26.965	316.1812	$C_{18}H_{24}N_2O_3$	297.1636	-
10	Ecgonine-methyl ester	Alkaloid	1.454	199.1195	$C_{10}H_{17}NO_3$	-	200.1268
11	2-Hydroxy-3-(4-methoxyethylphenoxy)-propanoic acid	Organic Acids	5.585	240.101	$C_{12}H_{16}O_5$	-	263.0902
12	Lomatin	Coumarins	6.222	246.878	$C_{14}H_{14}O_4$	-	247.0951
13	Marmesin	Coumarins	7.088	246.0903	$C_{14}H_{14}O_4$	-	269.0794
14	Purpurogallin	Natural Phenol	7.210	220.0361	$C_{11}H_8O_5$	-	269.0794
15	Atranorin	Polyphenol	7.363	196.0387	$C_9H_8O_5$	-	203.0328
16	Methyl 7-desoxypurpurogallin-7-carboxylate trimethyl ether	Natural Phenols	7.777	304.0935	$C_{16}H_{16}O_6$	-	287.0903
17	Dihydro-Obliquin	Coumarin	8.789	246.0904	$C_{14}H_{14}O_4$	-	269.0796
18	13-amino-tridecanoic acid	Fatty Acid	9.341	229.2031	$C_{13}H_{27}NO_2$	-	230.2104
19	Gummiferol	Fatty Acyl	9.417	286.0836	$C_{16}H_{14}O_5$	-	269.0803
20	Farnesyl pyrophosphate	Isoprenoid	9.786	382.128	$C_{15}H_{28}O_7P_2$	-	405.1171
21	Syringic acid	Natural Phenols	15.399	198.054	$C_9H_{10}O_5$	-	203.0326
22	Khayanthone	Polyphenols	18.482	570.2856	$C_{32}H_{42}O_9$	-	593.275

2.3. Antimicrobial Activities of D. flabellifolia Methanolic Extract

The antimicrobial activities of *D. flabellifolia* methanolic extract was assessed using disc diffusion and microdilution assays. Results summarized in Table 3 showed high diameter of growth inhibition zones (mGIZ $\pm$ SD) ranging from 10.33 $\pm$ 0.57 mm (*P. aeruginosa*; Environmental strain, pf8) to 14.67 $\pm$ 0.57 mm (*S. aureus* MDR, Clinical strain, 136). The highest mGIZ was recorded for *C. neoformans* (17.00 $\pm$ 1.73 mm) and *C. albicans* ATCC 10231 (16.33 $\pm$ 0.57 mm). However, *C. vaginalis* (6.00 $\pm$ 0 mm), *Candida* sp. (6.67 $\pm$ 0.57 mm), *A. fumigatus* ATCC 204305 (8.33 $\pm$ 1.15 mm), and *A. niger* (8.67 $\pm$ 0.57 mm) were the most resistant microorganisms. Using the MBC/MIC ratio, the tested extract showed bacteriostatic action against almost all tested bacterial strains (MBC/MIC ratio > 4) with the exception against *S. aureus* MDR (Clinical strain, 136), *S. paucimobilis* (Clinical strain, 144), and *A. baumannii* (Clinical strain, 146) with MBC/MIC ration lower than 4 highlighting a bactericidal action against these bacteria. Interestingly, *D. flabellifolia* methanolic extract exhibited fungicidal activity against the four yeast strains tested (MFC/MIC < 4).

2.4. Antioxidant Activities of D. flabellifolia Methanolic Extract

Total phenolic content (TPC), total flavonoids content (TFC), and total tannins content (TTC) were estimated, and the obtained results revealed a dominance of phenolic compounds (TPC = 38.85 $\pm$ 1.63 mg of gallic acid equivalent per gram of extract) followed by flavonoids (TFC = 17.06 $\pm$ 0.48 mg of quercetin equivalent per gram of dry extract), and condensed tannins (TTC = 7.80 $\pm$ 0.69 mg of tannic acid equivalent per gram of dry extract). In addition, *D. methanolic* extract was able to trap DPPH$^{\bullet}$ and ABTS$^{\bullet+}$ radicals with IC_{50} about 0.05 $\pm$ 0 mg/mL and 0.105 $\pm$ 0 mg/mL, respectively (Table 4).

Table 3. Growth inhibition zone, MICs, MBCs, and MFCs values obtained using disc diffusion and microdilution assays.

Code	Bacterial Strain	*D. flabellifolia* Methanolic Extract				Ampicillin Mean ± SD (mm)
		mGIZ ± SD (mm)	MIC [a]	MBC [b]	MBC/MIC Ratio	
B_1	*E. coli* ATCC 35218	12.66 ± 1.15 [bc]	12.50	200	16; bacteriostatic	7.00 ± 0.00 [d]
B_2	*P. aeruginosa* ATCC 27853	11.33 ± 0.57 [cde]	25	200	8; bacteriostatic	7.33 ± 0.57 [d]
B_3	*P. mirabilis* ATCC 29245	12.67 ± 0.57 [bc]	25	200	8; bacteriostatic	6.33 ± 0.57 [d]
B_4	*K. pneumoniae* ATCC 27736	14.33 ± 0.57 [a]	25	200	8; bacteriostatic	6.66 ± 0.57 [d]
B5	*P. mirabilis* (Environmental strain, 3)	12.67 ± 0.57 [bc]	25	200	8; bacteriostatic	21.00 ± 1.00 [a]
B6	*S. sciuri* (Environmental strain, 4)	11.33 ± 1.52 [cde]	25	200	8; bacteriostatic	7.00 ± 0.00 [d]
B7	*S. pyogens* (Clinical strain)	11.33 ± 1.15 [cde]	25	200	8; bacteriostatic	16.00 ± 1.73 [b]
B8	*P. aeruginosa* (Environmental strain, pf8)	10.33 ± 0.57 [e]	12.50	100	8; bacteriostatic	6.66 ± 0.57 [d]
B9	*S. aureus* MDR (Clinical strain, 136)	14.67 ± 0.57 [a]	12.50	50	4; bactericidal	7.33 ± 0.57 [d]
B10	*E. cloacae* (Clinical strain, 115)	14.33 ± 0.57 [a]	12.50	100	8; bacteriostatic	6.66 ± 0.57 [d]
B11	*S. paucimobilis* (Clinical strain, 144)	12.33 ± 0.57 [cd]	25	100	4; bactericidal	7.66 ± 0.57 [d]
B12	*A. baumannii* (Clinical strain, 146)	14.00 ± 0.00 [ab]	12.50	50	4; bactericidal	13.33 ± 0.57 [c]
Code	**Yeasts and molds**	**mGIZ±SD (mm)**	**MIC [a]**	**MFC [b]**	**MFC/MIC Ratio**	**Amphotericin B Mean ± SD (mm)**
Y_1	*C. albicans* ATCC 10231	16.33 ± 0.57 [a]	25	50	2; fungicidal	22.66 ± 1.15 [a]
Y_2	*C. neoformans* ATCC 14116	17.00 ± 1.73 [a]	6.25	12.50	2; fungicidal	15.33 ± 0.57 [b]
Y_3	*C. vaginalis* (Clinical strain)	6.00 ± 0.00 [d]	6.25	25	4; fungicidal	6.66 ± 0.57 [d]
Y_4	*Candida sp.* (Clinical strain)	6.67 ± 0.57 [cd]	25	100	4; fungicidal	12.33 ± 0.57 [c]
M_1	*A. fumigatus* ATCC 204305	8.33 ± 1.15 [bc]	-	-	-	15.00 ± 1.00 [b]
M_2	*A. niger*	8.67 ± 0.57 [b]	-	-	-	6.00 ± 0.00 [d]

Inhibition zone around the discs impregnated with *D. flabellifolia* methanolic extract (3 mg/disk) expressed as mean of three replicates (mm ± SD). SD: standard deviation. MIC: Minimal Inhibitory Concentration. MBC: Minimal Bactericidal Concentration. The letters (**a–e**) indicate a significant difference between the inhibition zones of the sample and amphotericin B against bacteria according to the Duncan test ($p < 0.05$).

Table 4. Antioxidant activities of *D. flabellifolia* methanolic extract as compared to standard molecules.

Tests	DPPH IC_{50} (mg/mL)	ABTS IC_{50} (mg/mL)	β-Carotene IC_{50} (mg/mL)
D. flabellifolia methanolic extract	0.05 ± 0 [a]	0.105 ± 0 [a]	5.00 ± 0.78 [a]
BHT (Butylated hydroxytoluene)	0.023 ± 0 [b]	0.018 ± 0 [b]	0.042 ± 0 [b]
Ascorbic Acid	0.022 ± 0 [b]	0.021 ± 0 [b]	0.017 ± 0 [b]

Letters (a,b) indicate a significant difference ($p < 0.005$) between *D. flabellifolia* methanolic extract and standard molecules.

Using the Duncan test, a significant difference ($p < 0.005$) was found between *D. flabellifolia* methanolic extract and the standards molecules used (BHT and AA) in all three antioxidants systems used. Using β-carotene/linoleic assay, our results indicated that high concentration from *D. flabellifolia* methanolic extract was needed for bleaching 50% of (IC_{50} = 5.00 ± 0.78 mg/mL) as compared to BHT (IC_{50} = 0.042 ± 0 mg/mL) and AA (IC_{50} = 0.017 ± 0 mg/mL).

2.5. Anticancer Activities of *D. flabellifolia* Methanolic Extract

The anticancer activity of *D. flabellifolia* methanolic extract was tested against breast (MCF-7), lung (A549), and colon (HCT-116) cancer cell lines using the MTT assay (Figure 1). Results showed an increase in cell viability inhibition in a concentration dependent manner. The highest percentages were recorded at 500 μg/mL for all cancer cell lines with IC_{50} about 240.56 μg/mL (A-549), 202.94 μg/mL (HCT-116), and 154.44 μg/mL (MCF-7).

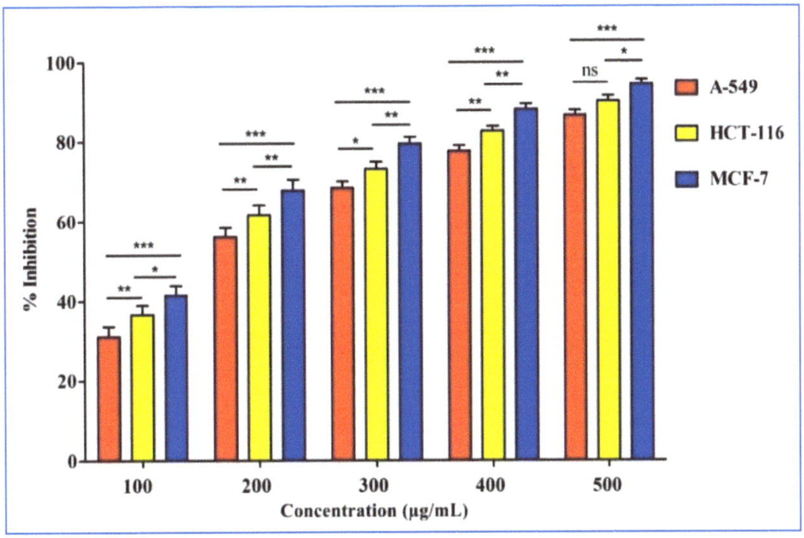

Figure 1. Effect of different concentrations from *D. flabellifolia* methanolic extract on breast (MCF-7), lung (A549), and colon (HCT-116) cancer cell lines. Error bars indicate SDs (± standard deviation) of three independent experiments. Significance; ns > 0.01, * $p < 0.01$, ** $p < 0.001$, *** $p < 0.0001$.

2.6. ADME Predictions

In drug development, it is imperative to study and investigate its safety and efficacy to know how a successful drug processes and reacts with the human body. For this, the major compounds were predicted for their absorption, distribution, metabolism, and excretion (ADME) properties tool to verify that the designed molecules are viable drugs. The selected phytocompounds showed good bioavailability score, high gastrointestinal absorption (GI) and some of them were predicted to be blood-brain-barrier (BBB) permeant. Except compound 9 (Ergoline-1,8-dimethanol, 10-methoxy-6-methyl-, (8b)-), the rest have been predicted to be not P-gp substrate, meaning that they are likely to have promising intestinal absorption and bioavailability. Simultaneously, predictive data showed that most of the selected compounds are not inhibitors of cytochrome P450 isoenzymes CYP 1A2, CYP2C19, CYP2C9, CYP2D6 and CYP3A4, meaning that they will not hamper the biotransformation of drugs metabolized by CYP450 enzymes. From the skin permeation (LogKP), we deduced acceptable values. All examined compounds were predicted to comply with Lipinski's rule-of-five, suggesting their good drug-likeness behavior. They also do not violate Ghose, Veber, Egan or Muegge (with some irregularities) filters.

The bioavailability radar of the selected analogues showed that the colored zone is the desired physicochemical space for good oral bioavailability in which the following properties were taken into account: flexibility, lipophilicity, saturation, size, polarity and solubility (Table 5). As shown in Figure 2, most of them fall entirely within the pink area, suggesting that they are suitable for better bioavailability.

Table 5. Pharmacokinetics, drug-likeness and medicinal chemistry of identified compounds according to SwissADME software.

Entry	Bioactive Compounds													
	2	**6**	**9**	**10**	**11**	**12**	**13**	**14**	**15**	**16**	**17**	**18**	**19**	**21**
Pharmacokinetics properties														
GI absorption	High	High	High	High	High	High	High	High	High	High	High	High	High	High
BBB permeant	Yes	Yes	No	No	No	Yes	Yes	No	No	Yes	Yes	Yes	No	No
P-gp substrate	No	No	Yes	No	No	No	No	No	No	No	No	No	No	No
CYP1A2 inhibitor	No	Yes	No	No	No	Yes	Yes	No	No	Yes	Yes	No	No	No
CYP2C19 inhibitor	No	Yes	No	No	No	No	No	No	No	Yes	Yes	No	No	No
CYP2C9 inhibitor	No	Yes	No	No	No	No	No	No	No	Yes	Yes	No	No	No
CYP2D6 inhibitor	No	No	Yes	No	No	No	No	No	No	Yes	No	No	No	No
CYP3A4 inhibitor	No	No	No	No	No	No	No	Yes	No	No	No	No	No	No
Log Kp (cm/s)	−5.15	−4.60	−7.73	−7.08	−7.09	−6.45	−6.45	−6.18	−5.51	−6.44	−5.66	−7.66	−8.02	−6.77
Druglikeness properties														
Lipinski	Yes	Yes	Yes	Yes	Yes	Yes	Yes	Yes	Yes	Yes	Yes	Yes	Yes	Yes
Ghose	Yes	Yes	Yes	Yes	Yes	Yes	Yes	Yes	Yes	Yes	Yes	Yes	Yes	Yes
Veber	Yes	Yes	Yes	Yes	Yes	Yes	Yes	Yes	Yes	Yes	Yes	Yes	Yes	Yes
Egan	Yes	Yes	Yes	Yes	Yes	Yes	Yes	Yes	Yes	Yes	Yes	Yes	Yes	Yes
Muegge	No	Yes	Yes	No	Yes	Yes	Yes	Yes	Yes	Yes	Yes	Yes	Yes	No
Bioavailability Score	0.55	0.85	0.55	0.55	0.56	0.55	0.55	0.55	0.55	0.55	0.55	0.55	0.55	0.56

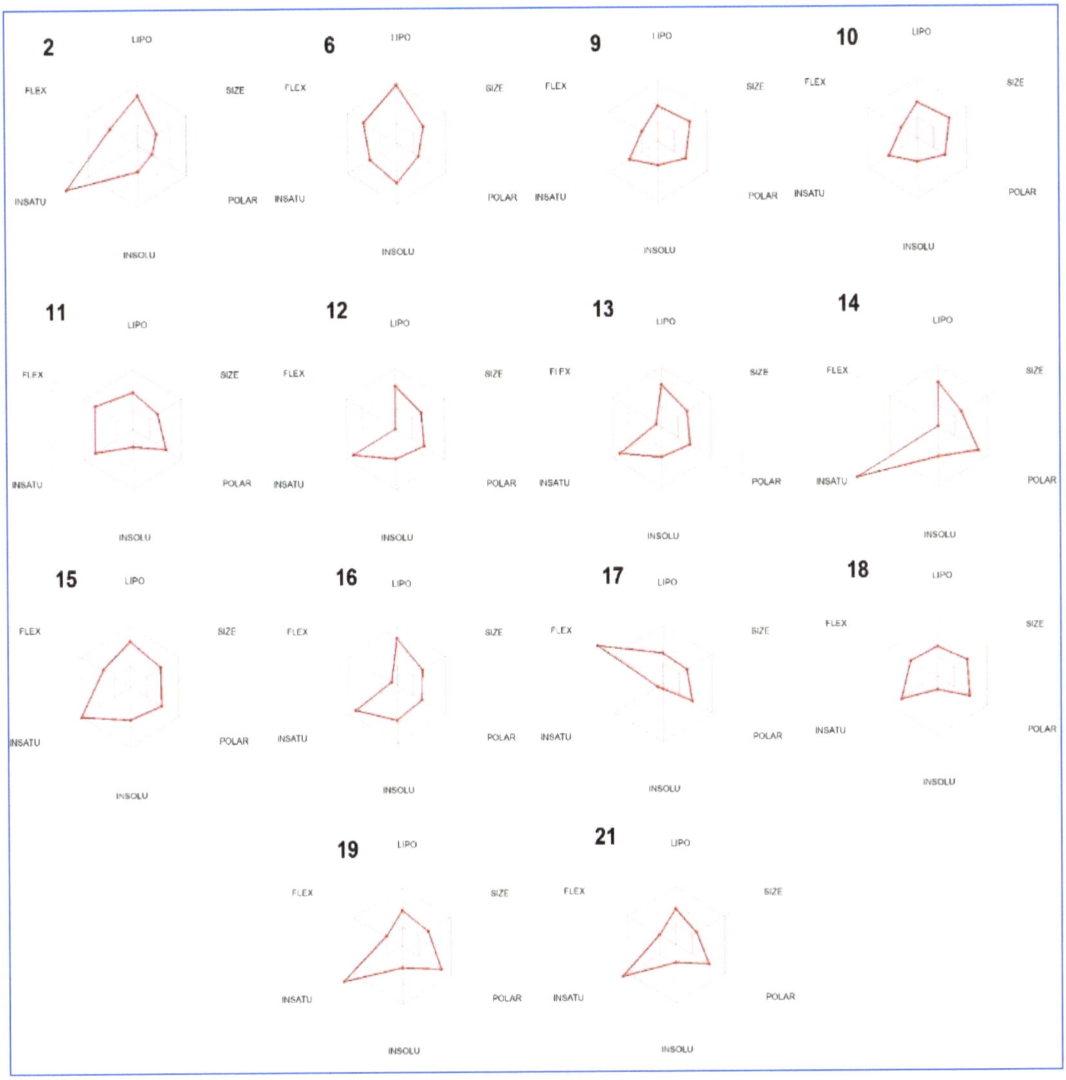

Figure 2. Radar plot of the selected peptides and phytoconstituents based on physicochemical indices ideal for oral bioavailability.

2.7. Molecular Docking Study

Table 6 and Table S2 showed that *D. flabellifolia* compounds bound the four targeted receptors with negative free binding energy but with different scores, except for Khayanthone for 1HD2 macromolecule. While the best binding score was predicted in the 3LN1-Harderoporphyrin complex, it ranged between –3.7 and –9.9 kcal/mol for the other complexes. Harderoporphyrin, which possessed the best binding score with 3LN1, was also predicted to interact with 1HD2, 2XCT and 2QZW receptors with acceptable binding scores of –6.4, –8.2 and –8.2 kcal/mol and included several different key residues. It was also deeply embedded into these targeted receptors and showed 2.047, 2.186, 1.804 and 1.898 Å only, respectively, for 1HD2, 2XCT, 2QZW, and 3LN1.

Table 6. Binding affinity, conventional hydrogen-bonding, the number of closest interacting residues and distance to closest interacting residue (Å) of some compounds with best scores with the different targeted receptors (1HD2, 2XCT, 2QZW, and 3LN1).

Complexes	Binding Affinity (kcalxmol^{-1})	Conventional H-Bonds	No. Closest Interacting Residues	Closest Interacting Residue	
				Residue	Distance (Å)
1HD2/2-Phenylaminoadenosine	−6.7	4	7	Arg86	1.822
1HD2/Atractyloside	−6.8	8	8	Gly92	2.153
2XCT/10-Hydroxyloganin	−7.6	6	4	Asp1105	1.904
2XCT/Harderoporphyrin	−8.2	6	5	Arg1377	2.186
2QZW/2-Phenylaminoadenosine	−8.3	6	9	Arg195	2.223
2QZW/Atractyloside	−8.3	6	7	Glu132	2.023
3LN1/Harderoporphyrin	−9.9	6	7	Ser160	1.898
3LN1/Dihydro-Obliquin	−9.2	1	9	Ser516	2.497

D. flabellifolia compounds were also found to be deeply embedded in all the studied receptors (1HD2, 2XCT, 2QZW, AND 3LN1). In this context, 1.476 Å only was reported in the compound (Khayanthone) while docked to the 1HD2 receptor. The bond network included H-bonds, which are commonly evaluated to assess the biological activities of the assessed compounds, associated several hydrophobic bonds: Pi-anion, Pi-cation, Pi-alkyl, and Pi-Pi T-shaped, as shown by the corresponding diagram of interactions (Figure 3).

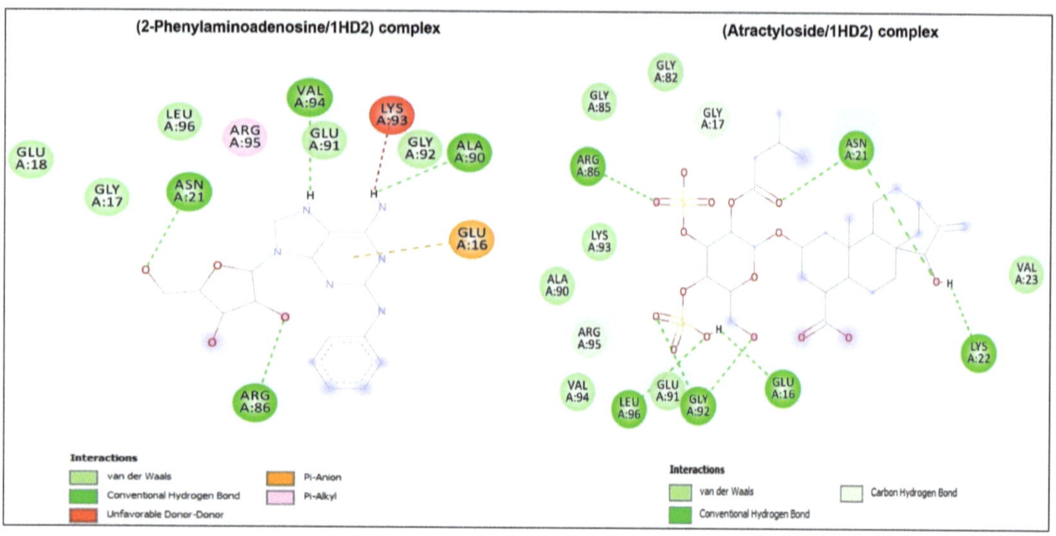

Figure 3. *Cont.*

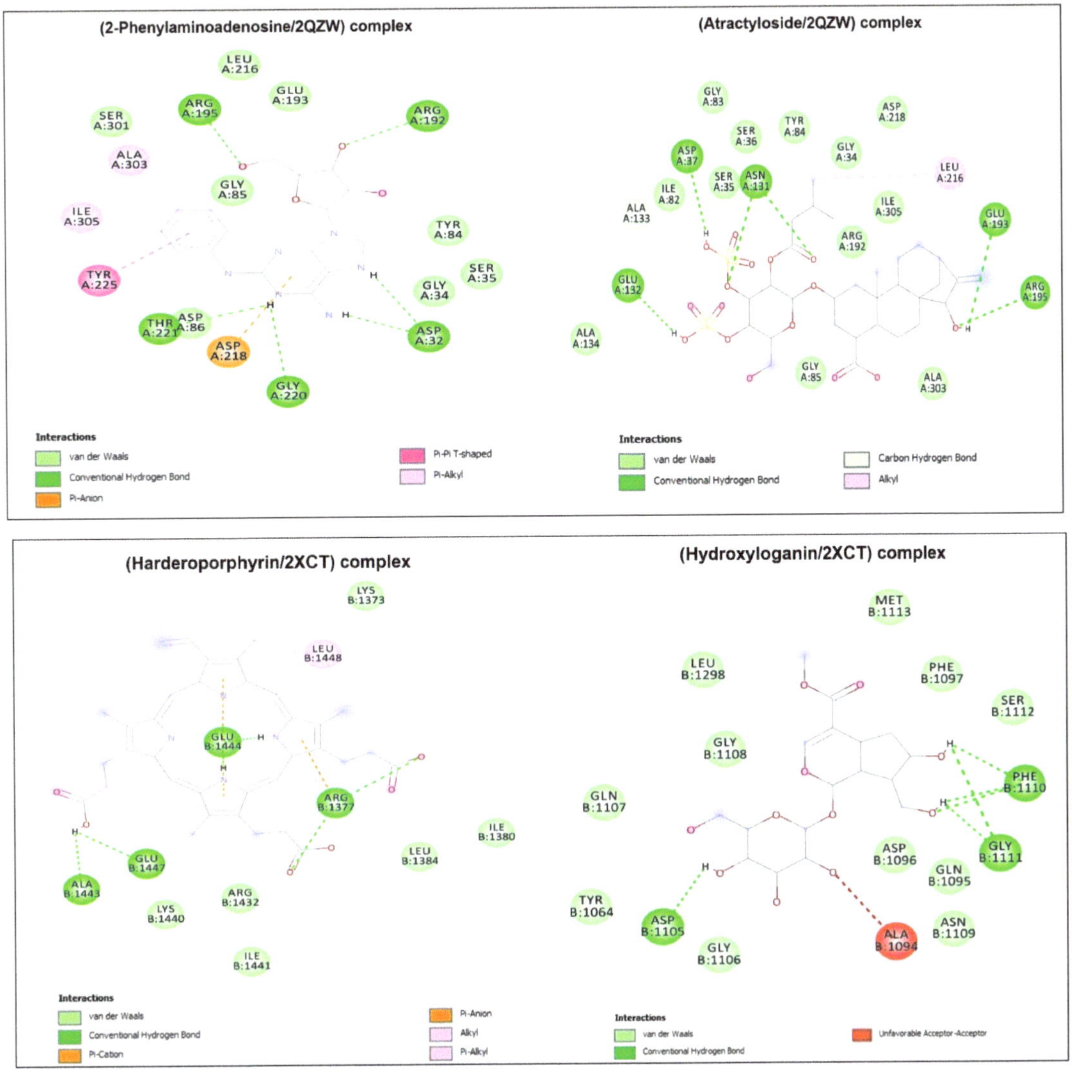

Figure 3. *Cont.*

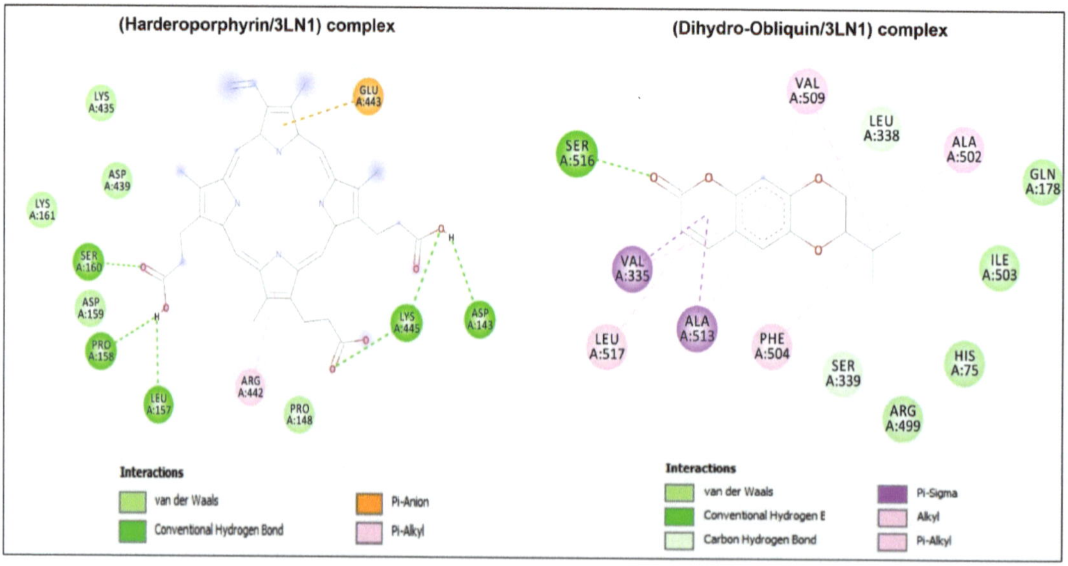

Figure 3. Three-dimensional (3D) residual interactions network of the best selective compounds with the active site of human peroxiredoxin 5 protein (PDB ID: 1HD2), *C. albicans* Sap1 (PDB ID: 2QZW), and *S. aureus* IIA topoisomerase (PDB ID: 2XCT).

3. Discussion

The yield of volatile oil was about 0.24 ± 0 mL/100 g of plant material, while the yield of methanolic extract was about 20.06 ± 0.19 g of dry extract/100 g of plant material. In this study, we report the identification of twenty-two bioactive compounds in the essential oil of *D. flabellifolia* aerial parts obtained by hydrodistillation. In fact, *D. flabellifolia* was dominated by decanal, dodecanal, 1-heptadecene, and α-pinene. Linear furanocoumarins have been previously identified as the major non-volatile components of the *Ducrosia* genus [20], whilst the main most volatile constituents were aliphatic hydrocarbons, with decanal (10.1–74.0%) and dodecanal (7.2–33.41%) as the major constituents [1]. Our results are in accordance with those reported by Al-Ghamdi et al. [21], who described the identification of 52 compounds in stems, leaves, and flowers of *D. flabellifolia* collected from northern border (Saudi Arabia). The same authors reported that the obtained essential oil was dominated by aldehyde hydrocarbons (stems 65.11%, leaves 65.39%, and flowers 67.30%), and decanal and dodecanal were the main identified compounds in the three tested organs. Similarly, Shahabipour and colleagues [3] reported that n-decanal (32.80%), dodecanal (32.6%), n-decanol (4.30%) and (2E)-tridecen-1-al (3.30%) were the dominant compounds identified in *D. flabellifolia* essential oil from Iran. The essential oils (fresh leaves and flowers) from Jordanian *D. flabellifolia* plant species obtained through hydrodistillation and solid phase microextraction (SPME) was a rich source of aliphatic compounds where n-decanal was the predominant compound obtained in *D. flabellifolia* fresh leaves (Hydrodistillation: 36.61%; SPME: 24.44%), while n-decanol was the dominant bioactive compound in the essential oils of dry leaves (27.88%) and fresh flowers' (11.49%) essential oils extracted by SPME technique [2].

In addition, as regards *D. flabellifolia* methanolic extract, we reported in this study the tentative identification of seventeen small peptides (tripeptides) and twenty-two phytochemical compounds (mainly farnesyl pyrophosphate, Methyl 7-desoxypurpurogallin-7-carboxylate trimethyl ether, Dihydro-Obliquin, Gummiferol, 2-Phenylaminoadenosine, and 2,4,6,8,10-dodecapentaenal) using HR-LCMS techniques. In fact, our team reported the identification of twenty-three bioactive compounds in *D. flabellifolia* hydroalcoholic extract dominated by (mg/Kg of crude extract) chlorogenic acid (5980.96 ± 73.12), ferrulic acid (180.58 ± 2.77), caffeic acid (70.90 ± 1.75), sinapic acid (61.74 ± 2.79), 2-5 dihydrobenzoic acid (59.74 ± 0.945), *p*-coumaric acid (55.11 ± 0.765), and 2-Hydroxycinnamic acid (31.28 ± 0.015) analyzed using liquid chromatography-electrospray tandem mass spectrometry [5]. Previous works reported the isolation of tetradecenol from leaves and fruit methanolic extract of *D. anethifolia* collected from the province Thadeq (180 km north of Riyadh) by using GC-MS technique [22]. Moreover, *D. anethifolia* is considered a good source of biologically active compounds especially coumarins and furanocoumarin [23].

We report also in this study that using Parveen scheme [24], *D. flabellifolia* methanolic extract was moderately to highly active against several Gram-positive and Gram-negative bacteria, yeast, and mold strains. In fact, on agar plates, the mean diameter of growth inhibition zone ranged from 10.33 ± 0.57 mm for *P. aeruginosa* to 14.67 ± 0.57 mm for *S. aureus* MDR strains. In addition, *C. albicans* ATCC 10231 and *C. neoformans* ATCC 14116 were the most sensitive yeast strains with mGIZ about 16.33 ± 0.57 mm and 17.00 ± 1.73 mm, respectively. Moreover, using the scheme proposed by Gatsing et al. [23] and Moroh et al. [25], *D. flabellifolia* methanolic extract exhibited bacteriostatic activity against almost all tested bacteria with the exception of *S. aureus*, *S. paucimobilis*, and *A. baumannii* (MBC/MIC = 4). Interestingly, *D. flabellifolia* exhibited fungicidal character against the four tested yeast strains (MFC/MIC $\leq$ 4). More recently, we reported that hydroalcoholic extract from *D. flabellifolia* collected from the Hail region (Saudi Arabia) was active against ESKAPE pathogens in a concentration dependent manner [5]. In addition, high concentrations (from 100 to 200 mg/mL) of *D. flabellifolia* methanol–water extract were needed to kill *C. utilis* ATCC 9255, *C. tropicalis* ATCC 1362, *C. guillermondii* ATCC 6260, and *C. albicans* ATCC 20402 [5].

Previous reports have demonstrated that extracts from *Ducrosia* plant species possessed good-to-moderate susceptibility against *S. epidermidis* ATCC 49461, *B. cereus* ATCC 10876, *S. aureus* clinical isolate and *S. aureus* ATCC 25923 [26]. Similarly, Alsaggaf [27] reported that *D. anethifolia* extract was able to inhibit the growth of both methicillin sensitive and methicillin resistant *S. aureus* strains with a diameter of growth inhibition zone ranging from (7.8 ± 0.4) mm for MRSA2 strain to (9.6 ± 0.6) mm, as compared to *S. aureus* ATCC 25923 (10.2 ± 0.5 mm). Al-Whibi and colleagues [22] studied the antimicrobial activities of *D. anethifolia* (leaves and fruit methanolic and acetone extracts) against *S. aureus* ATCC 25923, methicillin resistant *S. aureus* MRSA ATCC 12498, *B. subtilis* (ATCC 6633), *E. feacalis* ATCC 29122, *E. coli* ATCC 25966, *P. aeruginosa* ATCC 27853, *K. pneumoniae* ATCC 700603, *Salmonella* sp. and *Serratia* sp.

We also tested the antioxidant activities of *D. flabellifolia* methanolic extract using DPPH, ABTS, and β-carotene assays. Results obtained showed good ability to scavenge the three radicals at low IC_{50} values. Similar results were obtained with *D. flabellifolia* methanol/water extract (Table 7). In fact, low concentration from methanol/water extract was able to scavenge DPPH• free radicals (IC_{50}= 0.014 ± 0.045 mg/mL) as compared to methanolic extract (IC_{50} = 0.048 ± 0.004 mg/mL).

Table 7. Comparison between *D. flabellifolia* methanolic and methanol/water extract antioxidant activities.

***D. flabellifolia*/Tests**	**DPPH IC_{50} (mg/mL)**	**ABTS IC_{50} (mg/mL)**	**β-Carotene IC_{50} (mg/mL)**
Methanolic extract	0.05 ± 0	0.105 ± 0	5.00 ± 0.78
Methanol/Water extract *	0.014 ± 0.045	0.102 ± 0.024	7.80 ± 0.919

* Snoussi et al. [5].

Our results also indicated that *D. flabellifolia* methanolic extract exhibited anticancer activities against colon, lung, and breast cancer cell lines with IC_{50} about 240.56 μg/mL (A-549), 202.94 μg/mL (HCT-116), and 154.44 μg/mL (MCF-7). Previous reports have demonstrated that essential oil from *Ducrosia* members (*D. flabellifolia* and *D. anethifolia*) exhibited cytotoxic activity against human chronic myelogenous leukemia cell lines (K562), human colon adenocarcinoma (LS180), and human breast adenocarcinoma (MCF-7) [22]. The same authors reported that *D. flabellifolia* essential oil was active against K562, LS180, and MCF-7 cell lines with IC_{50} value about 304.0 ± 87.2 µg/mL, 286.9 ± 28.0 µg/mL, and 511.2 ± 133.2 µg/mL, respectively [22]. It has been also demonstrated *that D. ismaelis* essential oil (Decanal 40.6%, α-pinene 15.1%, and dodecanal 13.7%) exhibited anticancer activities against three cell lines, namely MCF-7 (IC_{50} 66.24 ± 1.26 µg/mL), LoVo (IC_{50} 102.53 ± 1.00 µg/mL), and HepG2 (IC_{50} 137.32 ± 2.48 µg/mL) [28].

The ADME prediction revealed that all the studied compounds possessed high GI absorption associated variable BBB permeation. Most of the selected compounds were also found to not inhibit the cytochrome P450 isoforms (1A2, 2C19, 2C9, 2D6 and 3A4), which indicates the safe use of these compounds and the absence of any disruption in drug distribution and metabolism [29–31]. It was noticed that bioavailability scores ranged between 0.55 and 0.85 for the major *D. flabellifolia* studied compounds, which indicate acceptable bioavailability associated potential physiological activity of these phytochemicals as previously reported by several studies on natural and synthesized compounds [29,32,33]. The acceptable bioavailability scores were confirmed by the polygons illustrations. In fact, most of the compounds physicochemical properties stayed in the pick areas that indicate the most suitable oral bioavailability.

The *D. flabellifolia* identified compounds were subjected to computational assay to assess their molecular interactions with some key receptors related to antimicrobial, antioxidant, and anticancer activities. Molecular docking results showed that all compounds bound the four targeted receptors with negative free binding energy but with different scores, except Khayanthone for 1HD2 macromolecule. The free binding energy ranged between −3.7 and −9.9 kcal/mol. It has been reported that variation in such score values is mainly linked to the 3D chemical structures of the ligands [29–33] The best binding score was predicted in 3LN1-Harderoporphyrin complex. The same compound (Harderoporphyrin) was also found to interact with each of 2XCT and 2QZW receptors with an interesting bound energy of −8.2 kcal/mol. The molecular interactions of *D. flabellifolia* compounds with the targeted receptors included up to twelve conventional H-bonds and involved several different key residues. In addition, *D. flabellifolia* compounds were also found to be deeply embedded in all the studied receptors. The lowest distance of 1.476 Å only was reported in the compound (Khayanthone) while docked to the 1HD2 macromolecule. Regardless of H-bonds that are commonly evaluated to assess the biological activities of the targeted compounds, a network of hydrophobic bonds was also found within the different studied complexes. This may contribute to the stability of the complexes as reported in several recent *in silico* studies [34–36]. Our results exhibit that all *D. flabellifolia* identified compounds established acceptable number of H-bonds. The corresponding diagram of interactions of the selected established complexes (Figure 4) showed involvement of several key residues and diversified bond network: Pi-anion, Pi-cation, Pi-alkyl, Pi-Pi T-shaped . . . , which support the H-bonds and contributed to the complex stability [32,33]. Interactions with key residues was found to promote biological activities including antimicrobial, antioxidant, and anticancer potential of the studied compounds. In this context, all *D. flabellifolia* compounds were found to be in close proximity of all the targeted receptor with distance less than 3 Å. Ligands deeply embedding were reported to enhance the biological activity [30,31].

Overall, the high antimicrobial, antioxidant, and anticancer activities of the tested *D. flabellifolia* methanolic extract can be attributed to its richness in bioactive compounds belonging to different chemical classes, such as alkaloids, coumarins, polyphenols, fatty acyls, and terpenoids. Docking results revealed that the high molecular interactions obtained justify that the antimicrobial, antioxidant, and anticancer potentials of the studied *D. flabellifolia* are thermodynamically possible and this could explain the results obtained in vitro.

Figure 4. Al-Haza plant species collected from Al-Mu'ayqilat (Hail region).

4. Materials and Methods

4.1. Plant Material Sampling

In this study, *D. flabellifolia* Boiss. locally known as Al-Hazaa (Figure 4), was collected from Al-Mu'ayqilat, 27°16′41.9″ N, 41°22′48.0″ E in October 2019. The plant material was air-dried at room temperature for one week. The methanolic extract was obtained using maceration technique (20 g of powdered aerial parts in 200 mL of pure methanol at room temperature for 72 h with low agitation). The filtrate was recuperated through lyophilization and kept at −4 °C until use.

4.2. Phytochemical Composition

4.2.1. Composition of the Essential Oil

The gas chromatography–mass spectrometry (GC–MS) analyses were performed on a gas chromatograph HP 6890 (II) interfaced with an HP 5973 mass spectrometer (Agilent Technologies, Palo Alto, CA, USA) with electron impact ionization (70 eV). The volatile compounds were identified by comparing their retention indices relative to (C7–C20) n-alkanes with those of literature and/or with those of authentic compounds available in our laboratory, and by matching their mass spectral fragmentation patterns with corresponding data (Wiley 275.L library) and other published mass spectra [37], as well as by comparison of their retention indices with data from the Mass Spectral Library.

4.2.2. Composition of the Methanolic Extract

The identification of phytoconstituents in the methanolic extract from *D. flabellifolia* methanolic extract was performed using High Resolution-Liquid Chromatography Mass Spectroscopy (HR-LCMS) as previously described by Noumi et al. [38]. MS data were provided in negative and positive ionization mode.

4.3. Antimicrobial Activities of *D. flabellifolia* Methanolic Extract

D. flabellifolia methanolic extract was tested for its ability to inhibit the growth of twelve clinical and environmental bacterial strains (*Escherichia coli* ATCC 35218, *Pseudomonas aeruginosa* ATCC 27853, *Proteus mirabilis* ATCC 29245, *Klebsiella pneumoniae* ATCC 27736, *P. mirabilis*, *Staphylococcus sciuri*, *Streptococcus pyogens*, *P. aeruginosa*, *S. aureus* MDR, *Enterobacter cloacae*, *Stenotrophomonas paucimobilis*, *Acinetobacter baumannii*), four yeasts (*Candida albicans* ATCC 10231, *Cryptococcus neoformans* ATCC 14116, *C. vaginalis*, *Candida* sp.), and two mold strains (*Aspergillus fumigatus* ATCC 204305 and *A. niger*). Disc diffusion assay was used (20 μL/disc) using the same protocol described by Snoussi et al. [5] for the determination of the diameter of growth inhibition zone estimated on agar medium (Mueller Hinton for bacteria and Sabouraud Chloramphenicol agar for fungi). Parveen et al. [24] was used to interpret the obtained. Ampicillin and Amphotericin B were used as control.

To estimate the minimal inhibitory concentrations (MICs values expressed in mg/mL) and minimal bactericidal/fungicidal concentration (MBCs and MFCs values), the obtained extract was serially diluted in DMSO-5% supplemented with Tween 80 (From 100 mg/mL to 0.097 mg/mL) in 96-well microtiter plates containing 95 μL of the microbial suspension and 95 μL of the enrichment broth (Lauria Bertani for bacteria and Sabouraud dextrose broth for fungal strains). To interpret the character of the tested extract, we used the ratios (MBC/MIC ratio and MFC/MIC ratio) described by Gatsing et al. [23] and Moroh et al. [17].

4.4. Antioxidant Activities

The ability of Al-Haza extract against DPPH-H was determined following the same method as Mseddi al. [39], and the method of Koleva et al. [40] for β-Carotene bleaching test. The radical scavenging activity against ABTS•+ (2,2′-azino-bis(3-ethylbenzothiazoline-6-sulfonic acid)) radical cations was measured using the same protocol described by Hamdi et al. [41].

4.5. Anticancer Activity

D. flabellifolia methanolic extract was tested against human lung (A549), breast (MCF-7), and colon (HCT-116) cancer cells at different concentrations. The percentage growth inhibition was calculated after subtracting the background and the blank, and the concentration of the test drug needed to inhibit cell growth by 50% (IC_{50}) was calculated from the dose–response curve for the respective cell line [42].

4.6. Computational Study

4.6.1. ADME Properties

The pharmacokinetic properties, drug-likeness, and medicinal properties of the identified bioactive molecules from *D. flabellifolia* methanolic extract were studied using the same recommendations described by Daina and colleagues [43]. Each structure was imported, and the structure SMILES was entered at the interface of the website (http://swissadme.ch/, accessed on 2 October 2022), a free web tool to assess the pharmacokinetics, drug-likeness and medicinal chemistry friendliness of small molecules. The SwissADME drug design study was run and the ADMET properties/parameters were generated.

4.6.2. Molecular Docking Study

Four different receptors, (PDB ID) 2XCT (*S. aureus* IIA topoisomerase), 2QZW (*C. albicans* Sap1), 1HD2 (human peroxiredoxin 5 protein) and 3LN1 (cyclooxygenase-2; COX-2), have been targeted to check the potential antimicrobial, antioxidant, and anticancer effect of the *D. flabellifolia* identified compounds. The crystalized structures of the selected receptors have been obtained from RCSB protein data bank. ChemDraw was used to obtain the chemical structures whenever needed, following the pre-processing of both ligands and receptors (removal of water molecules and addition of polar hydrogens and Koleman charges) using Autodock vina packages v1.2.3 [30,31]. The binding scores and calculation of embedding distances and bonding network were studied as previously described based on the CHARMM force field [32–35]. The reason behind the selection of these receptors is their involvement in antioxidant and anticancer pathways and the fact that they are commonly targeted in pharmaceutical and drug design approaches [33].

4.7. Statistical Analysis

Experiments were performed in triplicate and average values were calculated using the SPSS 25.0 statistical package for Windows. Duncan's multiple-range tests for means with a 95% confidence interval ($p \leq 0.05$) was used to calculate the differences in means.

5. Conclusions

In conclusion, our results indicated that fresh aerial parts of *D. flabellifolia* growing wild in the Hail region possess antimicrobial activity against several Gram-positive and Gram-negative bacteria, yeast, and molds with different degree. In fact, their phytochemical composition revealed the presence of various compounds with known biological properties in both essential oil and methanolic extract. The pharmacokinetic and ADMET properties of *D. flabellifolia* phytochemicals may explain the in vitro antimicrobial, antioxidant, and anticancer findings, which may result from the potential molecular interactions of these chemicals with the concerned receptors (1HD2, 2XCT, 2QZW, and 3LN1). These results support the benefits of this medicinal plant as a source of bioactive molecules for different ethnobotanical uses.

Supplementary Materials: The following supporting information can be downloaded at: https://www.mdpi.com/article/10.3390/metabo13010064/s1, Figure S1: HR-LCMS chromatograms (A: positive chromatogram, B: negative chromatogram) showing the main compounds identified in *D. flabellifolia* methanolic extract. Numbers on the chromatograms correspond to the compounds listed in Table 2; Table S1: Tripeptides identified by HR-LCMS technique in *D. flabellifolia* methanolic extract; Table S2: Binding affinity, conventional hydrogen-bonding, the number of closest interacting residues and distance to closest interacting residue (Å) of the compound with best scores with the different targeted receptors (1HD2, 2XCT, 2QZW, AND 3LN1).

Author Contributions: Conceptualization, M.S., V.D.F. and E.N.; methodology, R.B., M.A.A., M.A.-R., K.H., M.A. and F.P.; software, R.H.L., A.J.S., M.S.; validation, E.N. and M.S.; writing—original draft preparation, E.N., A.K., M.P., M.A., K.H. and M.S.; writing—review and editing, M.S., M.P., R.B., A.K. and F.P.; supervision, M.S.; project administration, R.H.L., V.D.F. and M.S.; funding acquisition, M.S. All authors have read and agreed to the published version of the manuscript.

Funding: This research received no external funding.

Institutional Review Board Statement: Not applicable.

Informed Consent Statement: Not applicable.

Data Availability Statement: Not applicable.

Acknowledgments: This research has been funded by the Scientific Research Deanship at University of Ha'il-Saudi Arabia through project number RD-21 040.

Conflicts of Interest: The authors declare no conflict of interest.

References

1. Mottaghipisheh, J.; Boveiri Dehsheikh, A.; Mahmoodi Sourestani, M.; Kiss, T.; Hohmann, J.; Csupor, D. *Ducrosia* spp., Rare Plants with Promising Phytochemical and Pharmacological Characteristics: An Updated Review. *Pharmaceuticals* **2020**, *13*, 175. [CrossRef]
2. Al-Shudiefat, M.; Al-Khalidi, K.; Abaza, I.; Afifi, F.U. Chemical composition analysis and antimicrobial screening of the essential oil of a rare plant from Jordan: *Ducrosia flabellifolia*. *Anal. Lett.* **2013**, *47*, 422–432. [CrossRef]
3. Shahabipour, S.; Firuzi, O.; Asadollahi, M.; Faghihmirzaei, E.; Javidnia, K. Essential oil composition and cytotoxic activity of *Ducrosia anethifolia* and *Ducrosia flabellifolia* from Iran. *J. Essent Oil Res.* **2013**, *25*, 160–163. [CrossRef]
4. Talib, W.H.; Issa, R.A.; Kherissat, F.; Mahasne, A.M. Jordanian *Ducrosia flabellifolia* inhibits proliferation of breast cancer cells by inducing apoptosis. *Br. J. Med. Med. Res.* **2013**, *3*, 771–783. [CrossRef] [PubMed]

5. Snoussi, M.; Ahmad, I.; Aljohani, A.M.A.; Patel, H.; Abdulhakeem, M.A.; Alhazmi, Y.S.; Tepe, B.; Adnan, M.; Siddiqui, A.J.; Sarikurkcu, C.; et al. Phytochemical Analysis, Antioxidant, and Antimicrobial Activities of *Ducrosia flabellifolia*: A Combined Experimental and Computational Approaches. *Antioxidants* **2022**, *11*, 2174. [CrossRef] [PubMed]
6. Siegel, R.L.; Miller, K.D.; Fuchs, H.E.; Jemal, A. Cancer statistics, 2022. *CA Cancer J. Clin.* **2022**, *72*, 7–33. [CrossRef]
7. Bergers, G.; Fendt, S.M. The metabolism of cancer cells during metastasis. *Nat. Rev. Cancer* **2021**, *21*, 162–180. [CrossRef]
8. World Health Organization. Cancer. 2018. Available online: https://www.who.int/health-topics/cancer#tab=tab_1 (accessed on 27 September 2022).
9. Mattiuzzi, C.; Lippi, G. Current cancer epidemiology. *J. Epidemiol. Glob. Health* **2019**, *9*, 217. [CrossRef]
10. Othman, I.M.M.; Gad-Elkareem, M.A.M.; Anouar, E.H.; Snoussi, M.; Aouadi, K.; Kadri, A. Novel fused pyridine derivatives containing pyrimidine moiety as prospective tyrosyl-tRNA synthetase inhibitors: Design, synthesis, pharmacokinetics, and molecular docking studies. *J. Mol. Struct.* **2020**, *1219*, 128651. [CrossRef]
11. Othman, I.M.M.; Gad-Elkareem, M.A.M.; Anouar, E.H.; Aouadi, K.; Kadri, A.; Snoussi, M. Design, synthesis ADMET and molecular docking of new imidazo[4,5-b]pyridine-5-thione derivatives as potential tyrosyl-tRNA synthetase inhibitors. *Bioorg. Chem.* **2020**, *102*, 104105. [CrossRef] [PubMed]
12. Othman, I.M.M.; Gad-Elkareem, M.A.M.; Radwan, H.A.; Badraoui, R.; Aouadi, K.; Snoussi, M.; Kadri, A. Synthesis, Structure-Activity Relationship and in silico Studies of Novel Pyrazolothiazole and Thiazolopyridine Derivatives as Prospective Antimicrobial and Anticancer Agents. *ChemistrySelect* **2021**, *6*, 7860–7872. [CrossRef]
13. Saha, S.K.; Lee, S.B.; Won, J.; Choi, H.Y.; Kim, K.; Yang, G.-M.; Dayem, A.A.; Cho, S.-G. Correlation between oxidative stress, nutrition, and cancer initiation. *Int. J. Mol. Sci.* **2017**, *18*, 1544. [CrossRef] [PubMed]
14. Patel, H.; Wu, Z.X.; Chen, Y.; Chen, Z.-S. Drug resistance: From bacteria to cancer. *Mol. Biomed.* **2021**, *2*, 27. [CrossRef] [PubMed]
15. Van den Boogaard, W.M.C.; Komninos, D.S.J.; Vermeij, W.P. Chemotherapy side-effects: Not all DNA damage is equal. *Cancers* **2022**, *14*, 627. [CrossRef]
16. Hajlaoui, H.; Arraouadi, S.; Mighri, H.; Ghannay, S.; Aouadi, K.; Adnan, M.; Kadri, A. HPLC-MS profiling, antioxidant, antimicrobial, antidiabetic, and cytotoxicity activities of *Arthrocnemum indicum* (Willd.) Moq. extracts. *Plants* **2022**, *11*, 232. [CrossRef]
17. Alminderej, F.; Bakari, S.; Almundarij, T.I.; Snoussi, M.; Aouadi, K.; Kadri, A. Antimicrobial and wound healing potential of a new chemotype from *Piper cubeba* L. essential oil and in silico study on *S. aureus* tyrosyl-tRNA Synthetase Protein. *Plants* **2021**, *10*, 205. [CrossRef]
18. Haddaji, F.; Papetti, A.; Noumi, E.; Colombo, R.; Deshpande, S.; Aouadi, K.; Adnan, M.; Kadri, A.; Selmi, B.; Snoussi, M. Bioactivities and in silico study of *Pergularia tomentosa* L. phytochemicals as potent antimicrobial agents targeting type IIA topoisomerase, TyrRS, and Sap1 virulence proteins. *Environ. Sci. Poll. Res.* **2021**, *28*, 25349–25367. [CrossRef]
19. Eddouks, M.; Chattopadhyay, D.; De Feo, V.; Cho, W.C. Medicinal plants in the prevention and treatment of chronic diseases. *Evid. Based Complement. Alternat. Med.* **2012**, *2012*, 458274. [CrossRef]
20. Mottaghipisheh, J.; Nové, M.; Spengler, G.; Kúsz, N.; Hohmann, J.; Csupora, D. Antiproliferative and cytotoxic activities of furocoumarins of *Ducrosia anethifolia*. *Pharm. Biol.* **2018**, *56*, 658–664. [CrossRef]
21. Al-Ghamdi, F.A.; Abdelwahab, A.T. Volatile oil composition from stems, leaves and flowers of Ducrosia flabellifolia Boiss. From northern border of Saudi Arabia. *Int. J. Appl. Biol. Pharm. Technol.* **2014**, *5*, 296–300.
22. Al-Whibi, M.; Moubayed, N.M.S.; Zahrani, H.; Mashhour, A. Antibacterial and cytotoxic activities of *Ducrosia anethifolia*: A potential biomedicine against selected human pathogens and cancer cell lines. *Biomedica* **2019**, *35*, 203–209.
23. Gatsing, D.; Tchakoute, V.; Ngamga, D.; Kuiate, J.R.; Tamokou, J.D.D.; Nji-Nkah, B.F.; Tchouanguep, F.M.; Fodouop, S.P.C. In vitro antibacterial activity of Crinum purpurascens Herb. leaf extract against the *Salmonella* species causing typhoid fever and its toxicological evaluation. *Iran. J. Med. Sci.* **2009**, *34*, 126–136.
24. Parveen, M.; Ghalib, R.M.; Khanam, Z.; Mehdi, S.H.; Ali, M. A Novel antimicrobial agent from the leaves of *Peltophorum vogelianum* (Benth.). *Nat. Prod. Res.* **2010**, *24*, 1268–1273. [CrossRef] [PubMed]
25. Moroh, J.L.; Bahi, C.; Dje, K.; Loukou, Y.G.; Guide Guina, F. Etude de l'activité antibactérienne de l'extrait acétatique de *Morinda morindoides* (Baker) Milne-Redheat (Rubiaceae) sur la croissance in vitro des souches d'*Escherichia coli*. *Bull. Soc. R. Sci. Liege* **2008**, *77*, 44–61.
26. Elsharkawy, E.R.; Abdallah, E.M.; Shiboob, M.H.; Alghanem, S. Phytochemical, antioxidant and antibacterial potential of *Ducrosia anethifolia* in Northern Border region of Saudi Arabia. *J. Pharm. Res. Int.* **2019**, *31*, 1–8. [CrossRef]
27. Alsaggaf, M.S. Application of wild Saudi plant extracts to control antibiotic resistant *Staphylococcus aureus*. *Egypt. J. Exp. Biol. (Bot.)* **2018**, *14*, 29–35. [CrossRef]
28. Mothana, R.A.; Nasr, F.A.; Khaled, J.M.; Noman, O.M.; Abutaha, N.; Al-Rehaily, A.J.; Almarfadi, O.M.; Kurkcuoglu, M. Ducrosia ismaelis Asch. essential oil: Chemical composition profile and anticancer, antimicrobial and antioxidant potential assessment. *Open Chem.* **2020**, *18*, 175–184. [CrossRef]
29. Mhadhbi, N.; Issaoui, N.; Hamadou, W.S.; Alam, J.M.; Elhadi, A.S.; Adnan, M.; Naïli, H.; Badraoui, R. Physico-Chemical Properties, Pharmacokinetics, Molecular Docking and In-Vitro Pharmacological Study of a Cobalt (II) Complex Based on 2-Aminopyridine. *ChemistrySelect* **2022**, *7*, e202103592. [CrossRef]

30. Badraoui, R.; Adnan, M.; Bardakci, F.; Alreshidi, M.M. Chloroquine and hydroxychloroquine interact differently with ACE2 domains reported to bind with the coronavirus spike protein: Mediation by ACE2 polymorphism. *Molecules* **2021**, *26*, 673. [CrossRef]
31. Badraoui, R.; Saeed, M.; Bouali, N.; Hamadou, W.S.; Elkahoui, S.; Alam, M.J.; Siddiqui, A.J.; Adnan, M.; Saoudi, M.; Rebai, T. Expression Profiling of Selected Immune Genes and Trabecular Microarchitecture in Breast Cancer Skeletal Metastases Model: Effect of α–Tocopherol Acetate Supplementation. *Calcif. Tissue Int.* **2022**, *110*, 475–488. [CrossRef]
32. Hamrita, B.; Noumi, B.; Papetti, A.; Badraoui, R.; Bouslama, L.; Ben Tekfa, M.I.; Hamdi, A.; Patel, M.; Elasbali, A.M.; Adnan, M.; et al. Phytochemical Analysis, Antioxidant, Antimicrobial, and Anti-Swarming Properties of Hibiscus sabdariffa L. Calyx Extracts: In Vitro and In Silico Modelling Approaches. *Evid. Based Complement. Alternat. Med.* **2022**, *2022*, 1252672. [CrossRef] [PubMed]
33. Jedli, O.; Ben-Nasr, H.; Zammel, N.; Rebai, T.; Saoudi, M.; Elkahoui, S.; Jamal, A.; Siddiqui, A.J.; Sulieman, A.E.; Alreshidi, M.M.; et al. Attenuation of ovalbumin-induced inflammation and lung oxidative injury in asthmatic rats by *Zingiber officinale* extract: Combined in silico and in vivo study on antioxidant potential, STAT6 and TNF-pathways. *3 Biotech.* **2022**, *12*, 191. [CrossRef] [PubMed]
34. Hchicha, K.; Korb, M.; Badraoui, R.; Naïli, H. A novel sulfate-bridged binuclear copper (II) complex: Structure, optical, ADMET and in vivo approach in a murine model of bone metastasis. *New J. Chem.* **2021**, *45*, 13775–13784. [CrossRef]
35. Zammel, N.; Saeed, M.; Bouali, N.; Elkahoui, S.; Alam, J.M.; Rebai, T.; Kausar, M.A.; Adnan, M.; Siddiqui, A.J.; Badraoui, R. Antioxidant and Anti-Inflammatory Effects of Zingiber officinale roscoe and Allium subhirsutum: In Silico, Biochemical and Histological Study. *Foods* **2021**, *10*, 1383. [CrossRef]
36. Akacha, A.; Badraoui, R.; Rebai, T.; Zourgui, L. Effect of *Opuntia ficus* indica extract on methotrexate-induced testicular injury: A biochemical, docking and histological study. *J. Biomol. Struct. Dyn.* **2022**, *40*, 4341–4351. [CrossRef] [PubMed]
37. Adams, R. *Identification of Essential Oil Components by Gas Chromatography/Quadrupole Mass Spectroscopy*; Allured: Carol Stream, IL, USA, 2001.
38. Noumi, E.; Snoussi, M.; Anouar, E.H.; Alreshidi, M.; Veettil, V.N.; Elkahoui, S.; Adnan, M.; Patel, M.; Kadri, A.; Aouadi, K.; et al. HR-LCMS-based metabolite profiling, antioxidant, and anticancer properties of *Teucrium polium* L. methanolic extract: Computational and in vitro study. *Antioxidants* **2020**, *9*, 1089. [CrossRef]
39. Mseddi, K.; Alimi, F.; Noumi, E.; Veettil, V.N.; Deshpande, S.; Adnan, M.; Hamdi, A.; Elkahoui, S.; Alghamdi, A.; Kadri, A.; et al. *Thymus musilii* Velen. as a promising source of potent bioactive compounds with its pharmacological properties: In vitro and in silico analysis. *Arab. J. Chem.* **2020**, *13*, 6782–6801. [CrossRef]
40. Koleva, I.I.; van Beek, T.A.; Linssen, J.P.; de Groot, A.; Evstatieva, L.N. Screening of plant extracts for antioxidant activity: A comparative study on three testing methods. *Phytochem. Anal.* **2002**, *13*, 8–17. [CrossRef]
41. Hamdi, A.; Viaene, J.; Mahjoub, M.A.; Majouli, K.; Gad, M.H.H.; Kharbach, M.; Demeyer, K.; Marzouk, Z.; Vander Heyden, Y. Polyphenolic contents, antioxidant activities and UPLC–ESI–MS analysis of *Haplophyllum tuberculatum* A. Juss leaves extracts. *Int. J. Biol. Macromol.* **2018**, *106*, 1071–1079. [CrossRef]
42. Hamadou, W.S.; Bouali, N.; Badraoui, R.; Hadj Lajimi, R.; Hamdi, A.; Alreshidi, M.; Patel, M.; Adnan, M.; Siddiqui, A.J.; Noumi, E.; et al. Chemical composition and the anticancer, antimicrobial, and antioxidant properties of acacia honey from the Hail region: The in vitro and in silico investigation. *Evid. Based Complement. Alternat. Med.* **2022**, *2022*, 1518511. [CrossRef]
43. Daina, A.; Michielin, O.; Zoete, V. SwissADME: A free web tool to evaluate pharmacokinetics, drug-likeness and medicinal chemistry friendliness of small molecules. *Sci. Rep.* **2017**, *7*, 42717. [CrossRef] [PubMed]

metabolites

Article

Prenylated Isoflavanones with Antimicrobial Potential from the Root Bark of *Dalbergia melanoxylon*

Duncan Mutiso Chalo [1,2,3], Katrin Franke [2,4,5,*], Vaderament-A. Nchiozem-Ngnitedem [6], Esezah Kakudidi [1], Hannington Origa-Oryem [1], Jane Namukobe [7], Florian Kloss [8], Abiy Yenesew [6,*] and Ludger A. Wessjohann [2,5,*]

1 Department of Plant Sciences, Microbiology and Biotechnology, Makerere University, Kampala P.O. Box 7062, Uganda; dunmutiso6@gmail.com (D.M.C.); esezahk@gmail.com (E.K.); horyemoriga@gmail.com (H.O.-O.)
2 Leibniz Institute of Plant Biochemistry, Weinberg 3, 06120 Halle (Saale), Germany
3 Department of Biology, University of Nairobi, Nairobi P.O. Box 30197-0100, Kenya
4 Institute of Biology/Geobotany and Botanical Garden, Martin Luther University Halle-Wittenberg, 06108 Halle (Saale), Germany
5 German Centre for Integrative Biodiversity Research (iDiv) Halle-Jena-Leipzig, 04103 Leipzig, Germany
6 Department of Chemistry, University of Nairobi, Nairobi P.O. Box 30197-0100, Kenya; n.vaderamentalexe@gmail.com
7 Department of Chemistry, Makerere University, Kampala P.O. Box 7062, Uganda; jnamukobe@gmail.com
8 Transfer Group Anti-Infectives, Leibniz Institute for Natural Product Research and Infection Biology, Leibniz-HKI, Beutenbergstr. 11a, 07745 Jena, Germany; florian.kloss@leibniz-hki.de
* Correspondence: kfranke@ipb-halle.de (K.F.); ayenesew@uonbi.ac.ke (A.Y.); wessjohann@ipb-halle.de (L.A.W.)

Abstract: *Dalbergia melanoxylon* Guill. & Perr (Fabaceae) is widely utilized in the traditional medicine of East Africa, showing effects against a variety of ailments including microbial infections. Phytochemical investigation of the root bark led to the isolation of six previously undescribed prenylated isoflavanones together with eight known secondary metabolites comprising isoflavanoids, neoflavones and an alkyl hydroxylcinnamate. Structures were elucidated based on HR-ESI-MS, 1- and 2-D NMR and ECD spectra. The crude extract and the isolated compounds of *D. melanoxylon* were tested for their antibacterial, antifungal, anthelmintic and cytotoxic properties, applying established model organisms non-pathogenic to humans. The crude extract exhibited significant antibacterial activity against Gram-positive *Bacillus subtilis* (97% inhibition at 50 µg/mL) and antifungal activity against the phytopathogens *Phytophthora infestans*, *Botrytis cinerea* and *Septoria tritici* (96, 89 and 73% at 125 µg/mL, respectively). Among the pure compounds tested, kenusanone H and (3*R*)-tomentosanol B exhibited, in a panel of partially human pathogenic bacteria and fungi, promising antibacterial activity against Gram-positive bacteria including methicillin-resistant *Staphylococcus aureus* (MRSA) and *Mycobacterium* showing MIC values between 0.8 and 6.2 µg/mL. The observed biological effects support the traditional use of *D. melanoxylon* and warrant detailed investigations of its prenylated isoflavanones as antibacterial lead compounds.

Keywords: *Dalbergia melanoxylon*; Fabaceae; prenylated isoflavanones; antibacterial; antifungal; anti-helminthic; cytotoxic activities

Citation: Chalo, D.M.; Franke, K.; Nchiozem-Ngnitedem, V.-A.; Kakudidi, E.; Origa-Oryem, H.; Namukobe, J.; Kloss, F.; Yenesew, A.; Wessjohann, L.A. Prenylated Isoflavanones with Antimicrobial Potential from the Root Bark of *Dalbergia melanoxylon*. *Metabolites* **2023**, *13*, 678. https://doi.org/10.3390/metabo13060678

Academic Editors: Ramona Paltinean and Irina Ielciu

Received: 21 April 2023
Revised: 17 May 2023
Accepted: 19 May 2023
Published: 23 May 2023

1. Introduction

The genus *Dalbergia* L.f. (Fabaceae) consists of approximately 274 species distributed in the tropics and subtropics. Among these, eight species naturally occur in Kenya [1,2]. Plants of this genus vary from shrubs and lianas to small trees referred to as rosewoods (e.g., *D. odorifera* T.C. Chen., *D. latifolia* Roxb. and *D. melanoxylon* Guill. & Perr.) due to their fine timber of high economic value [3]. The genus is extensively used in the treatment of various ailments in traditional medicine globally [2,4]. Furthermore, pharmacological activities such as anti-inflammatory, antiallergic [5], antibacterial, antifungal [6], antiplasmodial [7], larvicidal and mosquito repellant [8], antidiarrheal [9], anthelmintic [10], anticancer [11], antidiabetic [12] and antigiardial [13] activities have been reported. Consequently, the genus *Dalbergia* has attracted wide phytochemical interest and exhibits a broad spectrum of biologically active secondary metabolites such as flavonoids, isoflavonoids, neoflavones, sterols, quinones, cinnamyl phenols and triterpenes [2,14]. Flavonoids are quite abundant in nature while prenylated flavonoids are much less common and their distribution is mostly restricted to the family Fabaceae, but to a lesser extent are reported from Cannabaceae, Guttiferae, Hypericaceae, Moraceae, Rutaceae and Umbelliferae [15]. Rare 3-hydroxisoflavanones [16] and flavonoids having isoprene or geranyl units attached to B- and C-rings have been described in the genus *Dalbergia* and also occur in a few other related genera such as *Sophora* and *Echinosophora* [17,18]. Prenylation in flavonoids increases lipophilicity and thus affinity to biological membranes. Interaction with target proteins is improved and therefore antibacterial, antifungal, anti-inflammatory, antioxidant and cytotoxic activities are enhanced compared to the parent compound [19–21]. Moreover, isoflavonoids have continued to gain tremendous attention due to their phytoestrogenic effects leading to a correlation between their dietary consumption and health-advantageous effects towards osteoporosis, hormone-related cancer, cardiovascular diseases and menopausal symptoms. The dietary intake of legumes represents the main source of these so-called phytoestrogens which play a pivotal role for both humans and animals [22].

Dalbergia melanoxylon Guill. & Perr. (known as African Blackwood) occurs as a shrub or small tree with a wide ecological versatility in semi-arid, sub-humid and tropical lowland areas in Eastern Africa [1]. Traditionally, African Blackwood is widely applied in African communities. For instance, in Kenya, the leaves are boiled with goat soup to relieve joint pains [23], while the bark decoction is used in Zimbabwe for cleaning wounds [6]. Furthermore, *D. melanoxylon* is used indigenously in South Africa and Zambia as an emetic and aphrodisiac, respectively [6]. The ethnobotanical utilization of the roots in management of abdominal pain, helminths, gonorrhea, stomachache and as a mouth wash for toothache is documented in Kenya and Zimbabwe [6,24]. Crude extracts from different plant parts of *D. melanoxylon* have shown significant antibacterial and antifungal activity [6,25]. Besides the traditional and pharmacological applications, African blackwood has been used for decades for the manufacturing of musical instruments (oboe and clarinet) and ornamental objects (carvings, tables, sofas) ranking it among the most expensive timbers in the world [26]. Phytochemical investigations on the stem bark of this species led to the isolation of dihydrobenzofurans (melanoxin), neoflavones (*S*)(+)-3′-hydroxyl-4′,2,4,5-tetramethoxy-dalbergiquinol), a quinone (4-hydroxyl-4-methoxydalbergione) and isoflavanones (kenusanone F 7-methyl and sophoronol-7-methyl ether) [16,27–29]. A recent study also demonstrated the cardioprotective effects of several neoflavonoids from the heartwood of *D. melanoxylon* [30].

Because various tissues including the roots of *D. melanoxylon* are traditionally used to treat infection-related conditions, we hypothesized that phytochemical investigation of root bark might yield new secondary metabolites with antimicrobial properties. Although the plant species has attracted considerable attention from the scientific community, no phytochemical investigation or biological evaluation of *D. melanoxylon* root bark has been undertaken to date. The present study aims to fill this gap. Thus, the root bark of the target plant was investigated phytochemically leading to the isolation and identification of fourteen compounds, of which six are described for the first time. Herein, the isolation, structure elucidation, antibacterial, antifungal, anthelminthic and cytotoxic activity of these compounds are discussed.

2. Materials and Methods

2.1. General Experimental Procedures

Column chromatography was performed on silica gel (0.040–0.063 µm) (Merck, Darmstadt, Germany). Analytical TLC was conducted on silica gel plates 60 F254 (Merck). Spots were visualized using UV light at 254 and 366 nm or by spraying with vanillin-H_2SO_4. ^1H, ^{13}C NMR and 2D spectra were recorded on an Agilent DD2 400 NMR spectrometer at 399.915 and 100.569 MHz, a Bruker Avance NEO 500 NMR spectrometer equipped with a TXO cryogenic probe operating at 500 and 125 MHz and on an Agilent VNMRS 600 NMR spectrometer at 600 and 150 MHz, respectively. If not stated otherwise, the ^1H NMR chemical shifts are referenced to internal TMS (δ_H 0.00); ^{13}C NMR chemical shifts are referenced to internal methanol-d_4 (δ_C 49.0). The low-resolution ESI-MS spectra were obtained from a Sciex API-3200 instrument (Applied Biosystems, Concord, Ontario, Canada) combined with an HTC-XT autosampler (CTC Analytics, Zwingen, Switzerland). HRESIMS were obtained using Orbitrap Elite Mass spectrometer (Thermofisher Scientific, Bremen, Germany) and QTOF mass spectrometer Sciex TripleTOF 6600 LC-MS System (AB Sciex, Darmstadt, Germany). UV spectra were obtained on a JASCO V-560 UV/VIS spectrophotometer (JASCO Deutschland GmbH, Pfungstadt, Germany). Optical rotations were measured using a JASCO P-2000 digital polarimeter in methanol.

2.2. Plant Material

The root bark of *D. melanoxylon* Guill. & Perr. (Fabaceae) was collected from Muthetheni, Machakos County (S 1°28′60″, E 37°30′02″, El. 1200 masl) in Kenya, in December 2018. A voucher specimen (DMC/2018/001) was deposited at the University Herbarium (NAI), Department of Biology, University of Nairobi, Kenya. The species was identified by the curator of the herbarium, Mr. Patrick Mutiso.

2.3. Extraction and Isolation

The ground root bark of *D. melanoxylon* (1.6 kg) was macerated in a 1:1 mixture of CH_2Cl_2 and MeOH to yield a gummy extract (95.7 g). The crude extract was partitioned between CH_2Cl_2 and H_2O. After removal of the organic solvent 70.6 g CH_2Cl_2 extract was obtained. A portion of this extract (51.3 g) was subjected to column chromatography on silica gel (600 g, 80 × 4 cm) eluting with *n*-hexane containing increasing amounts of EtOAc.

A brownish precipitate obtained from the fraction eluted with 2% EtOAc in *n*-hexane was filtered and washed with methanol to afford a mixture of cinnamic acid esters with different chain lengths dominated by 3′,4′-dihydroxyl-trans-cinnamic acid octacosyl ester (**14**, 10.3 g). The precipitate obtained from the fraction eluted with 4% EtOAc in *n*-hexane was washed with methanol, giving sophoraisoflavone A (**10**, 12.7 g) as the major compound. The mother liquor was further purified by column chromatography on silica gel (eluted with 60% CH_2Cl_2 in *n*-hexane) to yield compound **2** (400.4 mg), compound **3** (6.3 mg), compound **4** (4.4 mg) and compound **5** (3.2 mg). The fraction eluted with 6% EtOAc in *n*-hexane, after concentration, was subjected to column chromatography on silica gel (eluted with 50% CH_2Cl_2 in *n*-hexane) followed by centrifugal thin-layer chromatography using a Chromatotron (CH_2Cl_2/MeOH 19:1) to yield methyl dalbergin (**11**, 5.1 mg), dalbergin (**12**, 4.7 mg) and mellanein (**13**, 3.5 mg). Chromatotron separation (CH_2Cl_2/MeOH 19:1) of the fraction eluted at 8% EtOAc in *n*-hexane resulted in the isolation of tomentosanol B (**9**, 400 mg). The fraction (7.5 g) eluted using 30% EtOAc in *n*-hexane was further purified by column chromatography applying a gradient of CH_2Cl_2 with increasing amount of MeOH to yield kenusanone F (**8**, 39.6 mg), kenusanone H (**7**, 202.3 mg) and compound **6** (45.4 mg). A white precipitate obtained from the fraction eluted with 60% EtOAc in *n*-hexane was filtered and washed with acetone to afford compound **1** (6.9 mg).

The physicochemical properties and spectroscopic data of new isolates (**1–6**) as well as previously not-reported data for compounds **7** and **9** are summarized below:

(3*S*)-3,4′,5,7-Tetrahydroxyl-2′-methoxy-3′-(4-hydroxyl-3-methylbut-2-enyl)-isoflavanone (**1**). White amorphous solid; $[\alpha]_D^{26}$ 123.4 (c 0.160, MeOH); CD (MeOH) $[\theta]_{215}$ − 33,894,

$[\theta]_{237}$ + 14,806, $[\theta]_{292}$ + 18,208, $[\theta]_{348}$ + 2614; UV (MeOH)λ_{max} (log ε) 291 (4.05); ^{1}H (referenced to methanol-d$_4$ solvent signal) and ^{13}C NMR data, see Tables 1 and 2; HRESIMS *m*/*z* 401.1234 [M − H]$^-$ (calcd. for $C_{21}H_{21}O_8$, 401.1236).

Table 1. ^{1}H NMR [δ_H (ppm), multiplicity (J in Hz)] data of compounds **1–6** in methanol-d$_4$.

Position	1 [a]	2 [a]	3 [b]	4 [b]	5 [b]	6 [c]
2A	4.70, d (11.8)	4.40, d (8.9)	4.39, d (8.9)	4.40, d (6.5)	4.40, m	4.53, t (10.8)
2B	4.03, d (11.8)	4.39, d (6.7)	4.38, d (6.8)	4.39, d (9.1)	4.38, m	4.41, dd (10.8, 5.5)
3		4.26, dd (8.9, 6.7)	4.24, dd (8.9, 6.8)	4.27, dd (9.1, 6.5)	4.27, m	4.17, dd (10.8, 5.5)
6	5.93, d (2.0)					5.94, s
8	5.88, d (2.0)	5.94, s	5.94, s	5.93, s	5.94, s	
3′						6.33, d (2.4)
5′	6.61, d (8.5)	6.55, d (8.4)	6.55, d (8.4)	6.55, d (8.4)	6.53, d (8.4)	6.26, dd (8.3, 2.4)
6′	7.27, d (8.5)	6.78, d (8.4)	6.78, d (8.4)	6.77, d (8.4)	6.88, d (8.4)	6.83, d (8.3)
1″A	3.40, dd (14.9, 6.4)	3.37, dd (14.4, 6.7)	3.36, dd (14.8, 7.1)	3.34, m	3.02, m	
1″B	3.33, dd (14.9, 6.4)	3.30 [d]	3.32 [d]		2.70, m	
2″	5.52, t (6.4)	5.24, t (6.7)	5.24, br t (7.2)	5.24 br t (7.2)	3.74, m	
4″	3.91, s	1.68, s	1.67, s	1.67, s	1.32/1.34, s	
5″	1.76, s	1.76, s	1.77, s	1.76, s	1.26/1.27, s	
1‴		3.22, d (7.2)	3.24, d (7.3)	3.23, br d (6.5)	3.22, d (7.2)	3.21, br d (7.1)
2‴		5.20, t (7.2)	5.25, br t (7.3)	5.24, br t (6.5)	5.19, br t (6.5)	5.20, br t (7.1)
4‴		1.95, m	2.65, br d (4.5)	1.97, m	1.94, d (7.6)	2.18, m
5‴		2.05, m	5.57, m	1.59, m	2.05, m	2.08, m
6‴		5.05, br t (6.8)	5.57, m	3.95, t (6.7)	5.05, m	5.16, m
7‴						
8‴		1.61, s	1.24, s	A 4.85 m B 4.76 m	1.61, s	1.67, s
9‴		1.74, s	1.74, s	1.77, s	1.74, s	1.65, s
10‴		1.55, s	1.24, s	1.67, s	1.56, s	1.61, s
2′-OCH$_3$	3.58, s	3.71, s	3.71, s	3.70, s	3.76/3.75, s	
5-OH	12.09, s	12.41, s	12.41, s	12.41, s		12.21, s

[a, b, c] recorded at 500, 600 and 400 MHz, respectively, [d] overlapping with solvent signal; s: singlet; d: doublet; t: triplet; m: multiplet; br s: broadened singlet; dd: doublet of doublets.

(3*R*)-6-Geranyl-4′,5,7-trihydroxyl-2′-methoxy-3′-prenylisoflavanone (**2**). White paste; $[\alpha]_D^{24}$ 0.77 (c 0.300, MeOH); CD (MeOH) $[\theta]_{197}$ − 15,079, $[\theta]_{210}$ +9341, $[\theta]_{285}$ − 3305, $[\theta]_{309}$ + 5596; UV (MeOH)λ_{max} (log ε) 294 (4.17); ^{1}H and ^{13}C NMR data, see Tables 1 and 2; HRESIMS *m*/*z* 505.2582 [M − H]$^-$ (calcd. for $C_{31}H_{37}O_6$, 505.2590).

6-((2*E*,5*E*)-7-Hydroxyl-3,7-dimethyl-octa-2,5-dienyl)-4′,5,7-trihydroxyl-2′-methoxy-3′-prenylisoflavanone (**3**). White amorphous solid; ^{1}H and ^{13}C NMR data, see Tables 1 and 2; HRESIMS *m*/*z* 521.2537 [M − H]$^-$ (calcd. for $C_{31}H_{37}O_6$, 521.2539).

(*E*)-6-(6-Hydroxyl-3,7-dimethylocta-2,7-dienyl)-4′,5,7-trihydroxyl-2′-methoxy-3′-prenylisoflavanone (**4**). White amorphous solid; ^{1}H and ^{13}C NMR data, see Tables 1 and 2; HRESIMS *m*/*z* 521.2521 [M − H]$^-$ (calcd. for $C_{31}H_{37}O_6$, 521.2539).

6-Geranyl-4′,5,7-trihydroxyl-2′-methoxy-3′-(2,3-epoxy-3-methyl-butyl)-isoflavanone (**5**). Colorless residue; ^{1}H and ^{13}C NMR data, see Tables 1 and 2; HRESIMS *m*/*z* 521.2530 [M − H]$^-$ (calcd. for $C_{31}H_{37}O_6$, 521.2539).

(*Z*)-2′,4′,5,7-Tetrahydroxyl-8-(3,7-dimethylocta-2,6-dienyl)-isoflavanone (**6**). Pale-yellow oil; ^{1}H and ^{13}C NMR data, see Tables 1 and 2; HRESIMS *m*/*z* 423.1848 [M − H]$^-$ (calcd. for $C_{25}H_{27}O_6$, 423.1808).

Kenusanone H (**7**). Yellow paste; $[\alpha]_D^{25}$ − 1.1 (c 0.490, MeOH); UV (MeOH) λ_{max} (log ε) 292 (4.32); ^{13}C NMR data, see Table 2; HRESIMS *m/z* 423.1837 [M − H]$^-$ (calcd. for $C_{25}H_{27}O_6$, 423.1808).

(3*R*)-Tomentosanol B (**9**). White paste $[\alpha]_D^{25}$ − 126.7 (c 0.300, MeOH)); CD (MeOH) $[\theta]_{206}$ − 71,898, $[\theta]_{236}$ −27,726, $[\theta]_{297}$ − 25,304, $[\theta]_{332}$ +4592; UV (MeOH)λ_{max} (log ε) 296 (4.32);

^{13}C NMR data, see Table 2; HRESIMS m/z 453.1913 [M − H]$^-$ (calcd. for $C_{26}H_{29}O_7$, 453.1908).

Table 2. ^{13}C NMR data (δ_C [ppm]) of isoflavanones **1–7** and **9** measured in methanol-d$_4$.

Position	1 [a]	2 [a]	3 [b]	4 [b]	5 [b]	6 [c]	7 [a]	9 [a]
2	75.8	72.5	72.0 [d]	72.2 [d]	72.0 [d]	71.4	71.5	75.6
3	75.4	47.0	46.8 [d]	46.8 [d]	46.9 [d]	47.9	47.9	75.7
4	197.1	199.4	199.6 [e]	199.2 [e]	199.1 [e]	199.8	199.8	197.3
5	166.4	162.8	162.9 [e]	165.6 [e]	162.8 [e]	163.5	163.5	163.3
6	97.3	109.8	109.8 [e]	109.4 [e]	109.8 [e]	96.4	96.4	109.9
7	168.3	166.0	166.1 [e]	165.5 [e]	166.0 [e]	165.8	165.8	165.9
8	96.1	95.3	94.8 [d]	95.0 [d]	95.3 [d]	108.9	108.9	95.4
9	164.5	162.8	162.9 [e]	162.5 [e]	162.8 [e]	161.8	161.8	162.1
10	102.1	103.8	103.9 [e]	103.5 [e]	103.7 [e]	103.9	103.8	102.0
1′	123.7	120.7	120.8 [e]	120.4 [e]	121.1 [e]	113.8	114.1	122.7
2′	157.7	159.2	159.3 [e]	158.8 [e]	158.8 [e]	157.7	157.6	157.7
3′	122.1	123.2	123.3 [e]	122.8 [e]	115.3 [e]	103.8	103.8	124.5
4′	158.7	157.5	157.7 [e]	157.2 [e]	155.2 [e]	159.1	159.1	158.6
5′	111.2	112.3	111.9 [d]	112.1 [d]	113.9[d]	107.8	107.8	111.3
6′	126.8	128.3	127.8 [d]	128.0 [d]	129.0 [d]	131.9	131.9	126.6
1″	24.4	24.4	24.2 [d]	24.1 [d]	27.6 [d]			4.8
2″	125.8	124.6	124.2 [d]	124.1 [d]	70.0 [d]			124.0
3″	135.8	131.7	131.4 [e]	131.4 [e]	77.8 [d]			132.2
4″	68.9	25.9	25.6 [d]	25.6 [d]	25.6 [d]			25.9
5″	14.0	18.0	17.8 [d]	17.7 [d]	20.8 [d]			17.9
1‴		21.8	21.7 [d]	21.6 [d]	21.5 [d]	22.1	22.2	22.0
2‴		124.0	124.2[d]	124.3 [d]	123.6 [d]	124.7	124.2	123.9
3‴		135.2	134.5 [e]	134.7 [e]	135.2 [e]	135.6	135.2	131.6
4‴		40.9	43.4 [d]	36.5 [d]	40.7 [d]	32.9	40.8	25.8
5‴		27.7	126.0 [d]	34.1 [d]	27.7 [d]	27.7	27.6	18.0
6‴		125.5	139.8 [d]	75.9 [d]	125.2 [d]	125.7	125.4	
7‴		132.0	71.3 [e]	148.4 [e]	132.0 [d]	131.9	132.1	
8‴		25.9	29.6 [d]	111.2 [d]	25.6 [d]	25.9	25.9	
9‴		16.2	16.0 [d]	16.0 [d]	16.0 [d]	23.7	16.2	
10‴		17.7	29.6 [d]	17.4 [d]	17.4 [d]	17.7	17.7	
2′-OCH_3	62.0	62.5	62.2 [d]	62.2 [d]	61.1 [d]			61.9

[a, b, c] recorded at 125, 150 and 100 MHz, respectively; [d, e] signals derived from HSQC and HMBC, respectively.

2.4. Biological Assays

2.4.1. Antibacterial Assays

The crude extracts of *Dalbergia melanoxylon* (50 and 500 µg/mL) and its isolated compounds (at 1 and 100 µM) were tested for their antibacterial activity against the Gram-negative *Aliivibrio fischeri* (DSM507) and the Gram-positive *Bacillus subtilis* (DSM 10) as described by dos Santos et al. [31]. Chloramphenicol (100 µM) was used as positive control and induced the complete inhibition of bacterial growth.

The results (mean value ± standard deviation, n = 6) were given as relative values (% inhibition) in comparison to the negative control (bacterial growth, 1% DMSO, without test compound). Negative values indicate an increase of bacterial growth. Calculations were performed applying the software Excel.

2.4.2. Antifungal Assays

The assays were performed according to the monitoring methods approved by the fungicide resistance action committee (FRAC) with minor modifications [32]. The phytopathogenic ascomycetes *Botrytis cinerea* Pers. and *Septoria tritici* Desm., and the oomycete *Phytophthora infestans* (Mont.) De Bary were used as test microorganisms. The crude extract and pure compounds were tested in 96-well microtiter plate assays at 125 and 42 µg/mL with DMSO used as negative control (max. concentration 2.5%), while epoxiconazole (100% inhibition at 42 μM) and terbinafine (67% inhibition at 42 μM) served as positive control. Five to seven days after inoculation, pathogen growth was evaluated by measurement of the optical density (OD) at λ 405 nm with a TecanGENios Pro microplate reader (5 measurements per well using multiple reads in a 3 × 3 square). Each experiment was carried out in triplicate.

2.4.3. Anthelmintic Assay

The anthelmintic bioassay was performed using the model organism *Caenorhabditis elegans* that previously was shown to correlate with anthelmintic activity against parasitic trematodes as described by Thomsen et al. [33]. The Bristol N2 wild-type strain of *C. elegans* was obtained from the Caenorhabditis Genetic Center (CGC), University of Minnesota, Minneapolis, USA. The nematodes were cultured on NGM (Nematode Growth Media) Petri plates using the uracil auxotroph *E. coli* strain OP50 as food source. In this assay, the solvent DMSO (2%) and the standard anthelmintic drug ivermectin (10 μg/mL, 100% dead worms after 30 min incubation) were used as negative and positive control, respectively.

2.4.4. Cytotoxicity Assay

Briefly, for the cytotoxicity assay, the human prostate cancer cell line PC-3 and the colon adenocarcinoma cancer cell line HT-29 (both from ATCC, Manassas, VA, USA) were used. The cell handling and assay techniques were in accordance with the method described by Khan et al. [34]. The extract was tested at the concentrations of 0.05 and 50 μg/mL. Anti-proliferative and cytotoxic effects of the extract were investigated by performing colorimetric MTT (3-(4,5-dimethylthiazol-2-yl)-2,5-diphenyltetrazolium bromide) and CV (crystal violet)-based cell viability assays (Sigma-Aldrich, Taufkirchen, Germany) after 48 h treatment time, respectively. The absorbance was measured with an automated microplate reader at 540 nm with a reference wavelength of 670 nm. Digitonin (125 μM) was used as positive control, which was set for data normalization to 0% cell viability. The results are presented as a percentage of control values obtained from untreated cultures.

2.4.5. Agar Diffusion Assay

The experiment was performed as previously published [35]. Briefly, test compounds were dissolved in dimethyl sulfoxide (DMSO) at a concentration of 1 mg/mL. Ciprofloxacin and amphotericin B (both positive control) were provided at 5 μg/mL and 10 μg/mL, respectively. The following test strains were used: *B. subtilis* (JMRC:STI:10880), *S. aureus* (JMRC:ST:10760 and JMRC:ST:33793 (MRSA)), *E. faecalis* (JMRC:ST:33700 (VRE)), *E. coli* (JMRC:ST:33699), *P. aeruginosa* (JMRC:ST:33772 and JMRC:ST:337721), *M. vaccae* (JMRC:STI:10670), *P. notatum* (JMRC:STI:50164), *C. albicans* (JMRC:STI:50163) and *S. salmonicolor* (JMRC:ST:35974). 2.4.6. MIC Assay

Minimal inhibitory concentrations were determined against *Mycobacterium vaccae* (JMRC:STI:10670), MRSA (JMRC:ST:33793) and *Enterococcus faecalis* (JMRC:ST:33700 (VRE)) by serial dilutions of the DMSO test item solutions of compounds **7**, **9** and **10** (1 mg/mL) Growth was inspected visually.

2.4.6. Cytotoxicity Testing (Compound 7)

HeLa cells (DSM ACC 57) were grown in RPMI 1640 medium supplemented with 10 mL/L ultraglutamine 1 (CAMBREX 17-605E/U1), 550 μL/L gentamicin sulfate (50 mg/mL, CAMBREX 17-518Z) and 10% heat inactivated fetal bovine serum (GIBCO Life Technologies 10270-106) at 37 °C in a 5% CO_2 atmosphere in high density polyethylene flasks (NUNC 156340). Cells were pre-incubated for 48 h in the absence of test substances. Subsequently, HeLa cells were incubated with serial dilutions of test compounds in 96 well microplates for 72 h at 37 °C in a humidified atmosphere and 5% CO_2. After incubation, the cytolytic effect of compounds was analyzed relative to the negative control (DMSO) using a colorimetric assay (methylene blue). The adherent HeLa cells were fixed by glutaraldehyde (MERCK

1.04239.0250) and stained with a 0.05% solution of methylene blue (SERVA 29198) for 15 min. After gentle rinsing, the stain was eluted through addition of 0.2 mL hydrochloric acid (0.33 M) to each well. The absorptions were measured at 660 nm in a SUNRISE microplate reader (TECAN). Four replicates were assayed for each substance. The half-cytotoxic concentration (CC_{50}) was defined as the test compound concentration required for 50% reduction of the viable cell count in the monolayer relative to the respective untreated control. All calculations of CC_{50} values were performed with the software Magellan (TECAN).

3. Results and Discussion

3.1. Isolation and Structure Elucidation

Chromatographic separation of the extract from the root bark of *D. melanoxylon* afforded six hitherto-undescribed isoflavanones (**1–6**) alongside eight known secondary metabolites comprising isoflavonoids (**7–10**), neoflavones (**11–13**) and alkyl hydroxylcinnamates (**14**) (Figure 1). Based on HRESIMS, NMR and ECD spectra and comparison to published data, the known compounds were identified as kenusanone H (**7**; $[\alpha]_D^{25} - 1.1$ (c 0.490, MeOH)), kenusanone F (**8**; $[\alpha]_D^{25} - 112.5$ (c 0.260, MeOH)) previously isolated from *Echinosophora koreensis* [36], tomentosanol B (**9**; $[\alpha]_D^{25} - 126.7$ (c 0.300, MeOH)) [18] and sophoraisoflavanone A (**10**) from *E. koreensis* [17], methyl dalbergin (**11**) from *Dalbergia sissoo* [37], dalbergin (**12**) from *D. odorifera* [38], melannein (**13**) from *D. melanoxylon* [28] and a mixture of cinnamic acid esters with the main compound being 3′,4′-dihydroxyl-trans-cinnamic acid octacosyl ester (**14**) known from *Gliricidia sepium* [39]. With the exception of compounds **12** and **13**, all known compounds were isolated for the first time from *D. melanoxylon*. For tomentosanol B (**9**) so far only the planar structure based on ^{1}H NMR data was described [18]. Herein we report its ^{13}C (Table 2) and 2D NMR data (Figures S9_3 and S9_4, Table S9). Based on ECD measurements (Figure S9_6) the configuration at C3 was determined as *R* and compound **9** thus elucidated as (3*R*)-6-prenyl-3,4′,5,7-trihydroxyl-2′-methoxy-3′-prenyl-isoflavanone (trivial name (3*R*)-tomentosanol B).

Compound **1** was purified as a white amorphous solid. It shows a deprotonated molecular ion in the HRESIMS at *m/z* 401.1234 $[M - H]^-$ (calcd. for $C_{21}H_{21}O_8$, 401.1236), corroborating the molecular formula $C_{21}H_{22}O_8$ (degree of unsaturation: 11 double bond equivalents (DBE)). Its ^{1}H [δ_H 4.70 (d, ^2J = 11.8 Hz, H-2A), 4.03 (d, ^2J = 11.8 Hz, H-2B)] and ^{13}C [δ_C 75.8 (C-2), 75.4 (C-3), 197.1 (C-4)] NMR spectral data displayed the signature of a 3-hydroxylisoflavanone core [16,40]. In addition, the NMR also exhibited signals of a methoxy (δ_H 3.58, δ_C 62.0) and a 3-hydroxylmethyl-3-methylbut-2-enyl [δ_H 3.40, 3.33 (dd, J = 14.9, 6.4 Hz, H-1″A/1B″), 5.52 (t, J = 6.4 Hz, H-2″), 3.91 (s, H-4″), 1.76 (s, H-5″)] substituent. Furthermore, compound **1** showed the typical pattern for meta-coupled protons of a 5,7-dioxygenated A-ring [δ_H 5.93 (d, J = 2.0 Hz, H-6) and 5.88 (d, J = 2.0 Hz, H-8)] alongside two ortho-coupled doublets of an AX spin system derived from B-ring protons [δ_H 7.27 (d, J = 8.5 Hz, H-6′), 6.61 (d, J = 8.5 Hz, H-5′)]. These observations were further supported with 2D spectra which showed cross-peaks in the ^{1}H-^{1}H COSY spectrum between H-6 and H-8 in the A-ring, and between H-5′ and H-6′ in the B-ring. Analysis of the ^{13}C-NMR spectrum indicated, in accordance with the molecular formula, the presence of 21 carbons with resonances ranging from δ_C 195.7 (sp^2 hybridized ketone) to 12.6 (sp^3 hybridized methyl unit). The ^{13}C-NMR chemical shift of the deshielded methoxy group signal (δ_C 62.0) indicated that it is di-ortho-substituted with two bulky groups, which is consistent with its placement at C-2′ [16,41]. This finding was further confirmed with NOESY correlations observed between 2′-OMe (δ_H 3.58) and H-2A (δ_H 4.70). HMBC correlations from H-1″A (δ_H 3.40) to C-2′ (δ_C 157.7), C-3′ (δ_C 122.1), C-4′ (δ_C 158.7) and C-2″ (δ_C 125.8) indicated the placement of the isoprenyl unit at C-3′. Analysis of the NMR spectroscopic data showed its structural similarity to kenusanone F 7-methyl ether ($C_{22}H_{24}O_7$) isolated previously from stem bark of D. melanoxylon [16] and to kenusanone F ($C_{21}H_{22}O_7$, **8**) obtained from stem bark of Erythrina brucei [42] and also isolated in this study. The difference is that compound **1** is missing one methyl group compared to kenusanone F 7-methyl ether while it possesses one more hydroxyl group than the two other compounds. The placement of the additional OH in the prenyl chain at C-4″ (δ_C 68.9) was supported by NOESY correlation between δ_H 5.52 (H-2″) and 3.91 (H-4″).

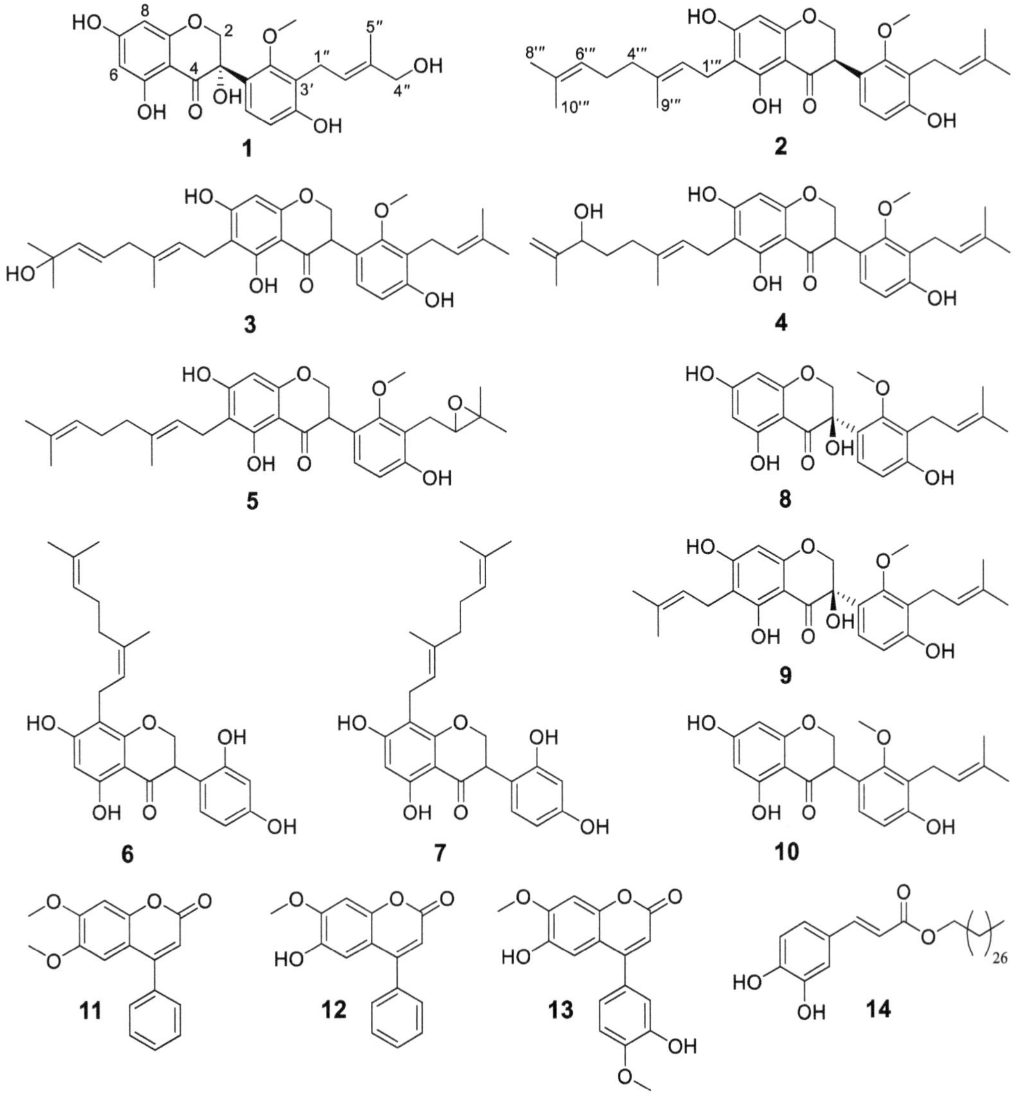

Figure 1. Structures of compounds **1–14** isolated from *Dalbergia melanoxylon*. Structures were elucidated based on HR-ESI-MS, 1- and 2-D NMR and ECD spectra and by comparison with literature values.

The absolute configuration of **1** was assigned by ECD spectroscopy. Usually, the octant rule modified for cyclic arylketones is applied to determine the stereochemistry of isoflavanones [43]. This predicts a positive Cotton effect (CE) for the n $\rightarrow$ π^* carbonyl transition between 320–352 nm for (3R)-isoflavanones with the B-ring in the favored equatorial position [16,43]. However, it should be kept in mind that the priority order according to the Cahn–Ingold–Prelog rules changes when hydrogen at C-3 in isoflavanones is replaced with a hydroxyl group in 3-hydroxylisoflavanones. Thus, (3R)-isoflavanones show the same spatial arrangement as (3S)-hydroxylisoflavanones. However, at least for 3-hydroxylisoflavones, the octant rule is not fully reliable and seems to be prone to misinterpretation. The ECD spectrum of **1** shows intense positive Cotton effects at 237, 292 and 348 nm, and a weak negative one around 330 nm (Figure 2). The weak CEs in the long wavelength region, around 330 (negative CE) or around 348 nm (positive CE), may not be reliable for the assignment of the absolute configuration of compound **1**. However, the ECD spectrum of **1** appears similar to the one calculated for (3*S*)-kenusanone F 7-methyl ether with a negative CE at 330 nm [42] and shows a mirror image to (3R)-kenusanone F (**8**, Figure 2; [42], hence it is consistent with (3*S*)-**1** configuration. This previously undescribed compound (**1**) was therefore characterized as (3*S*)-3,4′,5,7-tetrahydroxyl-2′-methoxy-3′-(4-hydroxylprenyl)isoflavanone.

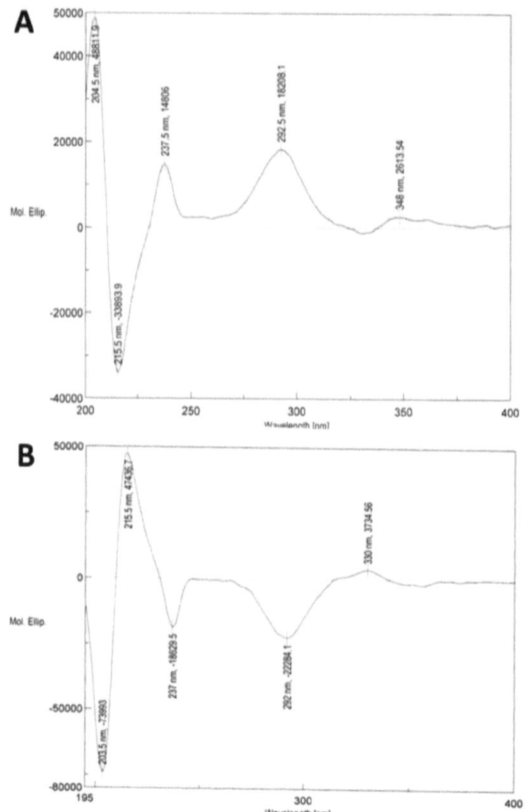

Figure 2. ECD spectra of (**A**) (3*S*)-3,4′,5,7-tetrahydroxyl-2′-methoxy-3′-(4-hydroxyl-3-methylbut-2-enyl)-isoflavanone (**1**) and (**B**) (3*R*)-kenusanone F (**8**).

Compound **2** was obtained as a white paste. Its molecular formula, $C_{31}H_{38}O_6$ (13 DBE) was established by means of HRESIMS (m/z 505.2582 $[M-H]^-$, (calcd. for $C_{31}H_{37}O_6$, 505.2590)) and NMR data. The ^{1}H [δ_H 4.40 (d, ^2J = 8.9 Hz, H-2A), 4.39 (d, ^2J = 6.7 Hz, H-2B), 4.26 (dd, J = 8.9, 6.7 Hz, H-3)] and ^{13}C NMR [δ_C 72.5 (C-2), 47.0 (C-3) and 199.4 (C-4)] spectral data confirmed that compound **2** possesses an isoflavanone skeleton [40]. The 1D and 2D NMR data of compound **2** were similar to those of sophoraisoflavanone A (**10**), isolated from *Erythrina droogmansiana* [44] and *Sophora tomentosa* [45], except for the presence of an additional geranyl group [δ_H 3.22 (d, J = 7.2 Hz, H-1‴), 5.20 (t, J = 7.2 Hz, H-2‴), 1.95

(m, H-4‴), 2.05 (m, H-5‴), 5.05 (br t, J = 6.8 Hz, H-6‴), 1.61 (s, H-8‴), 1.74 (s, H-9‴) and 1.55 (s, H-10‴); δ_C 21.8 (C-1‴), 124.0 (C-2‴), 135.2 (C-3‴), 40.9 (C-4‴), 27.7 (C-5‴), 125.5 (C-6‴), 132.0 (C-7‴), 25.9 (C-8‴), 16.2 (C-9‴) and 17.7 (C-10‴)] at C-6 in **2**. The tail-to-head linkage of the two isoprenyl moieties to form the geranyl group was further corroborated using ^{1}H-^{1}H COSY correlations between H-5‴/H-6‴ and H-5‴/H-4‴. HMBC cross-peaks from H-1‴ (δ_H 3.22) to C-5 (δ_C 162.8), C-6 (δ_C 109.8), C-7 (δ_C 166.0), C-2‴ (δ_C 124.0) and C-3‴ (δ_C 135.2) clearly establish the location of the geranyl substituent at C-6. A Cotton effect for n → π* transition was not observed in compound **2**, probably due to low concentration, and hence could not be used to determine absolute configuration. On the other hand, as in compound **1**, compound **2** showed a strong positive Cotton effect for π → π* transition at 309 nm, allowing the assignment of the same absolute configuration at C-3, but the designation is R (due to change in priority because of the absence of OH at C-3 in compound **2**). Thus, compound **2** was elucidated as (3R)-6-geranyl-4′,5,7-trihydroxyl-2′-methoxy-3′-prenylisoflavanone.

Compounds **3**, **4** and **5** were assigned the same molecular formula, $C_{31}H_{38}O_7$ (13 DBE) based on HRESIMS (m/z 521.2537 [M − H]$^-$, *m/z* 521.2521 [M − H]$^-$ and *m/z* 521.2582 [M − H]$^-$, respectively (calcd. for $C_{31}H_{37}O_6$, 521.2539)) combined with 1D (^{1}H and ^{13}C) and 2D (^{1}H-^{1}H COSY, HSQC and HMBC) NMR spectra. The molecular weights of **3**, **4** and **5** were 16 Dalton (Da) higher than that of **2** implying the presence of an additional oxygen atom in these compounds. Careful analyses of 1D and 2D NMR indicated that **3** and **4** had an isoflavanone skeleton similar to that of **2**, but with modifications in the geranyl residues. The terminal prenyl moiety was altered to a 3-hydroxyl-3-methyl-trans-but-1-enyl [δ_H 5.57 (m, H-5‴), 5.57 (m, H-6‴), 1.24 (s, H-8‴) and 1.24 (s, H-10‴); δ_C 126.0 (C-5‴), 139.8 (C-6‴), 71.3 (C-7‴), 29.6 (C-8‴) and 29.6 (C-10‴)] in **3** versus a 2-hydroxyl-3-methylbut-3-enyl [δ_H 1.59 (m, H-5‴), 3.95 (t, J = 6.7 Hz, H-6‴), 4.85 and 4.76 (m, H-8‴) and 1.67 (s, H-10‴); δ_C 34.1 (C-5‴), 75.9 (C-6‴), 148.4 (C-7‴), 111.2 (C-8‴) and 17.4 (C-10‴)] in **4**. The position of the hydroxyl group in compounds **3** and **4** was established using HMBC correlations from H3-8‴/H3-10‴ (δ_H 1.24) and H-6‴ (δ_H 5.57) to C-7‴ (δ_C 71.3) for **3** and from H-10‴ (δ_H 1.67) to C-6‴ (δ_C 75.9) for **4**. Hence, compounds **3** and **4** were characterized as 6-((2E,5E)-7-hydroxyl-3,7-dimethyl-octa-2,5-dienyl)-4′,5,7-trihydroxyl-2′-methoxy-3′-prenylisoflavanone (**3**) and (E)-6-(6-hydroxyl-3,7-dimethylocta-2,7-dienyl)-4′,5,7-trihydroxyl-2′-methoxy-3′-prenylisoflavanone (**4**), respectively.

Compound **5**, a colorless residue, possesses an isoflavanone scaffold as **2–4**, the major difference being in the isoprenyl substituent in ring B. The presence of an epoxyprenyl residue, formed through electrophilic addition of oxygen to the isoprenyl unit, was established from resonances at δ_H 3.02–2.70 (m, H-1″), 3.74 (m, H-2″), 1.32/1.34 (s, H-4″) and 1.26/1.27 (s, H-5″); δ_C 27.6 (C-1″), 70.0 (C-2″), 77.8 (C-3″), 25.6 (C-4″) and 20.8 (C-5″) as observed in **5**. Since the epoxidation seems not to be stereospecific, partly a double set of data are visible, especially for the methyl groups at position 4″ [(δ_H 1.32/1.34, s), 5″ ((δ_H 1.26/1.27, s) and 2′-OMe (δ_H 3.76/3.75, s)]. HMBC cross-peaks from H-1″ (δ_H 3.02) to C-2′ (δ_C 158.8) and C-4′ (δ_C 155.2) indicated that the epoxyprenyl moiety was located at C-3′. Hence, the planar structure of **5** was elucidated as 6-geranyl-4′,5,7-trihydroxyl-2′-methoxy-3′-(2,3-epoxy-3-methyl-butyl)-isoflavanone. Reliable optical rotation, UV and ECD spectra could not be generated for compounds **3–5**.

Compound **6** was obtained as pale-yellow oil. Its molecular formula was deduced as $C_{25}H_{28}O_6$ (12 unsaturation sites) based on HRESIMS (m/z 423.1813 [M − H]$^-$ (calcd. for $C_{25}H_{27}O_6$, 423.1808)) in conjunction with NMR data. The ^{1}H [δ_H 4.53 (t, J = 10.8 Hz, H-2A), 4.41 (dd, J = 10.8, 5.5 Hz, H-2B) and 4.17 (dd, J = 10.8, 5.5 Hz, H-3)] and ^{13}C NMR [δ_C 71.4 (C-2), 47.9 (C-3) and 199.8 (C-4)] data were consistent with an isoflavanone core similar to compounds **2–5**. In general, the NMR data of compound **6** were superimposable to 8-geranyl-2′,4′,5,7-tetrahydroxylisoflavanone (kenusanone H, **7**) reported from the roots of Echinosophora koreensis [36] and isolated in this study. Here we report for the first time ^{13}C (Table 2) and 2D NMR (Table S7) data for compound **7** which was isolated as a racemate. Nevertheless, signals observed at δ_C 32.9 (C-4‴) and 27.7 (C-5‴) in the ^{13}C NMR spectrum of compound **6** indicated that, unlike kenusanone H, the C10 unit is a neryl but not a geranyl group. These findings were further supported with NOESY correlation between H-1‴/H-4‴. The HMBC signals from H-6 (δ_H 5.94) to C-5 (δ_C 163.5), C-7 (δ_C 165.8), C-10 (δ_C 103.9) and C-8 (δ_C 108.9); from H-1‴ (δ_H 3.21) to C-7 (δ_C 165.8), C-8 (δ_C 108.9), C-9 (δ_C 161.8), C-2‴ (δ_C 124.7), C-3‴ (δ_C 135.6) verified the connectivity of the neryl side chain via C-8. The substitution at C-8 is further supported by the chemical shift of the H-bonded OH at position 5 (δ_H 12.21), which is shifted downfield to 12.41–12.43 ppm in compounds **2**–**4** bearing a prenyl chain at C-6 [18,46]. Hence, compound **6** was characterized as (Z)-2′,4′,5,7-tetrahydroxyl-8-(3,7-dimethylocta-2,6-dienyl)-isoflavanone. Nerylated flavonoids are very rare in nature, and compound **6** could have been formed from the geranylated analogue **7** through isomerization.

3.2. Biological Activity

Since *Dalbergia* species are known to exhibit a variety of biological activities, the partitioned crude extracts and the isolated compounds of *D. melanoxylon* were tested for their antibacterial, antifungal, anthelmintic and cytotoxic properties applying an established model organism non-pathogenic to humans (Table 3). The crude CH_2Cl_2 extract of the root bark induced nearly complete inhibition (97% ± 0%) of the Gram-positive bacterium *Bacillus subtilis* at the concentration of 50 µg/mL and complete inhibition (100% ± 0%) of the Gram-negative bacterium Aliivibrio fischeri at 500 µg/mL showing its potential especially against Gram-positive bacteria. The antifungal and anti-oomycetes activity was evaluated against the phytopathogens *Septoria tritici, Botrytis cinerea* and *Phytophthora infestans*, respectively. The extract showed promising activity against all phytopathogens at a concentration of 125 µg/mL. No anthelminthic activity against *Caenorhabditis elegans* could be detected at 500 µg/mL. Likewise, at low concentration (0.05 µg/mL) no antiproliferative or cytotoxic effects were observed against the human cancer cell lines PC3 and HT29 whereas a higher concentration (50 µg/mL) induced significant inhibition of cell growth and viability (Table S15). These results imply that the crude extract possesses moderate cytotoxic properties but might also show selective biological effects with focus on antibacterial and antifungal activities.

Table 3. Antibacterial (*Bacillus subtilis, Aliivibrio fischeri*) and antifungal (*Phytophthora infestans, Botrytis cinerea, Septoria tritici*) activities of the CH_2Cl_2 extract and isolated compounds from *D. melanoxylon* shown as growth inhibition [%] [a]. Data are presented as mean values ± standard deviation (n = 6 for antibacterial assays, n = 3 for antifungal assays.

	Antibacterial Assays				**Antifungal Assays**					
	B. subtilis		***A. fischeri***		***P. infestans***		***B. cinerea***		***S. tritici***	
Extract	500 µg/mL	50 µg/mL	500 µg/mL	50 µg/mL	125 µg/mL	42 µg/mL	125 µg/mL	42 µg/mL	125 µg/mL	42 µg/mL
CH_2Cl_2	77 ± 13	97 ± 0	100 ± 0	25 ± 2	96 ± 3	22 ± 10	89 ± 1	69 ± 8	73 ± 4	18 ± 19
Compounds	100 µM	1 µM	100 µM	1 µM	125 µg/mL	42 µg/mL	125 µg/mL	42 µg/mL	125 µg/mL	42 µg/mL
1	−7 ± 23	23 ± 34	−5 ± 26	3 ± 25	−68 ± 36	1 ± 19	−56 ± 40	−24 ± 11	6 ± 9	−1 ± 13
2	60 ± 2	−79 ± 17	24 ± 14	−25 ± 27	−32 ± 2	−14 ± 5	−38 ± 12	−20 ± 3	−24 ± 15	−1 ± 4
7	63 ± 2	−33 ± 73	−52 ± 23	−2 ± 28	81 ± 1	2 ± 3	95 ± 4	99 ± 0	88 ± 3	74 ± 10
9	96 ± 0	40 ± 2	67 ± 8	20 ± 18	71 ± 14	23 ± 12	0 ± 30	26 ± 5	76 ± 11	−16 ± 10
10	99 ± 0	n.d.	99 ± 0	8 ± 16	58 ± 12	86 ± 1	58 ± 13	56 ± 5	115 ± 6	13 ± 11
Positive control	100 µM chloramphenicol 100 ± 0		100 µM chloramphenicol 100 ± 0		42 µM terbinafine		42 µM epoxiconazole		42 µM epoxiconazole	
					87 ± 5	67 ± 8	99 ± 2	100 ± 0	97 ± 0	97 ± 0

[a] Negative values indicate an increase of bacterial growth in comparison to the negative control (0% inhibition).

Based on the results of the crude extracts, the isolated major compounds (**1**, **2**, **7**, **9**, **10**) were subjected to a preliminary biological screening in antibacterial and antifungal assays (Table 3). For the antibacterial assays, the compounds were tested at concentrations of 1 and 100 µM, and for the antifungal assays of 42 and 125 µg/mL. In both B. subtilis and A. fischeri assays, (3R)-tomentosanol B (**9**) and sophoraisoflavanone A (**10**) inhibited nearly 100% of bacterial growth at a concentration of 100 µM after 16 h incubation time. Both compounds had also a good antifungal activity against S. tritici at 125 µg/mL (corresponding to 0.28 and 0.34 mM, respectively) (Table 3). Furthermore, kenusanone H (**7**) at 42 µg/mL (0.1 mM) showed a promising growth inhibition of B. cinerea and S. tritici. Thus, these compounds were also tested against a panel of human pathogenic bacteria (Table 4) and fungi (Table S15). Kenusanone H (**7**), (3R)-tomentosanol B (**9**) and sophoraisoflavanone A (**10**) exhibited promising antibacterial activity against Gram-positive bacteria including MRSA as shown by the induction of significant inhibition zones in agar diffusion assays. Even more importantly, these compounds also inhibited the growth of Mycobacteria vaccae, a nonpathogenic member of the tuberculosis inducing the Mycobacteriaceae family. Indeed, previous docking studies indicated the potential binding of 3-hydroxylisoflavanones from *D. melanoxylon* to different mycobacterial target enzymes [16]. In the present study kensuanone H (**7**) displayed MIC values of 1.56, 1.56 and 0.78 µg/*mL* (3.7, 3.7 and 1.8 µM) against S. aureus (MRSA), *Enterococcus faecalis* and *Mycobacterium vaccae*, respectively, while tomentosanol B (**9**) inhibited the growth of these bacteria with MIC values of 3.12, 6.25 and 1.56 µg/mL corresponding to 6.9, 13.8 and 3.4 µM (Table 4), respectively. In addition, compound **7** also exhibited moderate antifungal effects against *Candida albicans*, *Penicillium notatum* and *Aspergillus fumigatus*, compound **9** against P. *notatum* and **10** against *S. salmicolor, C. albicans* and *P. notatum* (Table S15). Except for compounds **8**–**10**, the antimicrobial potential of the tested compounds is reported here for the first time. Nevertheless, in prior studies, kenusanone F (**8**), purified from *E. brucei* displayed moderate activity (MIC values ranging from 125 to 250 µg/mL) against four pathogenic test organisms, namely *S. aureus*, *B. cereus*, *B. megaterium* and *E. coli* [42], while tomentosanol B (**9**) showed antiplasmodial activity (IC_{50} = 25.3 µM) and virtually no in vitro cytotoxicity against the Chinese hamster ovarian (CHO) cell line (selectivity index = 5) [47]. Sophoraisoflavanone A (**10**) isolated from *Echinosophora koreensis* was already previously described as compound with strong antifungal (*C. albicans*, *S. cerevisiae*) and antibacterial activity (*E. coli*, *S. typhimurium*, *S. epidermis*, *S. aureus*) showing MIC values around 60 and 20 µg/mL, respectively [19]. In addition, this compound has proven toxic (IC_{50} = 22.1 µg/mL) to a human liver (HepG2) cell line [19]. Although we could not demonstrate anthelmintic activity for the crude extract of *D. melanoxylon*, mild anthelmintic effects of prenylated isoflavonoids have been reported [48]. Neoflavonoids (represented e.g., by methyl dalbergin (**11**) and dalbergin (**12**)) were not included in the biological testing in our study but were previously shown to possess osteogenic properties [37] whereas structurally related dalbergiones from *D. melanoxylon* exhibited anti-inflammatory effects [49].

Table 4. Antibacterial activity of compounds from *D. melanoxylon* against human pathogens determined by agar diffusion assay (inhibition zone [mm]) and microdilution (MIC [µg/mL]).

Compounds	*B. subtilis* 6633 B1	*S. aureus* 511 B3	*E. coli* 458 B4	*P. aeruginosa* SG137 B7	*P. aeruginosa* K799/61 B9	*S. aureus* (MRSA) 134/93 R9		*E. faecalis* (VRE) 1528 R10		*M. vaccae* 10670 M4	
(1 mg/mL)	[mm]	[mm]	[mm]	[mm]	[mm]	[mm]	[µg/mL]	[mm]	[µg/mL]	[mm]	[µg/mL]
1	10	11	13P	0	0	0/A	n.t.	0	n.t.	15p	n.t.
2	11/13p	10	0	0/A	0	11	n.t.	11	n.t.	12	n.t.
7	23	23	14P	0	0	24	1.56	22F	1.56	27	0.78
9	24	25	15P	0	0/A	26	3.12	18/23p/F	6.25	27/33p	1.56
10	20/23P	20/23p-P	14P	0	0	21	25	17/25p/F	25	20/32p	3.12
CIP [a]	29 EK	18	23/31p	25	28/35p	0	12.5	16F	0.78	20p	0.2
DMSO [b]	11P	13P	12P	12P	12P	11P	>100	12p-P	100	11P	100

[a] CIP: Ciprofloxacin, positive control tested at concentration of 5 µg/mL; [b] negative control; p = partial inhibition (few colonies visible within inhibition zone), P = partial inhibition (many colonies visible within inhibition zone), F = facilitation, A = indication of inhibition, n.t. = not tested.

Prenylated flavonoids and isoflavonoids play important roles in the defense strategy of plants by protecting them against diseases through a broad inhibition profile against bacteria and fungi [50]. At the same time, these compounds represent promising starting points for the development of new, natural therapeutics against MRSA and other Gram-positive bacteria [51]. Increased hydrophobicity and bioavailability (mediated by one or two prenyl groups) and electrostatic interactions are the main determinants for the anti-MRSA activity of prenylated isoflavonoids [51]. The effects might be mediated by damaging the membrane or cell wall function [19] whereby interaction with bacterial membranes reduces the fluidity of outer and inner membrane layers [52]. Specifically, prenylation at C-8, as present in kenusanone H (**7**), seems to be connected to strong biological activity, and also hydroxylation at C-3 in **10** plays a role for several biological effects [52]. In contrast, introduction of a hydroxyl group in the prenyl chain as in compound **1** seems to be connected to a reduction of activity. However, prenyl substitution increases antibacterial but also cytotoxic properties [52]. For the most promising candidate, kenusanone H (**7**), the cytotoxicity against HeLa cells was determined with a CC_{50} of 1.8 ± 1.4 μg/mL (4.2 μM). Prenylated flavonoids and isoflavonoids show moderate cytotoxic properties [19,50] which would have to be considered for potential applications or development.

Altogether, fourteen compounds including six new isoflavanones were isolated from the root bark of *D. melanoxylon*, a medicinal plant largely used for the treatment of infectious diseases. The crude CH_2Cl_2 extract of the root bark of *D. melanoxylon* induced in a concentration dependent manner different degrees of inhibition against the tested microorganisms. Among the tested compounds, **7** and **9** showed strong activities against several pathogenic microbes, while compound **10** was selective towards *M. vaccae* 10670 M4. It is worth noting that compounds **7** and **9** showed superior activity against *S. aureus* (MRSA) 134/93 R9 compared to the reference drug ciprofloxacin. Despite these activities, neither the crude extract, nor the tested compounds showed considerable anthelminthic and cytotoxic activities. Hence, the observed biological effects support the traditional use of *D. melanoxylon* against several conditions, which appear to be connected to bacterial or fungal infections [6,25]. The prenylated isoflavanone constituents proved to be of relevant bioactivity and are likely responsible for the activity of the roots of this plant, suggesting future investigations in terms of structure-activity-relationship, mode of action and in vivo experiments.

Supplementary Materials: The following supporting information can be downloaded at: https://www.mdpi.com/article/10.3390/metabo13060678/s1, Figure S1: NMR, HRMS, UV and CD spectra of compound **1**; Figure S2: NMR, HRMS, UV and CD spectra of compound **2**; Figure S3: NMR and HRMS spectra of compound **3**; Figure S4: NMR and HRMS spectra of compound **4**; Figure S5: NMR and HRMS spectra of compound **5**; Figure S6: NMR and HRMS spectra of compound **6**; Figure S7: NMR, UV and CD spectra of compound **7**; Figure S8: NMR, UV and CD spectra of compound **8**; Figure S9: NMR, HRMS, UV and CD spectra of compound **9**; Figure S10: NMR, HRMS, UV and CD spectra of compound **10**; Figure S11: NMR and HRMS spectra of compound **11**; Figure S12: NMR and HRMS spectra of compound **12**; Figure S13: NMR and HRMS spectra of compound **13**; Figure S14: NMR and HRMS spectra of compound **14**; Table S1: NMR data of compound **1**; Table S2: NMR data of compound **2**; Table S3: NMR data of compound **3**; Table S4: NMR data of compound **4**; Table S5: NMR data of compound **5**; Table S6: NMR data of compound **6**; Table S7: NMR data of compound **7**; Table S8: NMR data of compound **8**; Table S9: NMR data of compound **9**; Table S10: NMR data of compound **10**; Table S11: NMR data of compound **11**; Table S12: NMR data of compound **12**; Table S13: NMR data of compound **13**; Table S14: NMR data of compound **14**; Table S15: Cytotoxic activities of crude extract of *D. melanoxylon* against human cancer cell lines; Table S16: Antifungal activity of compounds from *D. melanoxylon* against human pathogens.

Author Contributions: Conceptualization, D.M.C. and L.A.W.; validation, D.M.C., K.F. and A.Y.; investigation, D.M.C., K.F. and F.K.; resources, F.K., A.Y. and L.A.W.; data curation, D.M.C., K.F., V.-A.N.-N., F.K. and A.Y.; writing—original draft preparation, D.M.C., K.F. and V-A.N-N.; writing—review and editing, D.M.C., K.F., V.-A.N.-N., E.K., H.O.-O., J.N., F.K., A.Y. and L.A.W.; visualization, D.M.C., K.F., V.-A.N.-N. and F.K.; project administration, L.A.W.; supervision, K.F., E.K., H.O.-O., J.N., A.Y. and L.A.W.; funding acquisition, D.M.C., A.Y., F.K. and L.A.W. All authors have read and agreed to the published version of the manuscript.

Funding: This research was funded by the German Academic Exchange Services (DAAD) with a doctoral scholarship to D.M.C. which was offered through the Natural Products Research Network for Eastern and Central Africa (NAPRECA) (No. 91635457). Additional support was provided by the Federal Ministry of Education and Research (BMBF) of Germany within the program InfectControl 2020 (FKZ 03ZZ0835A and 03ZZ0805A for F.K.).

Institutional Review Board Statement: Not applicable.

Informed Consent Statement: Not applicable.

Data Availability Statement: Additional information related to this manuscript can be found in the supporting information. Further data are available on request. Data is not publicly available due to privacy.

Acknowledgments: We thank Christiane Weigel for antimicrobial testing and Hans-Martin Dahse for cytotoxicity assays (HKI). The authors are also thankful to Andrea Porzel and Gudrun Hahn for the spectroscopic analysis, Andrej Frolov for the HRESIMS analysis and Anke Dettmer, Martina Brode, Dube Mthandanzo and Martina Lerbs for the antibacterial, antifungal, anthelminthic and cytotoxic assays, respectively (all IPB). Furthermore, we would like to thank Patrick B. Chalo Mutiso (Herbarium, University of Nairobi) for identification of the species.

Conflicts of Interest: The authors declare no conflict of interest. The funders had no role in the design of the study; in the collection, analyses, or interpretation of data; in the writing of the manuscript; or in the decision to publish the results.

References

1. Beentje, H.J. *Kenya Trees, Shrubs, and Lianas*; National Museums of Kenya: Nairobi, Kenya, 1994.
2. Saha, S.; Shilpi, J.A.; Mondal, H.; Hossain, F.; Anisuzzman, M.; Hasan, M.M.; Cordell, G.A. Ethnomedicinal, phytochemical, and pharmacological profile of the genus *Dalbergia* L. (Fabaceae). *Phytopharmacology* **2013**, *4*, 291–346.
3. Vatanparast, M.; Klitgård, B.B.; Adema, F.A.C.B.; Pennington, R.T.; Yahara, T.; Kajita, T. First molecular phylogeny of the pantropical genus *Dalbergia*: Implications for infrageneric circumscription and biogeography. *S. Afr. J. Bot.* **2013**, *89*, 143–149. [CrossRef]
4. Zheng, L.; Huang, X.; Wang, L.; Chen, Z. Physicochemical properties, chemical composition, and antioxidant activity of *Dalbergia odorifera* T. Chen seed oil. *J. Am. Oil Chem. Soc.* **2012**, *89*, 883–890.
5. Chan, S.C.; Chang, Y.S.; Wang, J.P.; Chen, S.C.; Kuo, S.C. Three new flavonoids and antiallergic, anti-inflammatory constituents from the heartwood of *Dalbergia odorifera*. *Planta Med.* **1998**, *64*, 153–158. [CrossRef]
6. Gundidza, M.; Gaza, N. Antimicrobial activity of *Dalbergia melanoxylon* extracts. *J. Ethnopharmacol.* **1993**, *40*, 127–130. [CrossRef]
7. Beldjoudi, N.; Mambu, L.; Labaïed, M.; Grellier, P.; Ramanitrahasimbola, D.; Rasoanaivo, P.; Martin, M.T.; Frappier, F. Flavonoids from Dalbergia louvelii and their antiplasmodial activity. *J. Nat. Prod.* **2003**, *66*, 1447–1450. [CrossRef]
8. Ansari, M.A.; Razdan, R.K.; Tandon, M.; Vasudevan, P. Larvicidal and repellent actions of *Dalbergia sissoo* Roxb. (*F. leguminosae*) oil against mosquitoes. *Bioresour. Technol.* **2000**, *73*, 207–211. [CrossRef]
9. Kalaskar, M.G.; Divekar, V.B.; Chaugule, P.D.; Surana, S.J.; Baheti, D.G. Studies on anti-diarrheal activity of *Dalberjia sissoo* Roxb. in experimental animals. *Pharmacologyonline* **2010**, *1*, 453–457.
10. Hood, M.M.; Tembhurne, S.V.; Sakarkar, D.M. Anthelmintic activity of different extracts of *Dalbergia sissoo* Roxb. on Indian adult earthworms. *Pharma Chem.* **2011**, *3*, 142–146.
11. Choi, C.W.; Choi, Y.H.; Cha, M.R.; Kim, Y.S.; Yon, G.H.; Kim, Y.K.; Choi, S.U.; Kim, Y.H.; Ryu, S.Y. Antitumor components isolated from the heartwood extract of *Dalbergia odorifera*. *J. Korean Soc. Appl. Biol. Chem.* **2009**, *52*, 375–379. [CrossRef]
12. Niranjan, P.S.; Singh, D.; Prajapati, K.; Jain, S.K. Antidiabetic activity of ethanolic extract of *Dalbergia sissoo* L. leaves in alloxan-induced diabetic rats. *Int. J. Curr. Pharm. Res.* **2010**, *2*, 24–27.
13. Khan, I.A.; Avery, M.A.; Burandt, C.L.; Goins, D.K.; Mikell, J.R.; Nash, T.E.; Azadegan, A.; Walker, L.A. Antigiardial activity of isoflavones from *Dalbergia frutescens* bark. *J. Nat. Prod.* **2000**, *63*, 1414–1416. [CrossRef]
14. Vasudeva, N.; Vats, M.; Sharma, S.K.; Sardana, S. Chemistry and biological activities of the genus *Dalbergia*—A review. *Pharmacogn. Rev.* **2009**, *3*, 307–319.
15. Henderson, M.C.; Miranda, C.L.; Stevens, J.F.; Deinzer, M.L.; Buhler, D.R. In vitro inhibition of human P450 enzymes by prenylated flavonoids from hops, *Humulus lupulus*. *Xenobiotica* **2000**, *30*, 235–251. [CrossRef]
16. Mutai, P.; Heydenreich, M.; Thoithi, G.; Mugumbate, G.; Chibale, K.; Yenesew, A. 3-Hydroxyisoflavanones from the stem bark of *Dalbergia melanoxylon*: Isolation, antimycobacterial evaluation and molecular docking studies. *Phytochem. Lett.* **2013**, *6*, 671–675. [CrossRef]
17. Iinuma, M.; Ohyama, M.; Tanaka, T.; Mizuno, M.; Soon-Keun, H. Three 2′,4′,6′-trioxygenated flavanones in roots of *Echinosophora koreensis*. *Phytochemistry* **1992**, *31*, 665–669. [CrossRef]
18. Tanaka, T.; Iinuma, M.; Asai, F.; Ohyama, M.; Burandt, C.L. Flavonoids from the root and stem of *Sophora tomentosa*. *Phytochemistry* **1997**, *46*, 1431–1437. [CrossRef]
19. Sohn, H.Y.; Son, K.H.; Kwon, C.S.; Kwon, G.S.; Kang, S.S. Antimicrobial and cytotoxic activity of 18 prenylated flavonoids isolated from medicinal plants: *Morus alba* L., *Morus mongolica* Schneider, *Broussnetia papyrifera* (L.) Vent, *Sophora flavescens* Ait and *Echinosophora koreensis* Nakai. *Phytomedicine* **2004**, *11*, 666–672. [CrossRef]
20. Wessjohann, L.; Zakharova, S.; Schulze, D.; Kufka, J.; Weber, R.; Bräuer, L.; Brandt, W. Enzymatic C–C-coupling prenylation: Bioinformatics–modelling–mechanism–protein-redesign–biocatalytic application. *Chimia* **2009**, *63*, 340–344. [CrossRef]

21. Wessjohann, L.; Schreckenbach, H.F.; Kaluđerović, G.N. Enzymatic C-alkylation of aromatic compounds. In *Biocatalysis in Organic Synthesis (Science of Synthesis)*; Faber, K., Fessner, W.-D., Turner, N.J., Eds.; Georg Thieme Verlag: Stuttgart, Germany, 2015; Volume 2, pp. 177–211.
22. Simons, R.; Gruppen, H.; Bovee, T.F.; Verbruggen, M.A.; Vincken, J.P. Prenylated isoflavonoids from plants as selective estrogen receptor modulators (phytoSERMs). *Food. Funct.* **2012**, *3*, 810–827. [CrossRef]
23. Kareru, P.G.; Gachanja, A.N.; Keriko, J.M.; Kenji, G.M. Antimicrobial activity of some medicinal plants used by herbalists in Eastern province, Kenya. *Afr. J. Tradit. Complement. Altern. Med.* **2008**, *5*, 51–55. [CrossRef] [PubMed]
24. Kokwaro, J.O. *Medicinal Plants of East Africa*, 3rd ed.; University of Nairobi Press: Nairobi, Kenya, 2009; pp. 1–310.
25. Amri, E.; Juma, S. Evaluation of antimicrobial activity and qualitative phytochemical screening of solvent extracts of *Dalbergia melanoxylon* (Guill. & Perr.). *Int. J. Curr. Microbiol. App. Sci.* **2016**, *5*, 412–423.
26. Amri, E.; Lyaruu, H.V.M.; Nyomora, A.S.; Kanyeka, Z.L. Vegetative propagation of African blackwood (*Dalbergia melanoxylon* Guill. & Perr.): Effects of age of donor plant, IBA treatment and cutting position on rooting ability of stem cuttings. *New. For.* **2010**, *39*, 183–194.
27. Donnelly, B.J.; Donnelly, D.M.X.; O'Sullivan, A.M.; Prendergast, J.P. *Dalbergia* Species—VII: The isolation and structure of melanoxin a new dihydrobenzofuran from *Dalbergia melanoxylon* Guill. and Perr. (Leguminoseae). *Tetrahedron* **1969**, *25*, 4409–4414. [CrossRef]
28. Dervilla, M.X.D.; O'Reilly, J.; Whalley, W.B. Neoflavanoids of *Dalbergia melanoxylon*. *Phytochemistry* **1975**, *14*, 2287–2290.
29. Lin, S.; Liu, R.H.; Ma, G.Q.; Mei, D.Y.; Shao, F.; Chen, L.Y. Two new compounds from the heartwood of *Dalbergia melanoxylon*. *Nat. Prod. Res.* **2020**, *34*, 2794–2801. [CrossRef]
30. Liu, Y.; Shu, J.C.; Wang, M.F.; Xu, Z.J.; Yang, L.; Meng, X.W.; Chen, L.Y. Melanoxylonin AG, neoflavonoids from the heartwood of *Dalbergia melanoxylon* and their cardioprotective effects. *Phytochemistry* **2021**, *189*, 112845. [CrossRef]
31. dos Santos, C.H.C.; de Carvalho, M.G.; Franke, K.; Wessjohann, L. Dammarane-type triterpenoids from the stem of *Ziziphus glaziovii* Warm. (Rhamnaceae). *Phytochemistry* **2019**, *162*, 250–259. [CrossRef]
32. Otto, A.; Porzel, A.; Schmidt, J.; Brandt, W.; Wessjohann, L.; Arnold, N. Structure and absolute configuration of pseudohygrophorones A12 and B12, alkyl cyclohexenone derivatives from *Hygrophorus abieticola* (Basidiomycetes). *J. Nat. Prod.* **2016**, *79*, 74–80. [CrossRef]
33. Thomsen, H.; Reider, K.; Franke, K.; Wessjohann, L.A.; Keiser, J.; Dagne, E.; Arnold, N. Characterization of constituents and anthelmintic properties of *Hagenia abyssinica*. *Sci. Pharm.* **2012**, *80*, 433–446. [CrossRef]
34. Khan, M.F.; Nasr, F.A.; Noman, O.M.; Alyhya, N.A.; Ali, I.; Saoud, M.; Rennert, R.; Dube, M.; Hussain, W.; Green, I.R.; et al. Cichorins D-F: Three new compounds from *Cichorium intybus* and their biological effects. *Molecules* **2020**, *25*, 4160. [CrossRef]
35. Krieg, R.; Jortzik, E.; Goetz, A.A.; Blandin, S.; Wittlin, S.; Elhabiri, M.; Rahbari, M.; Nuryyeva, S.; Voigt, K.; Dahse, H.M.; et al. Arylmethylamino steroids as antiparasitic agents. *Nat. Commun.* **2017**, *8*, 14478–14489. [CrossRef]
36. Iinuma, M.; Ohyama, M.; Tanaka, T.; Mizuno, M.; Soon-Keun, H. Five flavonoid compounds from *Echinosophora koreensis*. *Phytochemistry* **1993**, *33*, 1241–1245. [CrossRef]
37. Kumar, P.; Kushwaha, P.; Khedgikar, V.; Gautam, J.; Choudhary, D.; Singh, D.; Trivedi, R.; Maurya, R. Neoflavonoids as potential osteogenic agents from *Dalbergia sissoo* heartwood. *Bioorganic Med. Chem. Lett.* **2014**, *24*, 2664–2668. [CrossRef]
38. Chan, S.C.; Chang, Y.S.; Kuo, S.C. Neoflavonoids from *Dalbergia odorifera*. *Phytochemistry* **1997**, *46*, 947–949. [CrossRef]
39. Herath, H.M.T.B.; deSilva, S. New constituents from *Gliricidia sepium*. *Fitoterapia* **2000**, *71*, 722–724. [CrossRef]
40. Fan, J.R.; Kuang, Y.; Dong, Z.Y.; Yi, Y.; Zhou, Y.X.; Li, B.; Qiao, X.; Ye, M. Prenylated phenolic compounds from the aerial parts of *Glycyrrhiza uralensis* as PTP1B and α-glucosidase inhibitors. *J. Nat. Prod.* **2020**, *83*, 814–824. [CrossRef]
41. Yang, X.D.; Xu, L.Z.; Yang, S.L. Xanthones from the stems of *Securidaca inappendiculata*. *Phytochemistry* **2001**, *58*, 1245–1249. [CrossRef]
42. Gurmessa, G.T.; Kusari, S.; Laatsch, H.; Bojase, G.; Tatolo, G.; Masesane, I.B.; Spiteller, M.; Majinda, R.R.T. Chemical constituents from the stem bark of *Erythrina brucei*. *Phytochem. Lett.* **2018**, *25*, 37–42. [CrossRef]
43. Slade, D.; Ferreira, D.; Marais, J.P.J. Circular dichroism, a powerful tool for the assessment of absolute configuration of flavonoids. *Phytochemistry* **2005**, *66*, 2177–2215. [CrossRef]
44. Bedane, K.G.; Kusari, S.; Bullach, A.; Masesane, I.B.; Mihigo, S.O.; Spiteller, M.; Majinda, R.R.T. Chemical constituents of the root bark of *Erythrina droogmansiana*. *Phytochem. Lett.* **2017**, *20*, 84–88. [CrossRef]
45. Kinoshita, T.; Ichinose, K.; Takahashi, C.; Wu, J.-B.; Sankawa, U. Chemical studies on *Sophora tomentosa*: The isolation of a new class of isoflavonoid. *Chem. Pharm. Bull.* **1990**, *38*, 2756–2759. [CrossRef]
46. Fukai, T.; Nomura, T. Structure of 6- or 8-isoprenoid substituted flavanone: Chemical shift of the hydrogen-bonded hydroxyl group. *Heterocycles* **1990**, *31*, 1861–1872.
47. Zhang, G.P.; Xiao, Z.Y.; Rafique, J.; Arfan, M.; Smith, P.J.; Lategan, C.A.; Hu, L.H. Antiplasmodial isoflavanones from the roots of *Sophora mollis*. *J. Nat. Prod.* **2009**, *72*, 1265–1268. [CrossRef] [PubMed]
48. Xiang, W.; Li, R.T.; Mao, Y.L.; Zhang, H.J.; Li, S.H.; Song, Q.S.; Sun, H.D. Four new prenylated isoflavonoids in *Tadehagi triquetrum*. *J. Agric. Food Chem.* **2005**, *53*, 267–271. [CrossRef] [PubMed]
49. Shao, F.; Panahipour, L.; Omerbasic, A.; Tangm, F.; Gruber, R. Dalbergiones lower the inflammatory response in oral cells in vitro. *Clin. Oral Investig.* **2022**, *26*, 5419–5428. [CrossRef]

50. Chen, X.; Mukwaya, E.; Wong, M.S.; Zhang, Y. A systematic review on biological activities of prenylated flavonoids. *Pharm. Biol.* **2014**, *52*, 655–660. [CrossRef]
51. Kalli, S.; Araya-Cloutier, C.; Hageman, J.; Vincken, J.-P. Insights into the molecular properties underlying antibacterial activity of prenylated (iso)flavonoids against MRSA. *Sci. Rep.* **2021**, *11*, 14180. [CrossRef]
52. Boozari, M.; Soltani, S.; Iranshahi, M. Biologically active prenylated flavonoids from the genus *Sophora* and their structure-activity relationship—A review. *Phytother. Res.* **2019**, *33*, 546–560. [CrossRef]

metabolites

Article

Antioxidant Activity and Metabolite Profiling of *Xylocarpus granatum* Extracts Using Gas Chromatography–Mass Spectrometry

Rudi Heryanto [1,2,3], Cecep Abdurohman Putra [1], Munawar Khalil [4], Mohamad Rafi [1,2,3], Sastia Prama Putri [5], Alfi Hudatul Karomah [1] and Irmanida Batubara [1,2,*]

1 Department of Chemistry, Faculty of Mathematics and Natural Sciences, IPB University, Jalan Tanjung Kampus IPB Dramaga, Bogor 16680, Indonesia
2 Tropical Biopharmaca Research Center, Institute of Research and Community Services, IPB University, Jalan Taman Kencana No. 3 Kampus IPB Taman Kencana, Bogor 16128, Indonesia
3 Advanced Research Laboratory, Institute of Research and Community Services, IPB University, Jalan Palem Raya Kampus IPB Dramaga, Bogor 16680, Indonesia
4 Department of Chemistry, Faculty of Mathematics and Natural Sciences, Universitas Indonesia, Depok 16424, Indonesia
5 Department of Biotechnology, Graduate School of Engineering, Osaka University, 2-1 Yamadaoka, Suita 565-0871, Osaka, Japan
* Correspondence: ime@apps.ipb.ac.id

Abstract: The potential application of *Xylocarpus granatum*, a mangrove species, as traditional medicine has been widely linked to its high secondary metabolite and antioxidant contents. However, few studies have been reported to identify and classify active metabolites responsible for such excellent biological activities. Therefore, the aim of this work was to determine the antioxidant activity, identify the metabolite profiles, and predict the metabolites acting as antioxidants in *X. granatum* extract using a gas chromatography–mass spectrometry (GC-MS)-based metabolomics approach. The seeds, stems, fruit peel, pulp, leaves, and twigs of *X. granatum* were macerated with ethanol. Each extract was analyzed with GC-MS, and the data were processed using mass spectrometry data-independent analysis (MS-DIAL) software to identify the metabolites. The IC_{50} value of plant parts of *X. granatum* ranged from 7.73 to 295 ppm. A total of 153 metabolites were identified and confirmed in the *X. granatum* extracts. Among the identified metabolites, epicatechin and epigallocatechin were the two most abundant in the stem extracts and are expected to have the greatest potential as antioxidants. Principal component analysis (PCA) succeeded in grouping all parts of the plant into three groups based on the composition of the metabolites: group 1 (stems, fruit peel, and twigs), group 2 (seeds and pulp), and group 3 (leaves).

Keywords: antioxidant; GC-MS; metabolite; PCA; *Xylocarpus granatum*

Citation: Heryanto, R.; Putra, C.A.; Khalil, M.; Rafi, M.; Putri, S.P.; Karomah, A.H.; Batubara, I. Antioxidant Activity and Metabolite Profiling of *Xylocarpus granatum* Extracts Using Gas Chromatography–Mass Spectrometry. *Metabolites* **2023**, *13*, 156. https://doi.org/10.3390/metabo13020156

Academic Editors: Ramona Paltinean and Irina Ielciu

Received: 24 December 2022
Revised: 14 January 2023
Accepted: 16 January 2023
Published: 20 January 2023

1. Introduction

Mangrove is a type of plant that grows in coastal areas and is spread across Indonesia. Extracts and raw materials from mangrove have been widely utilized by coastal people for traditional medicines [1]. One such mangrove species widely used as material for traditional medicine is *X. granatum*. This species is a sea mangrove from the Meliaceae family with a majority of plants spread across Southeast Asia and along the Indian Ocean. In Indonesia, this plant can be found in the Kalimantan and Sulawesi regions [2]. Coastal people still use parts of *X. granatum* plants as traditional medicine, owing to their wide range of biological activities.

Parts of *X. granatum* such as leaves, stems, twigs, and fruit have been reported to contain several secondary metabolites such as alkaloids, flavonoids, monoterpenes, triterpenoids, tetratriterpenoids, limonoids, proanthocyanidins, and phenolic acids [3,4]. These compounds have the potential to exert anticancer, antihyperglycemic, antidyslipidemic, antidepressant, and neuroprotective activities [5–7]. Additionally, the extracts of the seeds, stems, fruit peel, and leaves of *X. granatum* can be used as antioxidants and antidiabetics, owing to their polyphenol contents [2,8,9]. The antioxidant activities of *X. granatum* extracts make this species a potential source of one of raw materials for cosmetic skin lightening, which work by inhibiting tyrosinase activity [10]. The biological activities of *X. granatum*, e.g., antioxidant activity, are greatly influenced by its contents of active compounds, which is a crucial factor affecting the quality of *X. granatum* as an antioxidant [2].

One method to identify the contents of active compounds in *X. granatum* extracts is by analyzing the metabolite profiles. Metabolite profiling is one method associated with metabolomic approaches that can be used to comprehensively identify primary or secondary metabolites in plants, both qualitatively and quantitatively, and is generally associated with specific metabolite pathways [11,12]. Metabolite profiling facilitates the efficient activity determination and use of active compounds utilization and can be used as a plant quality control process [13].

Comprehensive metabolite profile identification from a complex sample requires a high-resolution analysis method, such as liquid chromatography—mass spectrometry (LC-MS), gas chromatography—mass spectrometry (GC-MS), liquid chromatography—mass spectrometry—mass spectrometry (LC-MS/MS), or capillary electrophoresis—mass spectrometry (CE-MS) [14]. GC-MS analysis is often used for metabolite profiling, owing to its high sensitivity and high resolution, in addition to providing good reproducibility. Another advantage of the GC-MS technique is that it is easy to use and relatively inexpensive in terms of operational cost [15].

In this study, the metabolites in several parts of *X. granatum* plants, i.e., leaves, stems, twigs, fruit peel, pulp, and seeds, were identified using GC-MS. The resulting data were processed with the assistance of MS-DIAL version 4.20 software to identify metabolites; additionally, the compounds acting as antioxidants were predicted based on a comparison of the obtained profile to that of known antioxidant compounds reported in the literature. The result of GC-MS data processing include information on *m/z* values, retention time, retention index, area, and the peak intensity of the identified metabolites. The result of data normalization, which performed with MS-DIAL, was evaluated by PCA (principal component analysis) with SIMCA version 13 software (Umetrics, Umea, Sweden). PCA was used to discriminate every part of *X. granatum* based on similarities in metabolite content.

2. Materials and Methods

2.1. Samples and Instruments

The plant sample used in this study was *X. granatum* (leaves, stems, twigs, fruit peel, pulp, and seeds) from Togean Islands, Central Sulawesi. All parts of the plant were dried in an oven at 40 °C and ground before extraction.

The following instruments and software were used: GC-MS QP2010 Ultra (Shimadzu, Kyoto, Japan), MS-DIAL version 4.24 software, SIMCA version 13 software (Umetrics, Umea, Sweden), Abf Converter, AMDIS, MORPHEUS, and ChemSketch. The following chemicals were used: N-methyl-N-(trimethylsilyl) trifluoroacetamide (MSTFA), pyridine, alkane mixture (C10-C31), ribitol (internal standard), methanol, chloroform, Milli-Q water, methoxyamine HCl, 2,2-diphenyl-1-picrylhydrazyl (DPPH), and 2-(N-morpholino)ethanesulfonic acid (MES) buffer.

2.2. Extraction

The extraction method used was maceration with ethanol. Samples from each plant part (1 g) were soaked in 5 mL ethanol for 24 h, then filtered. The filtrate was concentrated using a rotary evaporator. This extract was used for antioxidant activity tests.

2.3. Antioxidant Activity

The antioxidant activity of every plant part of *X. granatum* was determined by DPPH method with referent to the method performed by Batubara et al. (2010) [10]. Each extract was diluted in ethanol to final concentrations of 1.67, 3.33, 6.67, 10.00, 13.33,

16.67, 33.33, 66.67, 100.00, 133.33, and 166.67 μg/mL. The sample aliquot, 100 μL 2-(N-morpholino)ethanesulfonic acid (MES) buffer (pH 7.4), and 100 μL DPPH solution (11.8 mg DPPH in 100 mL ethanol) were added to each well of a 96-well plate. The mixtures were incubated for 30 min; then, the absorbances were read at 514 nm. Vitamin C was used as positive controlm and ethanol was used as the blank. Inhibition activity was calculated using the following formula:

$$\text{Inhibition } (\%) = [1 - (\text{Asample} - \text{Acontrol})/(\text{Ablank} - \text{Acontrol})] \times 100\%$$

where Asample is the sample absorbance, Acontrol is the vitamin C absorbance as the control, and Ablank is the ethanol absorbance as the blank. The concentration of each sample and a positive control were tested in triplicate.

2.4. Sample Preparation for GC-MS Analysis

Firstr, 1 mL methanol:chloroform:water (5:2:2) and 100 μg/mL ribitol as an internal standard were added to 10 mg of ground herbal material of every part of *X. granatum* (stems, leaves, pulp, fruit peel, twigs, and seeds). The mixture was homogenized by vortexing and incubated in a shaker for 30 min, then centrifuged at 12,298× *g* at 4 °C for 3 min. The supernatant (600 μL) was transferred into a new tube and mixed with 300 μL Milli-Q water, then recentrifuged under the same conditions described above. Subsequently, 400 μL (sample) and 200 μL (QC sample) of the supernatant were transferred into separate new tubes and dried under vacuum using a centrifugal concentrator at room temperature for 2 h, then dried with a freeze dryer overnight (12 h) before derivatization. The QC (quality control) sample was a mixture of all samples. After drying, derivatization was performed by adding 100 μL methoxyamine HCl in pyridine (20 mg/mL), followed by incubation at 30 °C for 90 min, the addition of 50 μL MSTFA, and reincubation at 37 °C for 30 min in order to induce sylilation before being injected into the GC-MS. Sample derivatization was performed to improve the volatility of the compound for GC-MS analysis.

2.5. GC-MS Analysis

The GC-MS analysis performed in this study is a standard procedure used to analyze natural products and was conducted at Fukusaki Lab, Osaka University, Japan [16]. GC-MS QP2010 Ultra (Shimadzu, Japan) with ab InertCap 5 MS/NP column was used, with an injection temperature at 230 °C and an injection volume of 1 μL in split mode (25:1 *v/v*). The carrier gas (He) flow was 3.0 mL/minute, with a linear speed of 39 cm/s. The column temperature was held at 80 °C for 2 min, then increased at a rate of 15 °C/min to 330 °C and held for 6 min. The temperature of the transfer line and the ion source were 250 and 200 °C, respectively. The ions were generated by electron ionization (EI) at 1 kV, and the spectrum was recorded in the mass range of 85–500 *m/z*.

Then, 1 μL pyridine was injected into the GC-MS to check the background, and 1 μL of an alkane compound mixture was used to obtain the n-alkane retention time, which was further used for retention index calculation. Afterwards, pyridine was injected to clean the remaining n-alkane compounds, and 1 μL of the sample blank was injected to check for any contamination. Then, the samples or extracts from every part of *X. granatum* were analyzed. Every sixth sample injection was followed by a QC sample, a sample blank, and pyridine injections in order to check for any contamination and clean the column. For every plant part of *X. granatum*, injection was performed in quintuplicate, with 2 blank injections and 10 QC sample injections.

2.6. Data Analysis with MS-DIAL

The resulting data from GC-MS analysis were obtained as .QGD files. Peak conformation, peak filtering, and annotation were processed using MS-DIAL ver. 4.20 (Riken, Kanagawa, Japan). The output data from GC-MS were converted into CDF files, which were subsequently converted into Abf files with Abf Converter, then imported to MS-DIAL software. Data processing steps and compound identification in MS-DIAL included inputting data as a new project, setting some parameters before analysis (data collection, peak detection, deconvolution, identification, alignment, and filtering), and metabolite identification.

Compound identification at the peak was performed by comparing the conformity of the retention time (RT), retention index (RI), and mass spectrum values from the results of analysis with those of known metabolites in the available library from AllPublic-KovatsRI-VS2. MS-DIAL software identified the metabolites according to the predetermined parameters of analysis. The resulting accuracy of compound identification was confirmed by examining the total score of each compound in the compound search.

After identification was completed, data normalization was performed before the data were used in PCA analysis. Data normalization was performed by determining the ID number of a compound used as an internal standard (ribitol); then, the ID number of the standard compound was input in the ID column of the identified compound.

2.7. Data Analysis

All determinations were conducted in triplicate, and the data are reported as average values ± standard deviation (SD).

PCA analysis was achieved using SIMCA version 13 software (Umetrics, Umea, Sweden). The data were exported from MS-DIAL as .txt files, which were then converted into Microsoft Excel files. The Microsoft Excel data were then transposed. PCA classification was expected to be able to group every plant part of *X. granatum* extract based on the metabolite content.

3. Results

3.1. X. granatum Antioxidant Activity

In this study, the antioxidant activity of every part of *X. granatum* was determined using the DPPH method. As a free radical scavenging method, DPPH, has been widely used to evaluate the antioxidant activity of plant extracts, owing to its rapidity, sensitivity, simplicity, and reproducibility. The results of antioxidant activity are reported as IC_{50}, indicating the number of antioxidants needed to lower 50% of the initial concentration of DPPH. Higher IC_{50} values indicate lower antioxidant activity.

The IC_{50} values of every part of *X. granatum* are presented in Table 1. The antioxidant activities of each part of *X. granatum* vary, with IC_{50} values ranging from 7.73 to 295 ppm. In this study, vitamin C was used as a standard, with an IC_{50} value of 4.18 ppm. Among all plant parts, the stem extract showed the highest antioxidant potential, with an IC_{50} value of 7.73 ppm (Table 1).

Table 1. Antioxidant activity of *X. granatum* ethanol extract.

Plant Part	IC_{50} (ppm)
Seeds	104.64 [e]
Fruit peel	9.02 [c]
Stems	7.73 [b]
Pulp	44.48 [d]
Leaves	295.08 [f]
Twigs	9.86 [c]
Vitamin C	4.18 [a]

The same letter indicates non-significant differences at the 95% confidence interval.

3.2. X. granatum Metabolite Profiling Using GC-MS

The results of profiling were obtained after confirming the metabolites and performing data curation. One metabolite identified in MS-DIAL was confirmed by conforming the retention index value and the mass spectrum of the compound identified with a known compound found in the AllPublic-KovatsRI-VS2 library. Data curation involves reducing the data on the compounds with the same retention time values and the same compounds with different retention time values. The compound with the highest total score value (750 to 1000) was selected for profiling. The total score shows indicates the similarity of the identified compound relative to a compound in the database.

Metabolite identification from every part of *X. granatum* resulted in 153 identified compounds (Table 2). Every part of *X. granatum* is primarily composed of simple sugar (monosaccharides), as reflected by the relative area. The sugar compounds contained in *X. granatum* are sucrose (12.02%), glucose (7.59%), fructose (7.58%), and epicatechin (2.00%), with a chromatogram pattern as shown in Figure 1.

Table 2. Metabolites in the extract of *X. granatum*.

No	Rt (minutes)	RI	Ion (*m*/*z*)	Formula	Compound	Area (%)
1	4.33	1003	117	$C_3H_8O_2$	propylene glycol	0.028
2	4.60	1028	174	C_3H_9N	n-propylamine	0.006
3	4.71	1038	117	$C_4H_{10}O_2$	butane-2.3-diol	1.084
4	4.73	1039	152	C_5H_5NO	2-hydroxypyridine	0.035
5	4.94	1058	147	$C_3H_8O_2$	propane-1.3-diol	0.006
6	4.95	1058	130	$C_3H_4O_3$	Pyruvic acid	0.047
7	5.02	1064	147	$C_3H_6O_3$	Lactic acid	0.125
8	5.07	1069	131	$C_4H_8O_3$	2-Hydroxyisobutyric acid	0.341
9	5.18	1078	147	$C_2H_4O_3$	Glycolic acid	0.356
10	5.39	1097	174	$C_4H_{11}N$	n-Butylamine	0.003
11	5.55	1111	131	$C_4H_8O_3$	2-hydroxybutanoic acid	0.003
12	5.87	1141	152	C_5H_5NO	4-Hydroxypyridine	0.007
13	5.93	1146	147	$C_4H_8O_3$	3-hydroxybutyric acid	0.317
14	5.94	1147	116	$C_3H_7NO_2$	Sarcosine_2TMS	0.007
15	5.95	1148	142	C_4H_7NO	butyrolactam	0.036
16	6.68	1216	131	$C_4H_8O_3$	3-Hydroxyisovaleric acid	0.005
17	6.76	1224	144	$C_5H_{11}NO_2$	Valine_2TMS	0.063
18	6.89	1237	189	CH_4N_2O	Urea	0.005
19	6.91	1238	116	$C_4H_8O_3$	4-hydroxybutyric acid	0.011
20	6.95	1242	174	C_2H_7NO	Ethanolamine	0.008
21	7.01	1249	175	$C_6H_{10}O_4$	ethylsuccinate	0.007
22	7.04	1252	179	$C_7H_6O_2$	Benzoic acid	0.017
23	7.07	1255	147	$C_3H_6O_3$	Dihydroxyacetone	0.013
24	7.17	1264	116	$C_3H_7NO_3$	Serine_2TMS	0.003
25	7.29	1276	174	C_2H_7NO	2-Aminoethanol	0.070
26	7.33	1280	158	$C_6H_{13}NO_2$	Leucine_2TMS	0.033
27	7.35	1282	147	$C_3H_8O_3$	Glycerol	0.880
28	7.50	1297	180	$C_6H_5NO_2$	Nicotinic acid	0.008
29	7.53	1300	158	$C_6H_{13}NO_2$	Isoleucine_2TMS	0.033
30	7.69	1317	147	$C_4H_6O_4$	Succinic acid	0.068
31	7.80	1328	254	$C_6H_6O_2$	Catechol	0.007
32	7.91	1340	147	$C_3H_6O_4$	Glyceric acid	0.075
33	7.97	1347	241	$C_4H_4N_2O_2$	Uracil	0.010
34	7.99	1349	254	$C_4H_4O_4$	Fumaric acid	0.027
35	8.14	1365	240	$C_5H_5NO_2$	pyrrole-2-carboxylic acid	0.013
36	8.21	1372	188	$C_3H_7NO_2$	Alanine_3TMS	0.020
37	8.25	1376	156	$C_6H_{11}NO_2$	DL-Pipecolic acid	0.158
38	8.32	1384	147	$C_4H_6O_4$	erythronic acid lactone	0.032
39	8.46	1398	218	$C_4H_9NO_3$	Threonine_3TMS	0.007
40	8.55	1408	239	$C_6H_6O_2$	hydroquinone	0.021
41	8.55	1409	147	$C_5H_8O_4$	Glutaric acid	0.069
42	8.61	1415	255	$C_5H_6N_2O_2$	Thymine	0.001
43	8.66	1422	103	$C_4H_9NO_3$	homoserine	0.016
44	8.83	1440	233	$C_4H_8O_4$	2-deoxytetronic acid	0.004
45	8.86	1443	174	$C_4H_9NO_2$	3-aminoisobutyric acid	0.080

Table 2. *Cont.*

No	Rt (minutes)	RI	Ion (*m*/*z*)	Formula	Compound	Area (%)
46	8.96	1456	117	$C_{10}H_{20}O_2$	Decanoic acid	0.004
47	9.17	1480	158	$C_5H_9NO_3$	trans-4-hydroxy-L-proline	0.012
48	9.23	1487	247	$C_5H_8O_5$	Citramalic acid	0.003
49	9.33	1498	147	$C_4H_6O_5$	Malic acid	0.751
50	9.34	1499	223	$C_7H_6O_2$	p-Hydroxybenzal dehyde	0.011
51	9.43	1509	244	$C_4H_8N_2O_3$	asparagine dehydrated	0.004
52	9.45	1512	117	$C_4H_8O_5$	isothreonic acid	0.022
53	9.48	1516	147	$C_4H_{10}O_4$	Threitol	0.031
54	9.53	1521	174	$C_4H_{12}N_2$	putrescine 3tms	0.018
55	9.54	1523	267	$C_9H_8O_4$	Acetylsalicylic acid	0.014
56	9.55	1524	217	$C_4H_{10}O_4$	Meso erythritol	0.109
57	9.59	1528	232	$C_4H_7NO_4$	Aspartic acid_3TMS	0.010
58	9.63	1534	156	$C_5H_7NO_3$	Pyroglutamic acid_2TMS	0.168
59	9.68	1539	230	$C_5H_9NO_3$	Hydroxyproline	0.187
60	9.70	1542	174	$C_4H_9NO_2$	4-Aminobutyric acid	0.455
61	9.81	1556	263	$C_6H_6O_4$	5-hydroxymethyl-2-furoic acid	0.002
62	9.84	1560	239	$C_6H_6O_3$	Pyrogallol	0.004
63	9.96	1574	267	$C_7H_6O_3$	3-Hydroxybenzoate	0.007
64	9.99	1577	147	$C_4H_8O_5$	Threonic acid	0.097
65	10.01	1580	179	$C_8H_{10}O_2$	4-Hydroxyphenethyl alcohol	0.012
66	10.04	1584	129	$C_5H_8O_5$	2-hydroxyglutaric acid	0.015
67	10.12	1594	147	$C_7H_{12}O_5$	2-Isopropylmalic acid	0.124
68	10.28	1614	147	$C_6H_{10}O_5$	3-Hydroxy-3-methylglutarate	0.032
69	10.31	1617	217	$C_5H_8O_5$	xylonolactone	0.009
70	10.35	1622	117	$C_4H_6O_6$	L-(+)-Tartaric acid	0.034
71	10.41	1630	246	$C_5H_9NO_4$	Glutamic acid_3TMS	0.007
72	10.46	1636	267	$C_7H_6O_3$	4-Hydroxybenzoic acid	0.094
73	10.50	1642	200	$C_6H_{11}NO_2$	Pipecolic acid	0.037
74	10.55	1648	179	$C_8H_8O_3$	4-Hydroxyphenylacetic acid	0.004
75	10.74	1672	103	$C_5H_{10}O_5$	Xylose	0.013
76	10.82	1683	103	$C_5H_{10}O_5$	Lyxose	0.273
77	10.84	1686	271	$C_6H_6O_4$	Kojic acid	0.285
78	10.98	1703	103	$C_5H_{10}O_5$	Ribose	0.095
79	11.14	1726	204	$C_6H_{10}O_5$	1.6-Anhydroglucose	0.087
80	11.20	1734	217	$C_5H_{12}O_5$	Xylitol	0.016
81	11.22	1736	219	$C_6H_{14}O_5$	diglycerol	0.005
82	11.29	1746	117	$C_6H_{12}O_5$	Rhamnose	0.252
83	11.30	1746	217	$C_5H_{12}O_5$	Arabitol	1.275
84	11.34	1752	117	$C_6H_{14}O_5$	6-deoxyglucitol	0.971
85	11.48	1771	147	$C_6H_{12}O_5$	2-Deoxy-D-glucose	0.065
86	11.51	1775	297	$C_8H_8O_4$	vanillic acid	0.020
87	11.52	1777	231	$C_6H_{10}O_5$	3.6-anhydro-D-galactose	0.021
88	11.60	1787	156	$C_5H_{10}N_2O_3$	Glutamine_3TMS	0.007
89	11.62	1790	129	$C_6H_{14}O_5$	3-deoxyhexitol	0.077
90	11.65	1794	147	$C_5H_{10}O_6$	Xylonic acid	0.056
91	11.84	1821	204	$C_7H_{10}O_5$	Shikimic acid	0.434
92	11.93	1834	193	$C_7H_6O_4$	3.4-Dihydroxybenzoate	0.267
93	11.98	1841	204	$C_6H_{12}O_5$	1.5-Anhydro-D-glucitol	0.067
94	12.26	1881	159	$C_6H_{13}NO_5$	D-(+)-Galactosamine	0.268
95	12.29	1886	103	$C_6H_{12}O_6$	Psicose	0.118
96	12.30	1887	103	$C_6H_{12}O_6$	Tagatose	0.084

Table 2. *Cont.*

No	Rt (minutes)	RI	Ion (m/z)	Formula	Compound	Area (%)
97	12.33	1891	345	$C_7H_{12}O_6$	Quinic acid	1.910
98	12.46	1910	104	$C_6H_{10}O_6$	L-Gulcono-1.4-lactone	3.503
99	12.50	1916	103	$C_6H_{12}O_6$	Fructose	6.983
100	12.53	1921	319	$C_6H_{12}O_6$	Mannose	0.179
101	12.56	1925	204	$C_7H_{14}O_6$	1-methylgalactose	0.121
102	12.57	1927	319	$C_6H_{12}O_6$	Galactose	0.328
103	12.58	1928	217	$C_6H_{10}O_6$	Glucono-1.5-lactone	0.998
104	12.62	1935	319	$C_6H_{12}O_6$	Glucose	7.042
105	12.71	1947	203	$C_6H_{13}NO_5$	Glucosamine	0.023
106	12.74	1953	293	$C_9H_8O_3$	p-Coumaric acid	0.035
107	12.78	1959	218	$C_9H_{11}NO_3$	Tyrosine	0.086
108	12.80	1962	275	$C_6H_{12}O_7$	galactonic acid	0.014
109	12.82	1963	204	$C_6H_{12}O_6$	hexose	1.079
110	12.84	1968	319	$C_6H_{14}O_6$	Mannitol	0.181
111	12.89	1975	281	$C_7H_6O_5$	Gallic acid	0.034
112	12.90	1977	217	$C_6H_{14}O_6$	Galactitol	0.025
113	13.02	1994	318	$C_6H_{10}O_5$	conduritol-beta-expoxide	2.131
114	13.10	2006	147	$C_6H_{14}O_6$	hexitol	0.099
115	13.31	2040	353	$C_5H_4N_4O_2$	Xanthine	0.006
116	13.33	2043	147	$C_6H_{12}O_7$	Gluconic acid	0.154
117	13.35	2046	117	$C_{16}H_{32}O_2$	palmitic acid	0.103
118	13.44	2059	204	$C_8H_{15}NO_6$	N-Acetyl-D-glucosamine	0.363
119	13.48	2067	147	$C_6H_{12}O_6$	myo-inositol	0.019
120	13.53	2073	129	$C_8H_{15}NO_6$	N-acetyl-D-mannosamine	0.007
121	13.73	2105	338	$C_{10}H_{10}O_4$	Ferulic acid	0.016
122	13.84	2124	204	$C_8H_{15}NO_6$	n-acetyl-d-hexosamine	0.114
123	13.89	2132	217	$C_6H_{12}O_6$	Inositol	1.004
124	14.00	2149	219	$C_9H_8O_4$	Caffeic acid	0.011
125	14.07	2160	319	$C_8H_{15}NO_6$	N-Acetyl galactosamine	0.035
126	14.30	2198	204	$C_9H_{16}O_9$	beta-mannosylglycerate	0.027
127	14.57	2245	117	$C_{18}H_{36}O_2$	Stearic acid	0.196
128	14.61	2251	202	$C_{11}H_{12}N_2O_2$	Tryptophan_3TMS	0.011
129	14.69	2266	204	$C_9H_{18}O_8$	glycerol-3-galactoside	0.021
130	14.79	2282	361	$C_{10}H_{17}NO_9S_2$	Sinigrin	0.004
131	15.12	2341	217	$C_{11}H_{15}N_5O_4$	1-methyladenosine	0.013
132	15.18	2352	204	$C_6H_{13}O_9P$	galactose-6-phosphate	0.056
134	15.85	2474	217	$C_9H_{12}N_2O_6$	Uridine_4TMS	0.010
135	16.19	2538	204	$C_{12}H_{22}O_{11}$	Leucrose	0.130
136	16.89	2674	361	$C_{12}H_{22}O_{11}$	beta-Lactose	0.015
137	17.03	2702	361	$C_{12}H_{22}O_{11}$	Sucrose	5.839
138	17.05	2706	204	$C_{12}H_{22}O_{11}$	D-(+)-Maltose	0.236
139	17.16	2729	217	$C_{12}H_{22}O_{11}$	D-(+)-Turanose	0.361
140	17.36	2770	217	$C_{12}H_{24}O_{11}$	lactitol	0.105
141	17.53	2806	217	$C_{12}H_{22}O_{12}$	lactobionic acid	0.047
142	17.56	2814	361	$C_{12}H_{22}O_{11}$	Trehalose	0.137
143	17.76	2855	355	$C_{21}H_{22}O_{10}$	4.5-dihydroxy-7-glucosyloxyflavanone	0.012
144	17.92	2889	361	$C_{12}H_{24}O_{11}$	Palatinitol	0.009
145	17.94	2893	368	$C_{15}H_{14}O_6$	(−)-Epicatechin	0.422
146	18.08	2925	361	$C_{12}H_{22}O_{11}$	Gentiobiose	0.033
147	18.14	2937	204	$C_{12}H_{22}O_{11}$	Melibiose	0.051
148	18.31	2975	456	$C_{15}H_{14}O_7$	(−)-Epigallo catechin	0.032
149	18.74	3072	204	$C_{12}H_{22}O_{11}$	Galactinol	0.021
150	18.87	3100	219	$C_{16}H_{18}O_9$	Chlorogenic acid	0.008
151	18.93	3109	487	$C_{15}H_{10}O_6$	Kaempferol	0.005
152	20.61	3349	204	$C_{18}H_{32}O_{16}$	Maltotriose	0.003
153	20.93	3396	361	$C_{18}H_{32}O_{16}$	Kestose	0.010

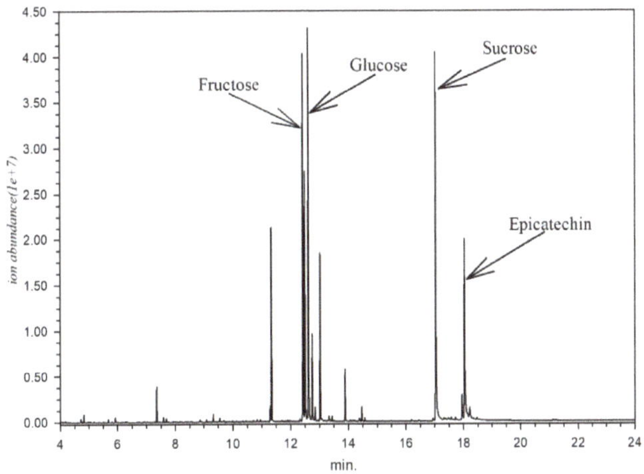

Figure 1. Chromatogram of *X. granatum* stem extract according to GC-MS analysis.

3.3. Prediction of Active Antioxidant Compounds in X. granatum

A total of 15 compounds with potential as antioxidants were identified in the various parts of *X. granatum*, as depicted in the form of a heat map in Figure 2. The heat map is intended to visualize the peak area percentage in a simpler way using the intensity of colors. In Figure 2, the intensity of red color indicates high compound content, whereas the intensity of green color shows indicates low compound content.

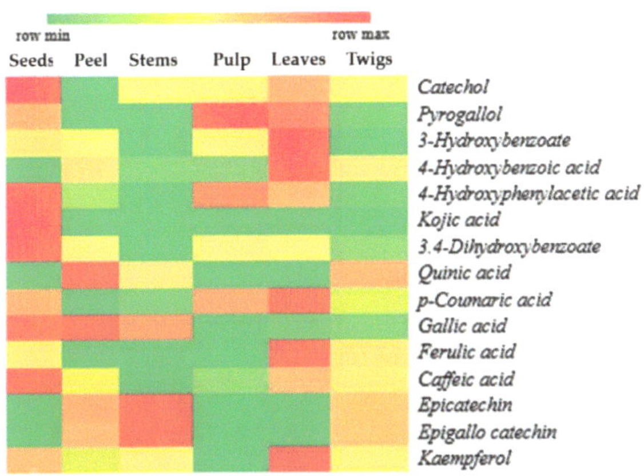

Figure 2. Distribution of antioxidant compounds in plant parts of *X. granatum*.

According to the results of antioxidant activity testing of *X. granatum* plant parts using the DPPH method, the ethanol extract of *X. granatum* stems has a very high potential compared to the other parts of the plant, as reflected by the lowest IC50 value (Table 1). This result indicates that the antioxidant compounds dominantly found in the stems have higher antioxidant activity compared to those found in the other parts of the plant. Figure 2 shows that the dominant compounds in the stems are epicatechin and epigallocatechin. Besides the stems, these two compounds can be found in the fruit peel and twigs, making the antioxidant activity in the fruit peel and twigs higher than that in the seeds, pulp, and leaves (Table 1).

3.4. *Discrimination of X. granatum Plant Parts with PCA Based on Metabolite Composition*

Discrimination is a process used to differentiate one sample from another by identifying similarities between samples. Samples with many similarities are grouped together; in other words, samples with many differences are separated into differerent groups [17]. The PCA technique is a method of analysis of double variables with the intention of simplifying the observed variables by reducing the dimensions to facilitate the visualization of data grouping and evaluation of similarities among the groups [18]. The application of PCA to the chromatogram enables a reduction in large-sized GC-MS data into several primary components (PCs) so that a two-dimension score plot can show severability among samples. The score plots of the first two components (PC1 and PC2) are usually used in the analysis because these PCs contain the most data variation.

The result of PCA analysis is considered acceptable if a small number of primary components can describe a large number of total variations. The result of PCA analysis was visualized as a score plot depicting the grouping of each plant part of *X. granatum* based on the metabolite composition contained in each plant part. Every point in the score plot represents a single sample, and samples with similarities are categorized in the same group [19]. Figure 3 shows the obtained score plot, which explains 91% of total variation (PC1 = 51%, PC = 40%). Samples with the same labels are grouped in adjacent positions. However, some samples did not provide a satisfactory grouping representation, so those data were considered outliers, such as teed 1, seed 5, and twig 3. Therefore, outlier reduction was performed in order to visualize a clearer and better data grouping.

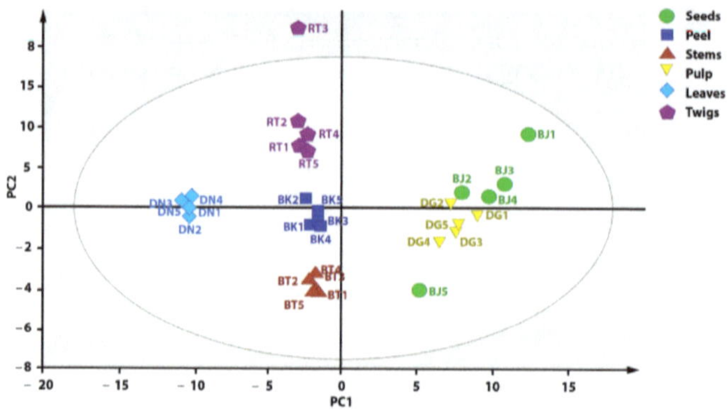

Figure 3. Score plot of PCA analysis before outlier reduction.

Outlier reduction resulted in the grouping presented in Figure 4, with an increase in the PC1 value of 1%. Figure 4 shows a plot score that can explain 92% of total variation (PC1 = 52%, PC = 40%). The result of score plot after outlier reduction shows that the metabolite profile of each *X. granatum* plant part can be differentiated into three groups based on metabolite composition, i.e., group 1 (stems, twigs, and fruit peel), group 2 (pulp and seeds), and group 3 (leaves). According to the PCA score plot (Figure 4), the metabolite characteristics of the fruit peel, stem, and twig parts of *X. granatum* are similar, as reflected by the antioxidant activities, with similar IC50 values: 9.02, 7.73, and 9.83 ppm, respectively (see, Table 2). The samples of the pulp and the seeds parts belong to one group, whereas the leaves are separated from the groups containing the other plant parts, showing that there are metabolite composition differences in the leaves. This is also supported by the antioxidant properties of the leaf extract, which has the highest IC50 value (295.08 ppm), showing that it has the lowest antioxidant activity. The association between the results of grouping and the antioxidant activity shows that the compounds with antioxidant activity play significant roles in the grouping of *X. granatum* plant parts.

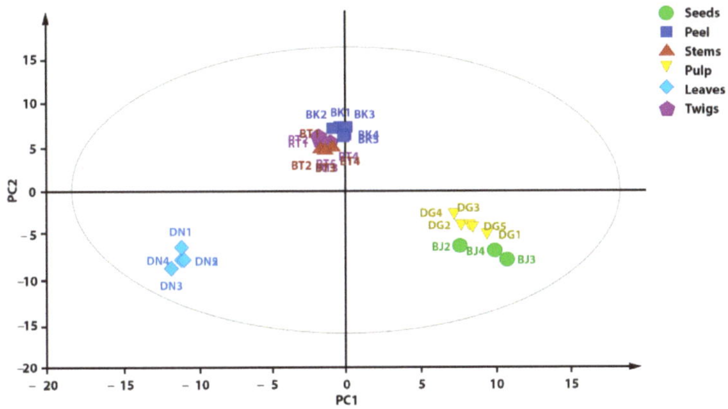

Figure 4. Distribution of antioxidant compounds in plant parts of *X. granatum*.

4. Discussion

Metabolite profiling is a technique of analysis used to determine the metabolite profile or the chromatogram pattern of chemical components of an extract with pharmacological activities or chemical components characterizing a plant with the objective of controling the quality [12]. Metabolite profiling and antioxidant compound prediction of *X. granatum* extracts consisted of four stages, i.e., information gathering through GC-MS data analysis using MSDIAL and antioxidant compound prediction through a literature study, data reduction and compilation to convert the spectrum into data that can be processed statistically, PCA multivariate analysis, and review and interpretation of the results of chemometric processes.

An antioxidant is an electron-donor compound that can lower the free radical level and help to reduce or prevent the impact of oxidative stress because of free radicals [20,21]. Common natural antioxidant compounds include vitamin C, vitamin E, carotenoids, phenolic compounds, and polyphenols, which can be grouped as flavonoids, cinnamate acid derivatives, coumarins, tocopherol, and polyfunctional organic acids, respectively [22]. Our literature review of several studies on the properties of antioxidant compounds, as well as the characteristics and applications thereof, indicated that natural metabolite compounds that are usually found in plants commonly include compounds in the group of polyphenols with active groups of hydroxy (-OH) and double-bonded carbons (-C=C-) and function as scavengers and inhibitors of free radicals reactions. Polyphenol secondary metabolites, such flavonoids, polyenes, and compounds containing many -OH groups, can react with free radicals as reducing agents, free radical scavengers, metal-chelating agents, and oxygen-singlet-forming suppressors [23–25].

According to our literature review of the properties and the general structures of antioxidant compounds, several compounds contained in *X. granatum* extracts have potential as antioxidants. These compounds mostly belong to the phenolics and polyphenols groups, including catechol [26], pyrogallol [24], 3-hydroxybenzoate, 4-hydroxybenzoic acid, 4-hydroxyphenylacetic acid, 3,4-dihydroxybenzoate [27], kojic acid [28], quinic acid [29], p-coumaric acid, gallic acid, ferulic acid, caffeic acid [30], epicatechin, epigallocatechin, and kaempferol [23].

Simple phenolic compounds such as catechol and pyrogallol are antioxidant compounds with the ability to lower the number of reactive oxygen species (ROS). Pyrogallol is a compound used widely to inhibit oxidation reactions in biodiesel [25]. On the other hand, a catechol-containing polyphenol, epicatechin, is a strong in vitro antioxidant compound, owing to its ability to rapidly lower the level of ROS, bind metallic ions, form inert complexes, and arrest the chain reaction of radical compounds [31].

Another phenolic compound identified as an antioxidant is phenolic acid. Phenolic acid is a strong antioxidant and exhibits antibacterial, antivirus, anticarcinogenic, and anti-inflammatory activities, as well as vasodilatory activity. Phenolic acid may further function as an anticancer agent and prevent heart diseases. Phenolic acid identified in *X. granatum* can be devided into two groups: benzoic acid derivatives and cinnamic acid derivatives. In this study, the identified compounds of benzoic acid derivatives were 3-hydroxybenzoate, 4-hydroxybenzoic acid, 4-hydroxyphenylacetic acid, and 3,4-dihydroxybenzoate, whereas the identified compounds of cinnamic acid derivatives were ferulic acid, gallic acid, p-coumaric acid, and caffeic acid. These compounds were reported to function as effective active antioxidants in radical scavenging of 2-azino-bis(3-ethylbenzthiazoline-6-sulfonic acid) (ABTS), 1,1-diphenyl-2-picryl-hydrazyl (DPPH), and superoxide anions, as well as metal chelation activity in iron ions [27,30,32]. Antioxidant activities of these phenolic acid compounds are influenced by the number of hydroxyl groups (OH-) in the phenyl ring. The length of conjugated double bonds, types of substituents, and the form of the molecules also contribute to the antioxidant activity [20].

In addition to the compounds in the phenolic group, other compounds were identified acting as antioxidants, such as quinic acid and kojic acid. Quinic acid is a carboxylic acid with many hydroxy groups (-OH), with potential as an antioxidant by inhibiting oral pathogens [29]. Kojic acid, on the other hand, is widely used in cosmetic products, especially as a skin-lightening agent [28,33].

5. Conclusions

The plant parts of *X. granatum* exhibit varying antioxidant activities. Metabolite profiling in *X. granatum* extracts using GC-MS succeeded in identifying 153 compounds. The compounds expected to have important roles in antioxidant activities are those compounds from the flavonol group, i.e., epicatechin and epigallocatechin, which are dominantly found in the stem of *X. granatum*. Multivariate analysis using PCA succeeded in grouping the plant parts of *X. granatum* into 3 groups based on metabolite composition: group 1 (stems, twigs, and fruit peel), group 2 (seeds and pulp), and group 3 (leaves). The grouping results of *X. granatum* plant parts using PCA can be associated with the antioxidant activities.

Author Contributions: Conceptualization, R.H., S.P.P., I.B.; methodology, R.H., C.A.P., M.K., M.R., S.P.P. and I.B.; software, R.H., C.A.P., M.R. and A.H.K.; validation, R.H., M.K., M.R. and S.P.P.; formal analysis, R.H. and C.A.P.; investigation, R.H. and I.B.; data curation, R.H. and C.A.P.; writing—original draft preparation, R.H., A.H.K. and I.B.; writing—review and editing, R.H., M.R., S.P.P., I.B. and M.K.; visualization, R.H.; project administration, I.B.; funding acquisition, I.B. All authors have read and agreed to the published version of the manuscript.

Funding: Directorate General of Strengten and Development Research Ministry of Research, Technology, and Higher Education, Republic of Indonesia No 3/E1/KP.PTNBH/2019.

Institutional Review Board Statement: Not applicable.

Informed Consent Statement: Not applicable.

Data Availability Statement: The data presented in this study are available in the main article.

Conflicts of Interest: The authors declare no conflict of interest.

References

1. Purnobasuki, H. Potensi mangrove sebagai tanaman obat. *J. Ilm. Ilmu-Ilmu Hayati* **2004**, *9*, 125–126. [CrossRef]
2. Zamani, N.P.; Ghazali, M.; Batubara, I. The study of tyrosinase and antioxidant activity of Xylocarpus granatum Koenig seed kernel extract toward evidence based indigenous knowledge from Togean Archipelago, Indonesia. *J. Mar. Sci: Res. Dev.* **2015**, *5*, 1–5. [CrossRef]
3. Das, S.K.; Samantaray, D.; Sahoo, S.K.; Pradhan, S.K.; Samanta, L.; Thatoi, H. Bioactivity guided isolation of antidiabetic and antioxidant compound from *Xylocarpus granatum* J. Koenig bark. *3 Biotech* **2019**, *9*, 198. [CrossRef] [PubMed]
4. Dey, D.; Quispe, C.; Hossain, R.; Jain, D.; Ahmed Khan, R.; Janmeda, P.; Islam, M.T.; Ansar Rasul Suleria, H.; Martorell, M.; Daştan, S.D.; et al. Ethnomedicinal use, phytochemistry, and pharmacology of *Xylocarpus granatum* J. Koenig. *Evid. Based Complement Altern. Med.* **2021**, *2021*, 8922196. [CrossRef]

5. Gao, Q.; Gao, Y.; Song, H.; Li, J.; Wu, Y.; Shi, X.; Shi, H.; Ma, Y. Cipadesin A, a bioactive ingredient of *Xylocarpus granatum*, produces antidepressant-like effects in adult mice. *Neurosci. Lett.* **2016**, *633*, 33–39. [CrossRef]
6. Pejin, B.; Glumac, M. New cytotoxic natural products from the mangrove biome: Covering the period 2007-2015. *J. Nat. Prod. Res.* **2018**, *33*, 1624–1628. [CrossRef]
7. Zhou, Z.F.; Kurtán, T.; Mándi, A.; Gu, Y.C.; Yao, L.G.; Xin, G.R.; Li, X.W.; Guo, Y.W. Novel and neuroprotective tetranortriterpenoids from Chinese mangrove *Xylocarpus granatum* Koenig. *Sci. Rep.* **2016**, *6*, 33908. [CrossRef]
8. Das, S.K.; Samanta, L.; Thatoi, H. In vitro antidiabetic and antioxidant potentials of leaf and stem bark extracts of a mangrove plant, *Xylocarpus granatum*. *J. Herbs Spices Med. Plants* **2016**, *22*, 105–117. [CrossRef]
9. Liao, M.; Pedpradab, P.; Wun, J. Thaixylograins A-H: Eight new limonoids from the thai mangrove, *Xylocarpus granatum*. *Phytochem. Lett.* **2017**, *19*, 126–131. [CrossRef]
10. Batubara, I.; Darusman, L.K.; Mitsunaga, T.; Rahminiwati, M.; Djauhari, E. Potency of Indonesian medicinal plants as tyrosinase inhibitor and antioxidant agent. *J. Bio. Sci.* **2010**, *10*, 138–144. [CrossRef]
11. Celano, R.; Docimo, T.; Piccinelli, A.L.; Rizzo, S.; Campone, L.; Di Sanzo, R.; Carabetta, S.; Rastrelli, L.; Russo, M. Specialized metabolite profiling of different *Glycyrrhiza glabra* organs by untargeted UHPLC-HRMS. *Ind. Crop. Prod.* **2021**, *170*, 113688. [CrossRef]
12. Feng, X.; Zhang, W.; Wu, W.; Bai, R.; Kuang, S.; Shi, B.; Li, D. Chemical composition and diversity of the essential oils of *Juniperus rigida* along the elevations in Helan and Changbai Mountains and correlation with the soil characteristics. *Ind. Crop. Prod.* **2021**, *159*, 113032. [CrossRef]
13. Dadwal, V.; Joshi, R.; Gupta, M. A comparative metabolomic investigation in fruit sections of *Citrus medica* L. and *Citrus maxima* L. detecting potential bioactive metabolites using UHPLC-QTOF-IMS. *Food Res. Int.* **2022**, *157*, 111486. [CrossRef] [PubMed]
14. Kang, M.J.; Suh, J.H. Metabolomics as a tool to evaluate nut quality and safety. *Trends Food Sci. Technol.* **2022**, *129*, 528–543. [CrossRef]
15. Lebanov, L.; Ghiasvand, A.; Paull, B. Data handling and data analysis in metabolomic studies of essential oils using GC-MS. *J. Chromatogr. A* **2021**, *1640*, 461896. [CrossRef]
16. Putri, S.P.; Fukusaki, E. *Mass Spectrometry-Based Metabolomics: A Practical Guide*; CRC Press: Boca Raton, FL, USA, 2014.
17. Niu, X.; Mi, S.; Jin, Q.; Sang, Y.; Wang, X. Characterization and discrimination of two varieties of eggplants using multi-element and metabolomics profiles coupled with chemometrics analysis. *Food Res. Int.* **2022**, *162*, 111976. [CrossRef]
18. Bro, L.; Smilde, A.K. Principal component analysis. *Anal. Methods* **2014**, *6*, 2812–2831. [CrossRef]
19. Feizi, N.; Hashemi-Nasab, F.S.; Golpelichi, F.; Saburouh, N.; Parastar, H. Recent trends in application of chemometric methods for GC-MS and GC×GC-MS-based metabolomic studies. *TrAC Trends Anal. Chem.* **2021**, *138*, 116239. [CrossRef]
20. Zduńska, K.; Dana, A.; Kolodziejczak, A. Antioxidant properties of ferulic acid and its possible application. *J. Skin Pharm. Physiol.* **2018**, *31*, 332–336. [CrossRef]
21. Singh, D.P.; Verma, S.; Prabha, R. Investigation on antioxidant potential of phenolic acid and flavonoid: The common phytochemical ingredients in plant. *J. Plant Biochem. Physiol.* **2018**, *6*, 219. [CrossRef]
22. Rahmawati, S.I.; Izzati, F.N.; Hapsari, Y.; Septiana, E.; Rachman, F.; Bustanussalam; Simanjuntak, P. Endophytic microbes and antioxidant activities of secondary metabolites from mangroves *Avicennia marina* and *Xylocarpus granatum*. *IOP Conf. Ser. Earth Env. Sci.* **2019**, *278*, 012065. [CrossRef]
23. Caban, M.; Lewandowska, U. Polyphenols and the potential mechanisms of their therapeutic benefits against inflammatory bowel diseases. *J. Funct. Foods* **2022**, *95*, 105181. [CrossRef]
24. Procházková, D.; Boušová, I.; Wilhelmová, N. Antioxidant and prooxidant properties of flavonoids. *Fitoterapia* **2011**, *82*, 513–523. [CrossRef] [PubMed]
25. Yan, Z.; Zhong, Y.; Duan, Y.; Chen, Q.; Li, F. Antioxidant mechanism of tea polyphenols and its impact on health benefits. *Anim. Nutr.* **2020**, *6*, 115–123. [CrossRef] [PubMed]
26. Li, H.; Lin, L.; Feng, Y.; Zhao, M.; Li, X.; Zhu, Q.; Xiao, Z. Enrichment of antioxidants from soy sauce using macroporous resin and identification of 4-ethylguaiacol, catechol, daidzein, and 4-ethylphenol as key small molecule antioxidants in soy sauce. *Food Chem.* **2018**, *240*, 885–892. [CrossRef] [PubMed]
27. Hang, D.T.N.; Hoa, N.T.; Bich, H.N.; Mechler, A.; Vo, Q.V. The hydroperoxyl radical scavenging activity of natural hydroxybenzoic acids in oil and aqueous environments: Insights into the mechanism and kinetics. *Phytochemistry* **2022**, *201*, 113281. [CrossRef]
28. Saeedi, M.; Eslamifar, M.; Khezri, K. Kojic acid application in cosmetic and pharmaceutical preparations. *J. Biomed Pharmacother* **2019**, *110*, 582–593. [CrossRef]
29. Nicolì, F.; Negro, C.; Vergine, M.; Aprile, A.; Nutricati, E.; Sabella, E.; Miceli, A.; Luvisi, A.; De Bellis, L. Evaluation of phytochemical and antioxidant properties of 15 Italian *Olea europaea* L. cultivar leaves. *Molecules* **2019**, *24*, 1998. [CrossRef]
30. Adeyemi, O.S.; Atolani, O.; Banerjee, P.; Arolasafe, G.; Preissner, R.; Etukudoh, P.; Ibraheem, O. Computational and experimental validation of antioxidant properties of synthesized biactive ferulic acid derivatives. *Int. J. Food Prop.* **2018**, *21*, 101. [CrossRef]
31. Prakash, M.; Basavaraj, B.V.; Chidambara Murthy, K.N. Biological functions of epicatechin: Plant cell to human cell health. *J. Funct. Foods* **2019**, *52*, 14–24. [CrossRef]

32. Spagnol, C.M.; de Assis, R.P.; Brunetti, I.L.; Isaac, V.L.B.; Salgado, H.R.N.; Corrêa, M.A. In vitro methods to determine the antioxidant activity of caffeic acid. *Spectrochim. Acta Part A Mol. Biomol. Spectrosc.* **2019**, *219*, 358–366. [CrossRef] [PubMed]
33. Phasha, V.; Senabe, J.; Ndzotoyi, P.; Okole, B.; Fouche, G.; Chuturgoon, A. Review on the use of kojic acid—A skin-lightening ingredient. *Cosmetics* **2022**, *9*, 64. [CrossRef]

metabolites

Article

Identification and Analysis of Antimicrobial Activities from a Model Moss *Ceratodon purpureus*

Ashley L. Dague [1], Lia R. Valeeva [2,3], Natalie M. McCann [2], Margarita R. Sharipova [3], Monica A. Valentovic [1], Lydia M. Bogomolnaya [1,*] and Eugene V. Shakirov [1,2,*]

1 Department of Biomedical Sciences, Joan C. Edwards School of Medicine, Marshall University, Huntington, WV 25755, USA
2 Department of Biological Sciences, College of Science, Marshall University, Huntington, WV 25701, USA
3 Institute of Fundamental Medicine and Biology, Kazan Federal University, Kazan 420008, Russia
* Correspondence: bogomolnaya@marshall.edu (L.M.B.); shakirov@marshall.edu (E.V.S.)

Abstract: The emergence of bacterial drug resistance is often viewed as the next great health crisis of our time. While more antimicrobial agents are urgently needed, very few new antibiotics are currently in the production pipeline. Here, we aim to identify and characterize novel antimicrobial natural products from a model dioicous moss, *Ceratodon purpureus*. We collected secreted moss exudate fractions from two *C. purpureus* strains, male R40 and female GG1. Exudates from the female *C. purpureus* strain GG1 did not exhibit inhibitory activity against any tested bacteria. However, exudates from the male moss strain R40 exhibited strong inhibitory properties against several species of Gram-positive bacteria, including *Staphylococcus aureus* and *Enterococcus faecium*, though they did not inhibit the growth of Gram-negative bacteria. Antibacterial activity levels in *C. purpureus* R40 exudates significantly increased over four weeks of moss cultivation in liquid culture. Size fractionation experiments indicated that the secreted bioactive compounds have a relatively low molecular weight of less than 1 kDa. Additionally, the R40 exudate compounds are thermostable and not sensitive to proteinase K treatment. Overall, our results suggest that the bioactive compounds present in *C. purpureus* R40 exudates can potentially add new options for treating infections caused by antibiotic-resistant Gram-positive bacteria.

Keywords: fire moss; bryophyte; antibacterial activity; exudate; moss; plant metabolite

check for updates

Citation: Dague, A.L.; Valeeva, L.R.; McCann, N.M.; Sharipova, M.R.; Valentovic, M.A.; Bogomolnaya, L.M.; Shakirov, E.V. Identification and Analysis of Antimicrobial Activities from a Model Moss *Ceratodon purpureus*. *Metabolites* **2023**, *13*, 350. https://doi.org/10.3390/metabo13030350

Academic Editors: Ramona Paltinean and Irina Ielciu

Received: 10 February 2023
Revised: 21 February 2023
Accepted: 24 February 2023
Published: 27 February 2023

1. Introduction

Bacterial infections lead to a number of human diseases and represent a leading cause of morbidity and mortality worldwide. A recent report by Antimicrobial Resistance Collaborators [1] indicated that the 33 most common bacterial pathogens caused up to 13.6% of all deaths in the world and 56.2% of all sepsis-related deaths in 2019. Among all bacterial pathogens, the Gram-positive *Staphylococcus aureus* was the leading bacterial cause of death in 135 countries and was also associated with the most deaths in individuals older than 15 years. Another Gram-positive pathogen, *Streptococcus pneumoniae*, was the bacterial species associated with the most deaths in children younger than 5 years. Overall, deaths associated with pathogenic bacteria ranked as the second leading cause of death globally in 2019 [1].

One of the main reasons bacterial infections continue to lead to high mortality throughout the world is the rapidly growing antibiotic resistance among many pathogens [2,3]. The emergence and spread of antibiotic resistance among pathogenic bacteria is a disturbing trend recently identified by the World Health Organization (WHO) as one of the ten biggest health challenges of our time (http://sdg.iisd.org/news/who-identifies-top-health-challenges-begins-five-year-health-plan/) (accessed on 30 January 2023). In its 2019 report, the Centers of Disease Control and Prevention categorized different antibiotic-resistant bacteria as urgent, serious, and concerning pathogens (https://www.cdc.gov/drugresistance/

biggest-threats.html) (accessed on 30 January 2023). Among these pathogens, several multidrug-resistant (MDR) Gram-positive bacteria are considered major healthcare problems [4]. These include methicillin-resistant *Staphylococcus aureus* (MRSA), which in 2017 caused 323,700 estimated cases in hospitalized patients and 10,600 estimated deaths, with USD 1.7 Billion estimated attributable healthcare costs. Also on the rise are infections caused by vancomycin-resistant *Enterococcus faecium* (VRE), which caused 54,500 estimated cases in hospitalized patients and 5400 estimated deaths in 2017. Another Gram-positive pathogen of concern is erythromycin-resistant Group A Streptococcus (GAS), which caused approximately 1 to 2.6 million cases of strep throat and invasive infections (cellulitis, pneumonia, flesh-eating infections, and sepsis), and up to 1900 deaths. These and other pathogens are becoming increasingly resistant to available antibiotics, raising concerns that the last remaining drugs to treat Gram-positive bacterial infections may become less effective. Thus, the rise of antimicrobial resistance demands increased efforts to discover new antibiotics.

Antibacterial drug development based on previously characterized chemical scaffolds is arguably reaching its technical limitations [5]. On the other hand, natural products produced by previously largely untapped groups of organisms, such as plants, could be a powerful source of new antimicrobials with potentially novel chemical structures [6]. Specifically, plant-derived natural antimicrobials have been historically used in traditional folk medicine, but not very commonly utilized by the modern pharmaceutical industry. Indeed, plants are known to produce a variety of potent secondary metabolites with important medical benefits, including well-characterized compounds such as acetylsalicylic acid, artemisinin, and paclitaxel [7]. In particular, members of the Bryophyte group (mosses, liverworts, and hornworts) are known to produce a plethora of biologically active secondary metabolites [8]. To date, a number of different natural compounds with antimicrobial, antioxidant, and ultraviolet-screening properties have been detected in bryophytes [9–11]. Specifically, metabolites from several mosses were shown to effectively inhibit the growth of Gram-negative or Gram-positive bacteria [12,13]. However, detailed studies aimed at chemical characterization of antibacterial secondary metabolites from Bryophytes are still lacking, especially in model moss species.

Even less is known about antibacterial properties of extracellular metabolites secreted by plants, many of which play a crucial role in communication with the root microbiota, including important herbivores and bacterial pathogens [14–16]. The model liverwort *Marchantia polymorpha*, for example, secretes specialized "oil bodies" containing sesquiterpenes and cyclic bisbibenzyls, which serve as a defense against arthropod herbivores [17,18]. The model moss *Physcomitrium patens* secretes a number of extracellular compounds and peptides with antibacterial properties [19,20]. For potential future biotechnological applications, such secreted bioactive metabolites may offer several technical advantages over intracellular compounds, as they are usually water-soluble and can be easier to biochemically isolate and purify.

Ceratodon purpureus (also known as the "fire moss") is a biogeographically diverse model moss species [21], known for its ability to grow on soils contaminated with heavy metals, in areas of high UV radiation, under salt and cold stress, and even in extremely cold conditions of Antarctica. *C. purpureus* is dioicous and harbors gene-rich UV sex chromosomes, enabling researchers to study how the presence of sex-linked genes affects plant sexual development and metabolism [22,23]. Recent data point to intriguing differences between male and female strains of *C. purpureus* in the amount and composition of produced volatile organic compounds (VOCs) that can affect the behavior of soil microarthropods and attract them to aid in moss fertilization processes [23,24]. Overall, due to its small size, ease of laboratory cultivation, sequenced and annotated genome, and rapidly developing metabolomics resources, *C. purpureus* is currently becoming a model organism of choice for many types of plant research [22,25,26].

Here, we aim to discover and characterize antibacterial metabolites secreted by the model moss *C. purpureus*. We detected an unexpected sexual dimorphism in the production of *C. purpureus* antibacterial metabolites: exudates from the male moss strain R40 contain potent antibacterial activity, while exudates from the female strain GG1 do not. Interestingly, R40 exudates are effective against several species of Gram-positive bacteria, but do not inhibit the growth of Gram-negative bacterial species. The mode of action of R40 exudate components appears to be bactericidal with a minimum inhibitory concentration of 6.25 mg/mL. Exudate stability and light sensitivity experiments indicate that secreted metabolites are relatively stable after boiling or multiple freezing/thawing cycles, in different light conditions and after treatment with Proteinase K, suggesting that the bioactive metabolites likely do not have a polypeptide structure. Furthermore, size fractionation experiments indicate that the metabolite with antibacterial activity has a low molecular weight of below 1 kDa. Taken together, our data suggest that *C. purpureus* antibacterial compounds represent secreted small metabolites with a bactericidal mode of action against Gram-positive bacteria. Overall, further characterization of bioactive *C. purpureus* compounds may lead to the development of novel antibacterial therapeutics.

2. Materials and Methods

2.1. Moss Strains, Growth Conditions and Exudate Collection

C. purpureus R40 (male) and GG1 (female) strains (a gift from Dr. Stuart McDaniel, University of Florida) were propagated on Petri plates with BCD nutrient agar medium as described before [20,27]. Moss was propagated weekly by homogenizing tissue with the IKA Ultra-Turrax T10 basic tissue dispenser, followed by plating the homogenized samples onto cellophane disks on top of solid BCD nutrient medium in Petri dishes. Moss plates were grown in a plant growth chamber (Model 7300, Caron Products) at 22 °C, 65% humidity, 880 lux light intensity, and 12/12 h light/dark conditions.

For secreted metabolite analysis, *C. purpureus* strains were grown in 250 mL flasks with 100 mL of the BCD liquid nutrient medium on the orbital shaker (at 150 rpm) at 22 °C, 65% humidity, 880 lux light intensity, and 12/12 h light/dark conditions. Exudate was collected and processed as described previously [20]. Frozen exudate samples were dried using a lyophilizer (model 7382021, Labconco, Kansas City, MO, USA) and stored at −80 °C until needed. For experimental analysis, dry samples were weighed and dissolved in sterile BCD medium with $CaCl_2$ in the final concentration of 100 mg/mL.

2.2. Tests for Antibacterial Activity

Antimicrobial activity of moss exudates was analyzed against Gram-positive (*Staphylococcus aureus* ATCC 25923, *Streptococcus pyogenes* ATCC 12344, and *Enterococcus faecium* ATCC 35667) and Gram-negative bacteria (*Serratia marcescens* SM6, *Salmonella enterica ser.* Typhimurium ATCC14028s). Gram-positive bacterial strains were grown in Tryptic Soy medium and Gram-negative bacterial strains were grown in LB medium.

Disk-Diffusion assays (DDM). Bacterial growth inhibition was tested first using the qualitative DDM method in LB or Tryptic Soy agar according to CLSI guidelines (www.clsi.org, accessed on 1 July 2022). Bacteria in liquid cultures were grown overnight (ON) at 35 °C and shaken at 200 rpm using an orbital incubator shaker (Excella E24 Incubator Shaker Series, New Brunswick Scientific). The bacterial inoculum (CFU (colony-forming units) = 1×10^7/plate) was prepared by dilution of 25 µL of ON culture in 5 mL of top agar (LB broth powder 25 g/L, 0.7% agar or Tryptic Soy broth powder 30 g/L, 0.7% agar), stirred and poured onto a Petri dish containing 20 mL of LB or Tryptic Soy agar. Next, 17.5 mg of moss exudate was dissolved in BCD medium and added onto sterile Whatman disks (disk diameter = 7 mm). Disks soaked with the same amount of sterile liquid BCD medium were used as the negative control and disks soaked with carbenicillin or chloramphenicol antibiotics were used as the positive control, as described previously [20]. Disks were placed on top of inoculated plates and incubated at 35 °C for 18 h. The diameter of the bacterial growth inhibition zone (halo) around each disk was then measured in mm and plotted. All experiments were carried out in triplicate on at least two separate occasions.

A broth microdilution method was used as a quantitative assay to determine the Minimal Inhibitory Concentration (MIC) of the moss exudate metabolites in a 96-well microtiter plate using the Synergy HTX BioTek plate reader spectrophotometer as described previously [20,28]. Moss exudates were added to bacterial cultures in twofold serial dilutions (100, 50, 25, 12.5, 6.25 mg/mL or less, depending on the experiment). The 96-well plate was sealed with a Breathe-Easy membrane (Diversified Biotech) to minimize evaporation, and the plate was incubated and continuously shaken at 35 °C for 18 h, with the optical density of the culture measured at 600 nm (OD600) every 15 min. MH broth with bacteria in the absence of exudates was used as the negative control, while a range of carbenicillin and chloramphenicol dilutions were used as the positive controls. MIC was defined as the lowest concentration of an antimicrobial agent that inhibits the visible growth of a bacterial culture. Each experiment was performed in four biological and three technical replicates.

2.3. *Evaluation of Minimal Bactericidal Concentration*

To determine MBC, bacterial cultures were prepared for MIC testing as described previously [20,28], and the inoculum aliquot was serially diluted and plated for CFU determination. The remaining bacterial cultures were grown overnight in the presence of different exudate concentrations in a 96-well microtiter plate using the Synergy HTX BioTek plate reader as described above for the MIC assays. After completion of this step, 50 µL of bacterial culture from each well (starting at the MIC concentration and higher) was plated onto LB plates and incubated overnight at 35 °C for 18 h. The number of growing bacterial colonies on each plate were counted to determine the CFU. The MBC was defined as the lowest concentration that demonstrates a 99.9% reduction in CFU compared to the inoculum [29].

2.4. *Size Fractionation of Extracellular Metabolites*

The Pall Corporation Macrosep Advance Centrifugal Devices (10 kDa, 3 kDa, and 1 kDa molecular weight cutoffs) were used to separate metabolite exudate components by molecular weight. Fractionation was performed following manufacturer instructions at 4 °C. Fractions of <1 kDa, 1–3 kDa, 3–10 kDa, and >10 kDa were collected, dried down, and analyzed in DDM or MIC assays.

2.5. *Exudate Metabolite Stability and Sensitivity Tests*

To determine light sensitivity, crude dried moss exudates were dissolved in BCD nutrient medium and either exposed to white light in a transparent microcentrifuge tube or kept in a similar microcentrifuge tube but covered with foil for 3 h at room temperature, followed by the MIC assay to determine residual activity. For thermostability assays, exudates were first subjected to size fractionation, and active <1 kDa fraction was subjected to two temperature treatments: boiling for ten minutes or repetitive thawing at 37 °C in a water bath and flash freezing in liquid nitrogen (three times total), followed by the MIC analysis to determine residual antibacterial activity. For Proteinase K sensitivity assay, 250 µg of active <1 kDa fraction was incubated with 32 µL (20 mg/mL) of proteinase K for 3 h at 37 °C, followed by the MIC analysis.

2.6. *Data Analysis*

Data are reported as mean ± standard deviation. Statistical significance was determined via an unpaired t-test with Welch's correction; $p < 0.05$. Analysis was performed using GraphPad Prism v.9.5.0.

3. Results

3.1. *Identification of Antibacterial Activity in C. purpureus exudates*

To test for the presence of antibacterial activity in *C. purpureus* exudates, R40 (male) and GG1 (female) moss strains were grown in BCD liquid cultures and their exudates were collected after 1, 2, and 4 weeks of growth. DDM assays detected distinct halos around filter disks containing R40 exudates when tested against Gram-positive *S. aureus* (Figure 1A), *S. pyogenes* (Figure 1B), and *E. faecium* (Figure 1C) bacteria, but not against Gram-negative *S. marcescens* or *S. enterica* (Table 1). Interestingly, significant levels of antibacterial activity were detected in R40 exudates after as little as 1 week of moss growth in liquid culture, with the highest activity detected at 4 weeks of growth, when the experiment was stopped (Figure 1D). These data suggest that moss cells continuously secreted bioactive metabolite/s when grown in liquid culture. Surprisingly, however, exudates of the *C. purpureus* female strain GG1 did not show antimicrobial activity in DDM assays against either Gram-positive or Gram-negative bacteria (Table 1). Collectively, these data indicate that the *C. purpureus* R40 strain produces extracellular metabolites with high inhibitory activity against a number of Gram-positive bacteria and also provide an intriguing example of sexual dimorphism in the production or exudation of antibacterial moss metabolites.

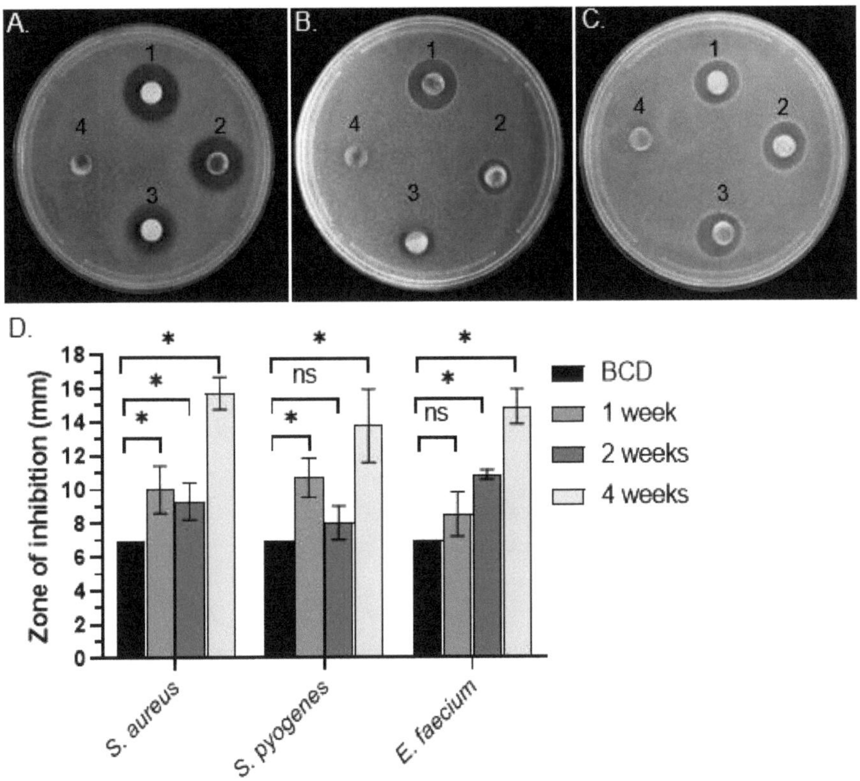

Figure 1. Bacterial growth inhibitory activity of extracellular metabolites of *C. purpureus* strains. (**A–C**) Representative pictures of qualitative DDM assays with filter disks soaked with exudates from four-week-old *C. purpureus* R40 strain and placed on top of *S. aureus* ATCC25923 (**A**), *S. pyogenes* ATCC12344 (**B**), and *E. faecium* ATCC35667 (**C**) bacterial lawns. (1–3) Disk with *C. purpureus* exudates; (4) negative control disks containing only BCD medium. (**D**) Diameter of *S. aureus* growth inhibition area (halo) around each cellulose disk containing secreted *C. purpureus* metabolites after one, two, and four weeks of moss growth was measured and plotted. Data represent the means from at least 3 independent experiments and a standard deviation. The asterisks indicate significance in an unpaired *t*-test; *—statistical significance $p \leq 0.05$; ns—not significant.

Table 1. Bacterial growth inhibition activity of *C. purpureus* exudates.

		Bacterial Growth Inhibition Zone in DDM Assays, in mm				MIC Values, mg/mL		MBC, mg/mL
C. purpureus strain	Bacteria	No Exudate Control [a]	1-Week-Old Moss Exudate	2-Week-Old Moss Exudate	4-Week-Old Moss Exudate	2-Week-Old Moss Exudate	4-Week-Old Moss Exudate	4-Week-Old Moss Exudate
R40	*S. aureus*	7	10.00 ± 1.41 *	9.31 ± 1.09 **	15.71 ± 0.95 **	12.5	6.25	6.25
	S. pyogenes	7	10.67 ± 1.16 *	8.00 ± 1.00	13.75 ± 2.17 **	-	50	-
	E. faecium	7	8.50 ± 1.32	11.00 ± 0.01 **	14.88 ± 1.05 **	-	12.5	-
	S. marcescens	7	7	7	7	-	-	-
	S. enterica ser. Typhimurium	7	7	7	7	-	-	-
GG1	*S. aureus*	7	7	7	7	50	50	-
	S. marcescens	7	7	7	7	-	-	-
	S. enterica ser. Typhimurium	7	7	7	7	-	-	-

[a] Cellulose disk diameter is 7 mm (no antibacterial activity). * Statistical significance $p \leq 0.05$; ** statistical significance $p \leq 0.01$; - not determined.

3.2. *Quantitative Analysis of Antimicrobial Activity from C. purpureus R40 Exudate*

To quantitatively measure the minimum inhibitory concentration (MIC) of *C. purpureus* exudates following the CLSI guidelines, we employed a broth microdilution method. Previous data indicated that BCD moss growth medium alone (negative control) does not inhibit the growth of *S. aureus* [20]. We serially diluted exudates from the male R40 strain to 50, 25, 12.5, 6.25, 3.125, 1.56, and 0.781 mg/mL concentrations, which were incubated with *S. aureus* cultures in MIC assays. Exudate from R40 strain grown for 2 weeks inhibited the growth of *S. aureus* at a concentration of 12.5 mg/mL (Figure 2A), while exudate from this strain grown for 4 weeks, as previously suggested by the DDM assays, had a stronger inhibitory activity and displayed a 2-fold lower MIC value of 6.25 mg/mL (Figure 2B). These data establish that longer growth times for the male R40 strain lead to higher antibacterial activity of its exudate.

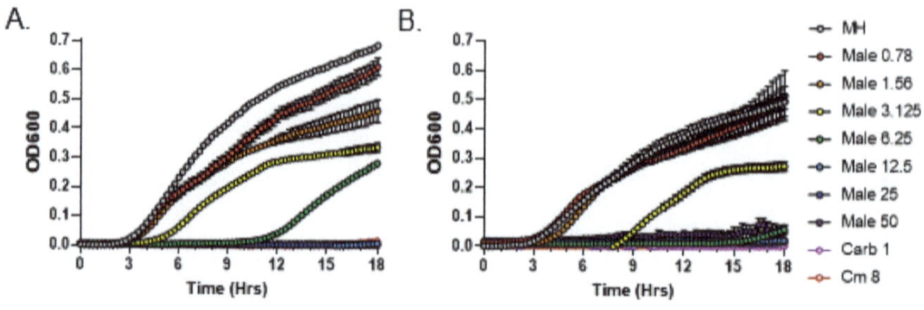

Figure 2. Broth microdilution method to determine the Minimal Inhibitory Concentration (MIC) of metabolites present in *C. purpureus* R40 exudates. Exudates from two-week-old (**A**) and four-week-old (**B**) *C. purpureus* R40 strains were tested in MIC assays against *S. aureus* ATCC25923. Growth curve of *S. aureus* cells was monitored in the presence of 0.781, 1.56, 3.125, 6.25, 12.5, 25, and 50 mg/mL of exudate solution. MH—negative control, no exudate added. Carbenicillin (Carb, 1 μg/mL) and chloramphenicol (Cm, 8 μg/mL) treatments were used as positive controls for *S. aureus* growth inhibition.

For comparison purposes, exudates from the female GG1 strain grown for 2 and 4 weeks were also tested in MIC assays. In both cases, bacterial growth inhibition was observed only at the highest exudate concentration of 50 mg/mL (Figure 3). Overall, these data confirm qualitative and quantitative differences in antibacterial potential between the male and female strains of *C. purpureus*.

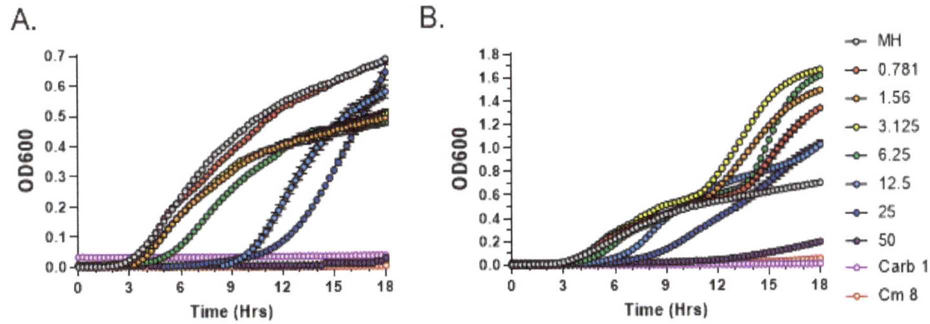

Figure 3. Broth microdilution method to determine the Minimal Inhibitory Concentration (MIC) of metabolites present in *C. purpureus* GG1 exudates. Exudates from two-week-old (**A**) and four-week-old (**B**) *C. purpureus* GG1 strains were tested in MIC assays against *S. aureus* ATCC25923. Growth curve of *S. aureus* cells was monitored in the presence of 0.781, 1.56, 3.125, 6.25, 12.5, 25, and 50 mg/mL of exudate solution. MH—negative control, no exudate added. Carbenicillin (Carb, 1 µg/mL) and chloramphenicol (Cm, 8 µg/mL) treatments were used as positive controls for *S. aureus* growth inhibition.

We next tested the most potent exudate from 4-week-old R40 strain in MIC assays against two other Gram-positive bacteria, *Streptococcus pyogenes* and *Enterococcus faecium*. While R40 exudates were indeed able to inhibit the growth of both bacteria, the MIC values were different. Specifically, the MIC value for *S. pyogenes* was 50 mg/mL (Figure 4A), while for *E. faecium* it was 4 times lower, 12.5 mg/mL (Figure 4B). These data indicate that both *S. aureus* and *E. faecium* are much more sensitive than *S. pyogenes* to the bioactive metabolites present in *C. purpureus* R40 exudates. Nevertheless, our quantitative MIC data confirm that antimicrobial compounds secreted by the *C. purpureus* R40 moss strain are effective against a spectrum of Gram-positive bacteria.

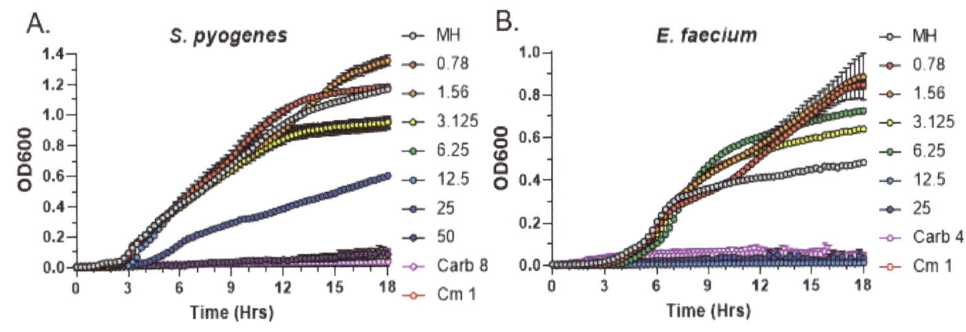

Figure 4. Analysis of the Minimal Inhibitory Concentration of *C. purpureus* R40 exudates against other Gram-positive bacteria. Exudates from four-week-old *C. purpureus* R40 strain were tested in MIC assays against *S. pyogenes* ATCC12344 (**A**) and *E. faecium* ATCC35667 (**B**). Growth curve of *S. pyogenes* and *E. faecium* cells was monitored in the presence of 0.781, 1.56, 3.125, 6.25, 12.5, 25, and 50 mg/mL of exudate solution. MH—negative control, no exudate added. Carbenicillin (Carb, 8 µg/mL or 4 µg/mL) and chloramphenicol (Cm, 1 µg/mL) treatments were used as positive controls for growth inhibition of *S. pyogenes* and *E. faecium*.

3.3. *C. purpureus* R40 Exudates Display Bactericidal Mode of Action

In general, antibiotics are divided into two groups based on whether they kill bacteria (bactericidal) or suppress bacterial growth (bacteriostatic). While both antibiotic types are often equally efficient in terms of clinical outcomes [30], understanding their mode of action is the first step in establishing their efficacy, especially since various antibiotics can be bacteriostatic for some pathogens and bactericidal for others. To test the mode of action for moss exudates, we followed the standard protocol for determining the ratio of minimum inhibitory concentration (MIC, the concentration that inhibits visible bacterial growth at 24 h of growth) to the minimum bactericidal concentration (MBC, the concentration of a compound that results in a 1000-fold reduction in bacterial CFU (colony-forming units) at 24 h of growth in the same specific conditions [31]. We monitored *S. aureus* growth dynamics for several *C. purpureus* exudate concentrations starting with 6.25 mg/mL and then plated serial dilutions of the growth cultures on LB plates to count the number of surviving colonies. Interestingly, no *S. aureus* colonies were detected in any plates containing bacterial cultures following incubation with all tested *C. purpureus* R40 exudate concentrations. Given that the calculated MIC value for *S. aureus* is 6.25 mg/mL (Figure 2), our data indicate that the ratio of MBC to MIC is equal to 1; thus, we conclude that the *C. purpureus* exudate displays a bactericidal mode of action.

3.4. Light Sensitivity of Antibacterial Compounds Present in C. purpureus Exudates

As the first step towards the characterization of the chemical nature of the moss metabolites with antibacterial activity, we tested bioactive compounds present in R40 exudates for stability at ambient temperature and for light sensitivity. Test tubes containing moss exudates were either covered in aluminum foil or exposed to direct sunlight for 3 h at room temperature and subsequently analyzed by MIC assays to test for any detrimental effects on antibacterial activity. Carbenicillin and chloramphenicol antibiotics in their MIC concentrations as established for the *S. aureus* strain were used as a positive control [20]. Neither treatment regimen changed MIC values of the exudates compared to the control with no treatment (Figure 5). These data suggest that the bioactive moss metabolites are not photo-sensitive and relatively stable at ambient temperature.

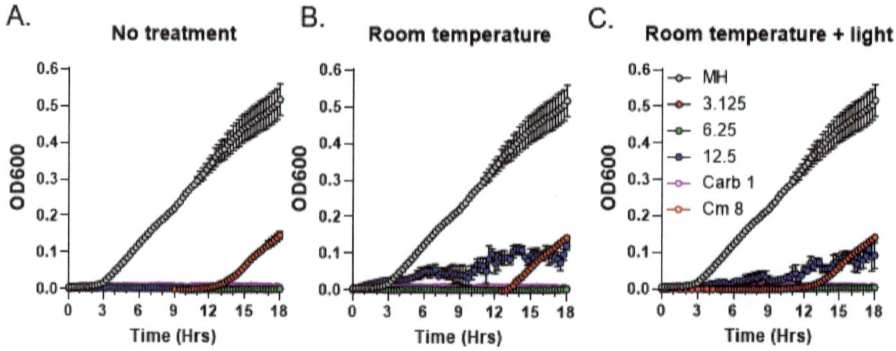

Figure 5. Residual antibacterial activity of *C. purpureus* R40 exudates after different treatments. Residual activity of exudate from four-week-old *C. purpureus* R40 culture was tested after treatments by MIC assays against *S. aureus*. (**A**) Samples were used in the assay immediately, without keeping them for 3 h at room temperature (untreated control). Samples were also kept at room temperature and either protected from light (**B**) or exposed to direct sunlight (**C**). Experiments were performed using 3.125, 6.25, and 12.5 mg/mL exudate concentrations. Carbenicillin (Carb, 1 μg/mL) and chloramphenicol (Cm, 8 μg/mL) were used as positive controls. MH, *S. aureus* growth in liquid MH medium without exudate addition.

3.5. Size Fractionation of Bioactive C. purpureus Exudate Components

To determine the approximate molecular weight range of the antimicrobial *C. purpureus* exudate compounds, we performed size fractionation using the Macrosep Advance centrifugal columns with molecular weight cutoffs of 10 kDa, 3 kDa, and 1 kDa. Following size fractionation, fractions were analyzed by the MIC assay. Antimicrobial activity was completely absent in three of the four size fractions: >10 kDa (Figure 6A), 3–10 kDa

(Figure 6B), and 1–3 kDa (Figure 6C). In contrast, the <1 kDa fraction harbored substantial antimicrobial activity, with MIC values for the fractionated exudate being similar to the unfractionated R40 samples, 6.25 mg/mL (Figure 6D). We conclude that the molecular weight of the bioactive *C. purpureus* compounds is relatively low, with the maximum upper limit not exceeding 1 kDa.

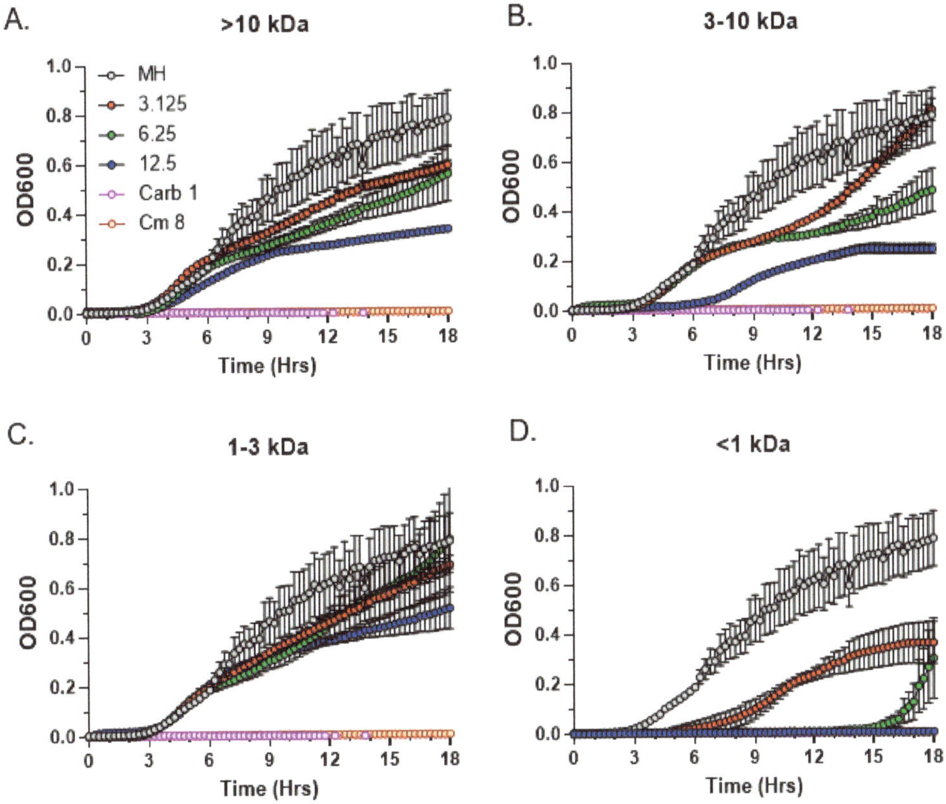

Figure 6. Size fractionation of *C. purpureus* exudates. Exudate from four-week-old *C. purpureus* R40 exudate was fractionated into four molecular weight fractions: >10 kDa (**A**), 3–10 kDa (**B**), 1–3 kDa (**C**), and <1 kDa (**D**). Each exudate fraction was analyzed by MIC assay at three different concentrations (3.125, 6.25, and 12.5 mg/mL) against *S. aureus* ATCC25923. MH medium without exudate addition was used as the negative control. Carbenicillin (Carb, 1 µg/mL) and chloramphenicol (Cm, 8 µg/mL) were used as positive controls.

3.6. Thermostability and Sensitivity to Proteinase K Treatment

After determining the approximate molecular weight range of the antibacterial *C. purpureus* R40 exudate components, we performed a series of thermostability and sensitivity to Proteinase K tests on the <1 kDa fraction containing the bioactive metabolites. Interestingly, similar to the untreated control (Figure 7A), antibacterial activity was not affected by repeated freezing and thawing cycles (Figure 7B) or by boiling (Figure 7C). Similarly, the exudate activity in the <1 kDa fraction was also not sensitive to the Proteinase K treatment (Figure 7D). These data indicate that the partially purified compounds are relatively thermostable and are unlikely to be proteinase-sensitive peptides or small proteins.

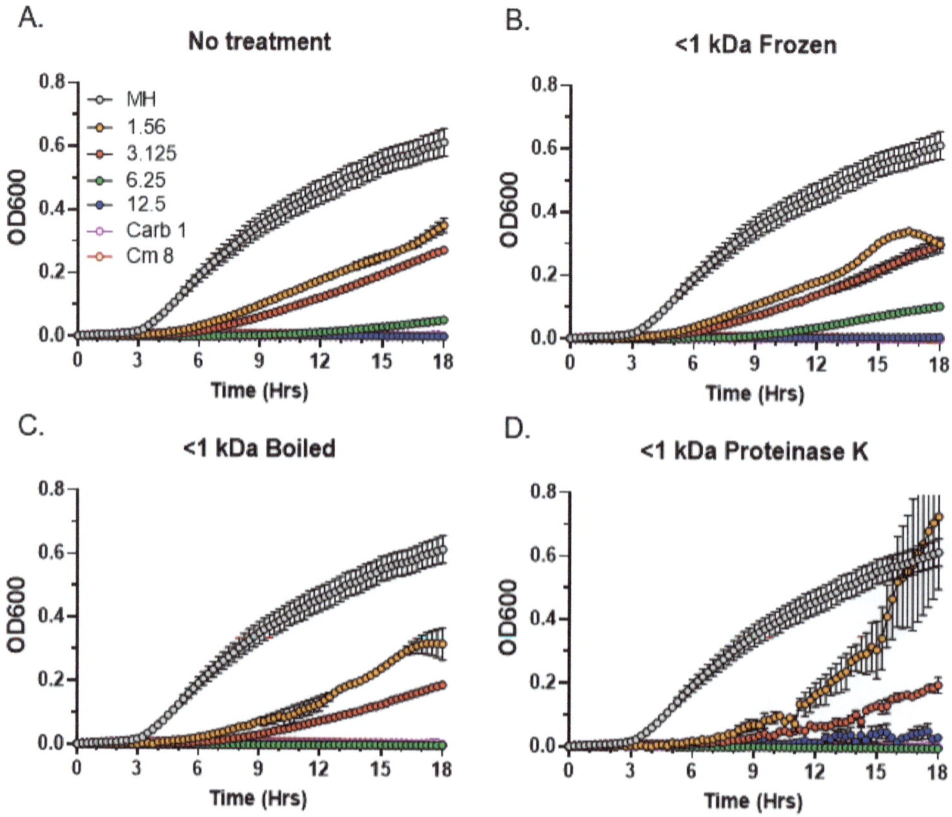

Figure 7. Residual antibacterial activity of *C. purpureus* R40 <1 kDa fraction after temperature and Proteinase K treatments. Residual activity of exudate from four-week-old *C. purpureus* R40 culture was tested after treatments by MIC assays against *S. aureus*. (**A**) Untreated control. (**B–D**) Samples were treated with repeated freezing and thawing (**B**), boiling (**C**) and incubation with Proteinase K (**D**). Samples were tested at exudate concentrations of 1.56, 3.125, 6.25, and 12.5 mg/mL. Carbenicillin (Carb, 1 µg/mL) and chloramphenicol (Cm, 8 µg/mL) were used as positive controls. MH, *S. aureus* growth in liquid MH medium without addition of exudate fraction.

4. Discussion

Flowering plants exude a number of phytochemicals into the rhizosphere that can influence soil characteristics, inhibit or stimulate root interactions with microorganisms, and promote plant growth [32]. While mosses do not have roots, they also secrete a number of complex compounds and peptides into the environment, with potential functions in antimicrobial defense and immune signaling [19,33]. We have previously detected potent antimicrobial activity in exudates from the model moss *Physcomitrium patens* and characterized its activity against *Staphylococcus aureus* ATCC25923 and several other Gram-positive bacteria [20]. Here, we extended our search for antimicrobial metabolites from model mosses and analyzed exudates from the dioicous moss *Ceratodon purpureus*.

Exudates of the *C. purpureus* R40 strain displayed a range of antimicrobial activities against *S. aureus* and two other Gram-positive bacteria, *E. faecium* and *S. pyogenes*, which represent close relatives of GAS and VRE bacteria from the CDC "Biggest Threats" list (https://www.cdc.gov/drugresistance/biggest-threats.html, accessed on 30 January 2023). Specifically, R40 exudates showed relatively high activity against *S. aureus* and *E. faecium* but low activity against *S. pyogenes*. These data are intriguing as Staphylococci and Streptococci are both classified as Gram-positive, non-motile, non-sporing, and facultative anaerobic cocci, yet R40 exudates appear to be effective against Staphylococci but not so much against Streptococci. These differences in activity coupled with known structural variations in cell wall or bacterial physiology can be further explored to characterize the specific molecular mechanisms of exudate's antibacterial action. Interestingly, the bioactive *C. purpureus* exudate compounds appear to have a bactericidal mode of action against *S. aureus* bacteria, as their MIC values are similar to the MBC values. As future research will focus on testing R40 exudates against a larger spectrum of Gram-positive pathogenic bacteria, a better understanding of their mode of antibacterial action will offer more meaningful prediction of their efficacy in vivo.

In contrast to the situation with Gram-positive bacteria, no antibacterial activity of *C. purpureus* exudates was observed against Gram-negative species *Salmonella* Typhimurium or *Serratia marcescens*. These data correlate well with our previous results on antibacterial activity of exudates from the moss *P. patens* [20] and suggest that the unique antimicrobial specificity towards Gram-positive bacteria is a unifying feature of exudates from two different model mosses that represent distinct subclasses of Bryopsida: Dicranidae (Ceratodon) and Funariidae (Physcomitrium). The biological significance of this observation remains to be established, but may involve implications for the type of bacteria mosses are often exposed to in their natural environment.

Several hundred different phytochemicals have previously been isolated from various mosses, with many of them possessing antimicrobial and antifungal activity [8,12]. However, most of these secondary plant metabolites have been detected in crude extracts of whole cells, making it more challenging to separate individual bioactive compounds from other components present in raw cellular mixtures. In contrast, exudates typically contain fewer components with more specific chemical structures that plants can easily secrete through the cell wall into the environment [14], making it potentially easier to isolate and further characterize the water-soluble bioactive compounds of interest. Furthermore, our time course data indicate that while antibacterial activity can be detected in R40 exudates as early as after 1 week of moss growth, it reaches maximum levels at four weeks, suggesting that the antimicrobial compounds are relatively stable in plant growth medium under standard conditions. This conclusion is further corroborated by our thermostability and light sensitivity assays, by apparent low molecular weight and by complete insensitivity to Proteinase K treatment. Collectively, these conclusions are encouraging, indicating that the chemical structure of *C. purpureus* R40 antibacterial compounds may be positively identified and characterized in the future.

Interestingly, only exudates from the male *C. purpureus* R40 strain, but not the female GG1 strain, exhibited strong activity against *S. aureus*, implying intriguing sexual dimorphism for the presence of metabolites with antibacterial activity in this dioicous moss species. The underlying mechanism of such dimorphism is currently unknown and will require identification of the chemical nature of the secreted metabolites from the male R40 strain. Nevertheless, sexual dimorphism in *C. purpureus* strains has previously been reported for a number of traits, with females having larger leaves and generally greater values for photosynthetic parameters [34]. Particularly compelling is the example of sex-specific compounds that influence moss fertilization and overall fitness. *C. purpureus* strains emit complex volatile scents, whose chemical composition and abundance are sex-specific, with moss-dwelling microarthropods being preferentially attracted to the female-produced volatile cues [24]. As the presence of sperm-dispersing microarthropods increases reproductive rates for *C. purpureus* moss strains [35], these data suggest that moss compounds released into the environment can indeed lead to a substantial fitness benefit. It would be similarly interesting to analyze the effects of sexual dimorphism in secreted antibacterial compounds on stress response, fitness, or physiological differences in *C. purpureus* male and female strains. Another implication of our study is that if genetic variation in secreted antibacterial compounds can also be demonstrated for *C. purpureus* male and female strains in their natural habitats, R40- and GG1-associated microbiomes could also be different, as was recently demonstrated for roots and rhizosphere soils of the dioecious flowering plant *Carica papaya* [36].

Further experiments, such as mass spectrometry analysis, will be necessary to establish both the nature of R40 antibacterial exudate components and their biological role in the *C. purpureus* life cycle. Currently, metabolomic, proteomics, and transcriptomics assays are well-established for *P. patens* [19,33,37,38], and similarly powerful tools are also being developed for *C. purpureus* [25]. Generally, the metabolome of *C. purpureus* appears to be diverse, with a number of biflavonoids, phospholipids, disaccharides, long-chain fatty acids, carotenoids, and antioxidants that change in abundance depending on the environmental conditions [26]. Some of the identified intracellular or cell-wall-bound biflavonoids display antioxidant and UV-protective activity [9]. However, little is known about natural metabolites secreted by *C. purpureus* into the environment. Future progress in characterizing secretome and exudate components will be instrumental in helping to identify new potent antimicrobial compounds from this model moss species.

Author Contributions: All authors contributed significantly to this work. L.M.B. and E.V.S. designed the experiments. A.L.D., L.R.V., N.M.M., M.A.V. and L.M.B. performed the experiments. A.L.D., M.R.S., L.M.B. and E.V.S. analyzed data and prepared figures. A.L.D., L.R.V., L.M.B. and E.V.S. wrote the paper with contributions from all other authors. Correspondence and requests for materials should be addressed to L.M.B. or E.V.S. All authors have read and agreed to the published version of the manuscript.

Funding: This work was supported in part by the NASA West Virginia Space Grant Consortium Training Grant #NNX15AI01H to D.A.L., by the WV Higher Education Policy Commission, Division of Science and Research Grant number dsr.20.1698-001 to N.M.M., and by the Kazan Federal University Strategic Academic Leadership Program (Priority-2030). Research in the Valentovic lab was supported by NIH grants P20GM103434, 2R15CA161491-03, R15AI15197-01, and R15HL145573-01. Moss metabolite research in the Shakirov lab was supported by the WV-INBRE Center for Natural Products Research, funded by NIGMS grant P20GM103434.

Institutional Review Board Statement: Not applicable.

Informed Consent Statement: Not applicable.

Data Availability Statement: The authors confirm that the data supporting the findings of this study are available within the article.

Acknowledgments: We thank Stuart McDaniel (University of Florida) for sharing *C. purpureus* strains, Tim Long (Marshall University, School of Pharmacy), and members of our labs for fruitful discussions.

Conflicts of Interest: The authors declare no conflict of interest.

References

1. GBD 2019 Antimicrobial Resistance Collaborators. Global mortality associated with 33 bacterial pathogens in 2019: A systematic analysis for the Global Burden of Disease Study 2019. *Lancet* **2022**, *400*, 2221–2248. [CrossRef] [PubMed]
2. Patel, J.; Harant, A.; Fernandes, G.; Mwamelo, A.J.; Hein, W.; Dekker, D.; Sridhar, D. Measuring the global response to antimicrobial resistance, 2020–2021: A systematic governance analysis of 114 countries. *Lancet Infect. Dis.* **2023**, *202*, 1–13. [CrossRef]
3. Antimicrobial Resistance Collaborators. Global burden of bacterial antimicrobial resistance in 2019: A systematic analysis. *Lancet* **2022**, *399*, 629–655. [CrossRef] [PubMed]
4. Asokan, G.V.; Ramadhan, T.; Ahmed, E.; Sanad, H. WHO Global Priority Pathogens List: A Bibliometric Analysis of Medline-PubMed for Knowledge Mobilization to Infection Prevention and Control Practices in Bahrain. *Oman Med. J.* **2019**, *34*, 184–193. [CrossRef] [PubMed]
5. Mattingly, J.M.; Dunham, C.M. ESKAPE velocity: Total synthesis platforms promise to increase the pace and diversity of antibiotic development. *Nat. Struct. Mol. Biol.* **2022**, *29*, 3–9. [CrossRef]
6. Khameneh, B.; Iranshahy, M.; Soheili, V.; Bazzaz, B.S.F. Review on plant antimicrobials: A mechanistic viewpoint. Antimicrobial Resistance and Infection Control. *BMC* **2019**, *8*, 118. [CrossRef]
7. Shen, B. A New Golden Age of Natural Products Drug Discovery. *Cell* **2015**, *163*, 1297–1300. [CrossRef] [PubMed]

8. Horn, A.; Pascal, A.; Lončarević, I.; Marques, R.; Lu, Y.; Miguel, S.; Bourgaud, F.; Thorsteinsdóttir, M.; Cronberg, N.; Becker, J.D.; et al. Natural Products from Bryophytes: From Basic Biology to Biotechnological Applications. *Crit. Rev. Plant Sci.* **2021**, *40*, 191–217. [CrossRef]
9. Waterman, M.J.; Nugraha, A.S.; Hendra, R.; Ball, G.E.; Robinson, S.A.; Keller, P.A. Antarctic Moss Biflavonoids Show High Antioxidant and Ultraviolet-Screening Activity. *J. Nat. Prod.* **2017**, *80*, 2224–2231. [CrossRef]
10. Kang, S.J.; Kim, S.H.; Liu, P.; Jovel, E.; Towers, G.H. Antibacterial activities of some mosses including *Hylocomium splendens* from South Western British Columbia. *Fitoterapia* **2007**, *78*, 373–376. [CrossRef]
11. Wolski, G.J.; Sadowska, B.; Fol, M.; Podsędek, A.; Kajszczak, D.; Kobylińska, A. Cytotoxicity, antimicrobial and antioxidant activities of mosses obtained from open habitats. *PLoS ONE* **2021**, *16*, e0257479. [CrossRef]
12. Olofin, T.A.; Akande, A.O.; Oyetayo, V.O. Assessment of the antimicrobial properties of fractions obtained from bryophytes. *J. Microbiol. Antimicrob.* **2013**, *5*, 50–54. [CrossRef]
13. Mishra, R.; Pandey, V.K.; Chandra, R. Potential of Bryophytes as therapeutics. *IJPSR* **2014**, *5*, 3584–3593.
14. Seitz, V.A.; McGivern, B.B.; Daly, R.A.; Chaparro, J.M.; Borton, M.A.; Sheflin, A.M.; Kresovich, S.; Shields, L.; Schipanski, M.E.; Wrighton, K.C.; et al. Variation in Root Exudate Composition Influences Soil Microbiome Membership and Function. *Appl. Environ. Microbiol.* **2022**, *88*, e00226-22. [CrossRef] [PubMed]
15. Hu, L.; Robert, C.A.M.; Cadot, S.; Zhang, X.; Ye, M.; Li, B.; Manzo, D.; Chervet, N.; Steinger, T.; van der Heijden, M.G.A.; et al. Root exudate metabolites drive plant-soil feedbacks on growth and defense by shaping the rhizosphere microbiota. *Nat. Comm.* **2022**, *9*, 2738. [CrossRef] [PubMed]
16. Samaddar, S.; Karp, D.S.; Schmidt, R.; Devarajan, N.; McGarvey, J.A.; Pires, A.F.A.; Scow, K. Role of soil in the regulation of human and plant pathogens: Soils' contributions to people. *Phil. Trans. R. Soc. B* **2021**, *376*, 20200179. [CrossRef] [PubMed]
17. Ichino, T.; Yazaki, K. Modes of secretion of plant lipophilic metabolites via ABCG transporter-dependent transport and vesicle-mediated trafficking. *Curr. Opin. Plant Biol.* **2022**, *66*, 102184. [CrossRef]
18. Romani, F.; Banić, E.; Florent, S.N.; Kanazawa, T.; Goodger, J.Q.D.; Mentink, R.A.; Dierschke, T.; Zachgo, S.; Ueda, T.; Bowman, J.L.; et al. Oil body formation in *Marchantia polymorpha* is controlled by MpC1HDZ and serves as a defense against arthropod herbivores. *Curr. Biol.* **2020**, *30*, 2815–2828. [CrossRef]
19. Fesenko, I.; Azarkina, R.; Kirov, I.; Kniazev, A.; Filippova, A.; Grafskaia, E.; Lazarev, V.; Zgoda, V.; Butenko, I.; Bukato, O.; et al. Phytohormone treatment induces generation of cryptic peptides with antimicrobial activity in the Moss *Physcomitrella Patens*. *BMC Plant Biol.* **2019**, *19*, 9. [CrossRef]
20. Valeeva, L.R.; Dague, A.L.; Hall, M.H.; Tikhonova, A.E.; Sharipova, M.R.; Valentovic, M.A.; Bogomolnaya, L.M.; Shakirov, E.V. Antimicrobial Activities of Secondary Metabolites from Model Mosses. *Antibiotics* **2022**, *11*, 1004. [CrossRef]
21. Biersma, E.M.; Convey, P.; Wyber, R.; Robinson, S.A.; Dowton, M.; van de Vijver, B.; Linse, K.; Griffiths, H.; Jackson, J.A. Latitudinal Biogeographic Structuring in the Globally Distributed Moss *Ceratodon Purpureus*. *Front. Plant Sci.* **2020**, *11*, 502359. [CrossRef]
22. Carey, S.B.; Jenkins, J.; Lovell, J.T.; Maumus, F.; Sreedasyam, A.; Payton, A.C.; Shu, S.; Tiley, G.P.; Fernandez-Pozo, N.; Healey, A.; et al. Gene-rich UV sex chromosomes harbor conserved regulators of sexual development. *Sci. Adv.* **2021**, *7*, eabh2488. [CrossRef]
23. Kollar, L.M.; Kiel, S.; James, A.J.; Carnley, C.T.; Scola, D.N.; Clark, T.N.; Khanal, T.; Rosenstiel, T.N.; Gall, E.T.; Grieshop, K.; et al. The genetic architecture of sexual dimorphism in the moss *Ceratodon purpureus*. *Proc. Biol. Sci.* **2021**, *288*, 20202908. [CrossRef]
24. Rosenstiel, T.N.; Shortlidge, E.E.; Melnychenko, A.N.; Pankow, J.F.; Eppley, S.M. Sex-specific volatile compounds influence microarthropod-mediated fertilization of moss. *Nature* **2012**, *489*, 431–433. [CrossRef]
25. Brennan, D.L.; Kollar, L.M.; Kiel, S.; Deakova, T.; Laguerre, A.; McDaniel, S.F.; Eppley, S.M.; Gall, E.T.; Rosenstiel, T.N. Measuring volatile emissions from moss gametophytes: A review of methodologies and new applications. *Appl. Plant Sci.* **2022**, *10*, e11468. [CrossRef] [PubMed]
26. Sala-Carvalho, W.R.; Montessi-Amaral, F.P.; Esposito, M.P.; Campestrini, R.; Rossi, M.; Peralta, D.F.; Furlan, C.M. Metabolome of *Ceratodon purpureus* (Hedw.) Brid.; a cosmopolitan moss: The influence of seasonality. *Planta* **2022**, *255*, 77. [CrossRef]
27. Ashton, N.V.; Cove, D.J. The Isolation and Preliminary Characterization of Auxotrophic and Analogue Resistant Mutants of the Moss, *Physcomitrella patens*. *Molec. Gen. Genet.* **1977**, *154*, 87–95. [CrossRef]
28. Shirshikova, T.V.; Sierra-Bakhshi, C.G.; Kamaletdinova, L.K.; Matrosova, L.E.; Khabipova, N.N.; Evtugyn, V.G.; Khilyas, I.V.; Danilova, I.V.; Mardanova, A.M.; Sharipova, M.R.; et al. The ABC-Type Efflux Pump MacAB is Involved in Protection of *Serratia marcescens* against Aminoglycoside Antibiotics, Polymyxins, and Oxidative Stress. *mSphere* **2021**, *6*, e00033-21. [CrossRef] [PubMed]
29. Pankey, G.A.; Sabath, L.D. Clinical relevance of bacteriostatic versus bactericidal mechanisms of action in the treatment of Gram-positive bacterial infections. *Clin. Infect. Dis.* **2004**, *38*, 864–870. [CrossRef] [PubMed]
30. Wald-Dicker, N.; Holtom, P.; Spellberg, B. Busting the myth of "static vs. cidal": A systemic literature review. *Clin. Infect. Dis.* **2018**, *66*, 1470–1474. [CrossRef]
31. Motyl, M.; Dorso, K.; Barrett, J.; Giacobbe, R. Basic Microbiological Techniques for Antibacterial Drug Discovery. *Curr. Protoc. Pharmacol.* **2003**, *13A.3* (Suppl. S31), 1–22. [CrossRef] [PubMed]
32. Badri, D.V.; Vivanco, J.M. Regulation and function of root exudates. *Plant Cell Environ.* **2009**, *32*, 666–681. [CrossRef]

33. Lyapina, I.; Filippova, A.; Kovalchuk, S.; Ziganshin, R.; Mamaeva, A.; Lazarev, V.; Latsis, I.; Mikhalchik, E.; Panasenko, O.; Ivanov, O.; et al. Possible role of small secreted peptides (SSPs) in immune signaling in bryophytes. *Plant Mol. Biol.* **2021**, *106*, 123–143. [CrossRef] [PubMed]
34. Slate, M.L.; Rosenstiel, T.N.; Eppley, S.M. Sex-specific morphological and physiological differences in the moss *Ceratodon purpureus* (Dicranales). *Ann. Bot.* **2017**, *120*, 845–854. [CrossRef]
35. Shortlidge, E.E.; Carey, S.B.; Payton, A.C.; McDaniel, S.F.; Rosenstiel, T.N.; Eppley, S.M. Microarthropod contributions to fitness variation in the common moss *Ceratodon purpureus*. *Proc. Biol. Sci.* **2021**, *288*, 20210119. [CrossRef] [PubMed]
36. Zhou, Y.; Pang, Z.; Yuan, Z.; Fallah, N.; Jia, H.; Ming, R. Sex-based metabolic and microbiota differences in roots and rhizosphere soils of dioecious papaya (*Carica papaya* L.). *Front. Plant Sci.* **2022**, *13*, 991114. [CrossRef]
37. Fesenko, I.; Shabalina, S.A.; Mamaeva, A.; Knyazev, A.; Glushkevich, A.; Lyapina, I.; Ziganshin, R.; Kovalchuk, S.; Kharlampieva, D.; Lazarev, V.; et al. A vast pool of lineage-specific microproteins encoded by long non-coding RNAs in plants. *Nucleic Acids Res.* **2021**, *49*, 10328–10346. [CrossRef]
38. Fesenko, I.; Khazigaleeva, R.; Kirov, I.; Kniazev, A.; Glushenko, O.; Babalyan, K.; Arapidi, G.; Shashkova, T.; Butenko, I.; Zgoda, V.; et al. Alternative splicing shapes transcriptome but not proteome diversity in *Physcomitrella patens*. *Sci. Rep.* **2017**, *7*, 2698. [CrossRef]

metabolites

MDPI

Article

Moringa oleifera and Propolis in Cattle Nutrition: Characterization of Metabolic Activities in the Rumen In Vitro

Mubarik Mahmood [1], Hasan Ersin Samli [2], Arife Sener-Aydemir [3], Suchitra Sharma [3], Qendrim Zebeli [3] and Ratchaneewan Khiaosa-ard [3,*]

1 Animal Nutrition Section, Department of Animal Sciences, University of Veterinary and Animal Sciences, Lahore, Sub Campus Jhang, 12 km Chiniot Road, Jhang 35200, Pakistan
2 Department of Animal Science, Faculty of Agriculture, Tekirdag Namik Kemal University, Degirmenalti Campus, Tekirdag 59030, Turkey
3 Institute of Animal Nutrition and Functional Plant Compounds, Department for Farm Animals and Veterinary Public Health, University of Veterinary Medicine Vienna, Veterinaerplatz 1, 1210 Vienna, Austria
* Correspondence: ratchaneewan.khiaosa-ard@vetmeduni.ac.at

Abstract: *Moringa oleifera* by-products such as seed cake and leaves are protein-rich ingredients, while raw propolis has the potential to influence ruminal protein metabolism. These substances are also known to be sources of functional compounds. With these properties, they could modulate ruminal fermentation activities. Using the rumen simulation technique, we investigated ruminal fermentation and the antioxidant properties of four dietary treatments. These included a control diet (CON) without supplementation; the CON diet top-dressed on a dry matter (DM) basis, either with moringa seed cake (MSC, containing 49% crude protein (CP)), moringa leaf powder (ML, containing 28% CP), or raw propolis (PRO, 3% CP). MSC, ML, and PRO accounted for 3.8, 7.4, and 0.1% of the total diet DM, respectively. Both ML and MSC resulted in 14 and 27% more ammonia concentration, respectively than CON and PRO ($p < 0.05$). MSC increased the propionate percentage at the expense of acetate ($p < 0.05$). Both ML and MSC decreased methane percentages by 7 and 10%, respectively, compared to CON ($p < 0.05$). The antioxidant capacity of the moringa seed cake, moringa leaf powder, and raw propolis were 1.14, 0.56, and 8.56 mg Trolox/g DM, respectively. However, such differences were not evident in the fermentation fluid. In conclusion, the supplementation of moringa seed cake desirably modulates rumen microbial activities related to protein and carbohydrate metabolism.

Keywords: moringa seed cake; moringa leaf; propolis; functional compounds; ruminal fermentation metabolite; methane

check for updates

Citation: Mahmood, M.; Samli, H.E.; Sener-Aydemir, A.; Sharma, S.; Zebeli, Q.; Khiaosa-ard, R. *Moringa oleifera* and Propolis in Cattle Nutrition: Characterization of Metabolic Activities in the Rumen In Vitro. *Metabolites* **2022**, *12*, 1237. https://doi.org/10.3390/metabo12121237

Academic Editors: Ramona Paltinean and Irina Ielciu

Received: 16 November 2022
Accepted: 5 December 2022
Published: 9 December 2022

Publisher's Note: MDPI stays neutral with regard to jurisdictional claims in published maps and institutional affiliations.

1. Introduction

With the help of microbial fermentation taking place in their reticulorumen, ruminants can remarkably utilize low-quality feed ingredients that are not suitable for human consumption. The feedstuff is degraded to various metabolites in the rumen, which fulfill the majority of the nutrient demand of ruminant animals [1]. Suboptimal protein feeding is prevalent in tropical and subtropical areas of the globe, where most of the livestock follows semi-intensive or non-intensive feeding practices [2]. Low dietary protein levels not only limit the availability of the microbial protein to the host animal but also negatively impact their activities to ferment carbohydrates, and thus, suppress the production of total short-chain fatty acids (SCFA) [3] and consequently the productivity of the host animal. Including protein-rich alternatives as well as increased efficiency of nitrogen metabolism in the rumen are options to fill the animal requirement gap.

The *Moringa oleifera* plant is native to tropical and subtropical regions and is a storehouse for a variety of nutrients in sufficient quantities [4]. The best-known benefit is its high crude protein (CP) contents, which are comparable to those of other common protein fodders such as soybeans and alfalfa [5]. A rising trend in the cultivation of moringa plants worldwide during recent years [6] has generated more foliage and seed portions. Moringa seeds are used to extract moringa oil, leaving behind the seed cake that is richer in protein than those of the leaves, seeds, and pods of the parent plant [7,8]. Besides being nutritious, moringa products are also known for harboring a variety of functional compounds, including myricetin, quercetin, moringyne, vanillin, rutins, tannins, gallic acid, and kaempferol [9]. Many compounds present in moringa hold anti-inflammatory, antibiotic, and, importantly, antioxidant properties [10]. However, the composition and quantity of the functional compounds vary greatly among leaf and seed portions of the plant [9]. For instance, moringa leaves contain slightly higher tannins (1.19 mg/g) in comparison to moringa seeds (0.89 mg/g) [11,12]. The role of moringa in cattle nutrition is mainly seen as protein supplements; notably, studies have indicated that moringa by-products may also have functional potential, for instance, in decreasing methanogenesis [13,14], but these are short-term studies. Another functional substance of interest when feeding low-protein diets is raw propolis, which is a non-proteinaceous substance prepared by honeybees from materials of plant origin. Its ammonia (NH_3)-reducing effect in vitro was reported previously [15]. They suspected that the effect was possibly associated with the reduced deamination of amino acids. This suggests that propolis supplementation might be beneficial in reducing N loss via ammonia production. In addition, propolis has been shown to have antioxidant, antibacterial, and anti-inflammatory effects [16–18]. However, due to limited data on moringa and even more lacking raw propolis, their roles as functional feeds for cattle cannot be generalized. In the present study, we aimed to characterize the metabolic activities of ruminal microbiota in response to the supplementation of different moringa by-products and raw propolis. We hypothesized that incorporating these by-products into low-protein diets would improve the ruminal protein metabolism and modulate carbohydrate fermentation, which may suggest their functional effects in cattle nutrition. Using the in vitro rumen simulation technique (RUSITEC), we evaluated the effects of dried moringa leaf powder, moringa seed cake, and raw propolis when supplemented with a low protein diet on ruminal fermentation characteristics and antioxidant capacity.

2. Materials and Methods

2.1. Experimental Design and Treatments

The trial was performed using two RUSITEC systems, each consisting of six fermenters. Each fermenter had an effective volume capacity of 800 mL. The RUSITEC systems simulated the ruminal conditions by maintaining an anaerobic condition, a temperature of 39.5 °C , and a continuous infusion of a salivary buffer throughout the trial. The trial consisted of two experimental runs in a changeover design [19]. Each run lasted 10 days, with five days of adaptation and system equilibration and the last five days of the sampling period. In each experimental run, we tested four dietary treatments in triplicates: the control (CON), moringa seed cake (MSC), moringa leaves (ML), and propolis (PRO). The CON diet contained hay and a grain mix (50:50 on a dry matter (DM) basis) and was without any supplementation. ML and MSC diets were the CON diet top-dressed with dried moringa leaf powder (containing 28% CP on a DM basis) and moringa seed cake (containing 49% CP on a DM basis), respectively, resulting in the inclusion level of 7.4 and 3.8% of the total diet DM, respectively (Table 1).

The chosen supplementation levels were aimed at providing similar dietary CP content boosting 1.2–1.4% units from that of CON to reach the target CP content of around 11–12%, which was shown to be adequate for milk production under 20 kg/day [20]. These scenarios are typical in tropical and subtropical regions utilizing extensive farming. The PRO diet was a CON diet top-dressed with raw propolis at the rate of 0.1% of the diet DM. The moringa by-products were products of Nicaragua origin and were provided by a private supplier (see Acknowledgments). Propolis, which was a brown type, was obtained from a local honeybee-keeping supplier in Tekirdağ, Turkey. To our knowledge, there was no reference dosage of raw propolis in the cattle reported in the literature. Therefore, the test dosage used in our study was adapted from Santos et al. [21], who used a dried propolis extract plus excipient. The ingredients and chemical composition of all four dietary treatments are shown in Table 1. Before use, the hay and concentrate were ground with a Wiley mill (Pulverisette 25/19; Fritsch GmbH, Idar-Oberstein, Germany) to pass through a 6 mm sieve.

Table 1. Ingredient and chemical composition of control and experimental diets (g/kg DM) *.

Item	CON	PRO	ML	MSC
Ingredients				
Meadow hay	503	502	465	483
Concentrate [1]	497	497	461	478
Propolis	0.0	1.0	0.0	0.0
Moringa leaf powder	0.0	0.0	74	0.0
Moringa seed cake	0.0	0.0	0.0	38
Chemical composition				
Dry Matter	894	894	897	895
Organic Matter	915	915	914	916
Crude Protein	99	99	112	114
Ash	85	85	86	84
Neutral detergent fiber	475	474	457	464
Ether extract (crude fat)	17	17	20	21
Non-fiber carbohydrates	325	325	324	318

* All fermenters were supplied with the same basal diet containing 50:50 hay and concentrate on a dry matter basis. The respective test ingredient was top-dressed. A diet containing concentrate at 500 g/kg of diet dry matter without supplementation (CON) or top dressed in propolis (PRO), dried moringa leaves (ML) or moringa seed cake (MSC). [1] Contained 216 barley; 216 wheat; 517 maize; and 52 vitamin and mineral supplement on dry matter basis (g/kg) (Rindavit TMR 11 ASS-CO + ATG; H. Wilhelm Schaumann GmbH & Co KG, Brunn/Gebirge, Austria).

2.2. RUSITEC Procedure

On the first day of each experimental run, all fermenters were inoculated with rumen fluid, and solid digesta obtained from 2 ruminally cannulated non-pregnant dry cows (one Holstein cow and one Brown Swiss cow) kept at the ruminant clinic of the University of Veterinary Medicine (Vetmeduni), Vienna, Austria. The cows were fed hay ad libitum with a daily allowance of 0.5 kg of commercial concentrates (KuhKorn PLUS Energie, Garant-Tiernahrung GmbH, Pölchlarn, Austria). They were maintained according to the Austrian guidelines for animal welfare [22]. The inoculation protocol was followed as previously described by Mahmood et al. [19]. Specifically, inoculum from both donor cows was prepared by straining through 4 layers of medical gauze, which were pooled into one batch before inoculation. Subsequently, a total of 600 mL of pooled rumen fluid was transferred into each fermenter already containing 100 mL of McDougall's buffer ($NaHCO_3$, $Na_2HPO_4{\cdot}2H_2O$, NaCl, KCl, $CaCl_2{\cdot}2H_2O$, and $MgCl_2{\cdot}6H_2O$ at 95.1, 23.6, 8.04, 7.64, 0.37, and 0.63 mmol/L, respectively). Equal amounts of the solid digesta from both donor cows were taken, pooled, and used to inoculate the fermenter. The pooled solid digesta and the respective diet containing 12 g DM were packed into separate nylon bags and placed into the respective fermenter. The dimensions of each nylon bag were 120 × 70 mm with a pore size of 70 μm (Linker Industrie-Technik GmbH, Kassel, Germany). Each fermenter was connected to a gas-tight bag for the collection of fermentation gases (TecoBag 8 L, Tesseraux-Spezialverpackungen GmbH, Bürstadt, Germany), and a glass bottle for collecting the outflow was constantly kept cool at 1 °C to prevent further fermentation. After inoculation and the placement of feed bags, each fermenter was closed and flushed with a stream of nitrogen gas for 3 min to establish an anaerobic condition. Throughout the trial, the McDougall's buffer was continuously infused into each fermenter using a multi-channel peristaltic pump (model ISM932, Ismatec, Idex Health & Science GmbH, Wertheim, Germany) at a flow rate of 375 mL/day. On the next day, the nylon bag with solid rumen digesta was replaced by a new nylon bag containing the respective diet. Before removal, the bag was rinsed and squeezed with 40 mL of a pre-warmed McDougall's buffer. Before the opening of the fermenter, nitrogen gas was flushed for 30 s to collect all the entrapped fermentation gases into the gas bag, followed by gas bag exchange. The associated effluent bottle was emptied and reconnected. Finally, the fermenter was again made air-tight, and nitrogen gas was flushed for 3 min to re-establish anaerobic conditions. The procedure was performed daily, and each feed bag, which was incubated for 48 h, was daily replaced with a new feed bag of the same treatment.

2.3. Sampling, Daily Measurements and Laboratory Analyses

During sampling days, the incubation fluid was collected daily from each fermenter for measurements and analyses. One portion of the aliquot was immediately measured for pH, and the redox potential using a pH meter (Seven Multi TM; Mettler-Toledo GmbH, Schwerzenbach, Switzerland) furnished with separate electrodes: InLab Expert Pro-ISM for pH and Pt 4805-DPA-SC-S8/120 for redox (Mettler-Toledo GmbH, Vienna, Austria). Additionally, another portion of the aliquot was preserved at −20 °C for the later analysis of SCFA and NH_3. Feed bags taken on the sampling days were rinsed using a machine wash for 30 min with cold water, a gentle cycle mode, and no spinning. The washed bags were manually squeezed to remove excess water and then stored at −20 °C for later chemical analysis.

For the chemical analysis of the incubated feed samples, the feed bags collected across the last 5 days were freeze-dried, pooled per fermenter, and then ground, passing through a 0.75 mm sieve prior to analysis. The ground material was used for analyzing the chemical composition, including the DM, organic matter (OM), CP, ether extract (EE, i.e., crude fat), neutral detergent fiber (NDF), ash, and non-fiber carbohydrates (NFC) and using previously described protocols [23]. Shortly, DM was determined after oven drying at 103 °C and ash after combustion at 580 °C overnight. EE was analyzed using a soxhlet extractor (Extraction System B-811, Buchi, Flawil, Switzerland) and CP using Kjeldahl's method. The amylase-treated NDF was determined using Fiber Therm FT 12 (Gerhardt GmbH & Co. KG, Königswinter, Germany). The OM was calculated based on the ash percentage. NFC calculation was estimated as follows: NFC = 100 − (CP + ash + EE + NDF). The same chemical analysis was performed on the original diets. The nutrient degradation (% of supply) was based on the apparent nutrient disappearances and was estimated from the differences between the nutrient concentrations before (original diet) and after incubation (feed residue) relative to the supply amount in the original diet times 100.

The analysis of SCFA concentration and the profile of incubation fluid was performed using gas chromatography (GC) and a GC apparatus (Shimadzu GC 2010-Plus, Shimadzu, Kyoto, Japan) equipped with a flame-ionization detector and a 30 m × 0.53 mm i.d. × 0.53 μm capillary column (Trace TR Wax, Thermo Fisher Scientific, Waltham, MA, USA). The quantification of the identified SCFA was conducted using an internal standard (4-methylvaleric acid, Sigma-Aldrich, St. Louis, MO, USA). Helium was used as a carrier gas and was maintained at a flow rate of 6 mL/min. The injector temperature was set at 170 °C while that of the detector was at 220 °C. The indophenol reaction method [24] was used to determine the daily NH_3 concentrations of the incubation fluid. Accordingly, the preserved samples were thawed at room temperature prior to centrifugation at 15,115× g for 10 min. The supernatant was diluted with deionized water to obtain the concentration range within the standard calibration curve. Sodium hydroxide was used to oxidize the phenol and NH_3 in the presence of dichloroisocyanuric acid and sodium nitroprusside. The absorbance of the treated samples was measured at 655 nm with a spectrophotometer U3000 (INULA GmbH, Vienna, Austria).

The volume of the fermentation gas was estimated by a water replacement method, as described by Soliva and Hess [25]. The composition of the fermentation gases (CH_4 and carbon dioxide (CO_2)) was determined with the help of an infrared detector machine (ATEX Biogas Monitor Check BM 2000, Ansyco, Karlsruhe, Germany). Afterward, the absolute production of CH_4 and CO_2 (mL/day) was calculated.

The ferric reducing antioxidant power (FRAP) assay was performed on original materials (moringa leaf powder, moringa seed cake, and raw propolis) and daily samples of the incubation fluid followed the procedure of Benzie and Strain [26] with minor modifications. Shortly, 24 μL each of the blank, standard, and sample were transferred in duplicates into a 96-well plate, followed by the addition of 180 μL of the pre-warmed (37 °C) working reagent. The working reagent consisted of 25 mL of acetic acid buffer, 2.5 mL of TPTZ (2,4,6-tripyridl-s-triazin) solution, and 2.5 mL of $FeCl_3 \cdot 6H_2O$. The absorbance was measured at 490 nm after 5 min of reaction time using a thermostat spectrophotometer (xMark, Bio-Rad). A calibration curve with an increasing Trolox concentration in the range of 0–9.6 μg/24 μL was used for the quantification to express the results in Trolox equivalents.

2.4. Statistical Analysis

Statistical analysis was performed using the MIXED procedure of SAS (version 9.4, SAS Institute Inc., Cary, NC, USA). There were two kinds of data: daily data in the case of fermentation characteristics and fermentation gas formation and one-time data (pooled feed bags) for nutrient degradation. For the daily data, repeated measures of ANOVA were used to compare the fixed effect of the treatments on the fermentation characteristics and fermentation gas formation. The variation among the experimental runs was regarded as a random effect. The measurement day was the repeated measure factor, and compound symmetry was the variance-covariance structure. For the one-time data, one-way ANOVA was used to compare the effect of the treatments on nutrient degradation. The variation between experimental runs was regarded as a random effect. Pairwise comparisons between the treatments were carried out using Tukey's test. The significance was declared at $p \leq 0.05$, whereas the tendency of an effect was observed at $0.05 \leq p \leq 0.1$.

3. Results

The nutrient disappearances are summarized in Table 2. Overall, no difference among the treatments was detected for the degradation of DM, OM, EE, CP, NDF, and NFC. Only the treatment ML lowered the ash disappearance compared to the other groups ($p < 0.05$).

Table 2. Ruminal nutrient disappearance (percentage of supply) as affected by dietary treatment *.

Item	CON	PRO	ML	MSC	SEM	*p*-Value
Dry matter	42.9	43.0	41.7	42.8	2.0	0.735
Organic matter	40.5	40.6	39.7	40.3	1.9	0.909
Crude protein	43.8	42.9	41.5	43.4	2.4	0.881
Ash	69.13 [a]	69.30 [a]	64.11 [b]	69.45 [a]	2.69	0.019
Neutral detergent fiber	20.86	19.93	19.71	20.00	3.29	0.277

SEM: standard error of the mean. * A diet containing concentrate at 500 g/kg of diet dry matter without supplementation (CON) or top dressed in propolis (PRO), dried moringa leaves (ML), or moringa seed cake (MSC). [ab] The values within the same row with different superscripts indicate a significant difference ($p < 0.05$) according to Tukey's test.

Table 3 illustrates the ruminal fermentation characteristics as affected by the treatment. While the pH of the incubation fluid was unaffected, both ML and MSC groups lowered their redox potential compared to CON and PRO ($p < 0.05$). There was an increase in the NH_3 concentration (mmol/mL) with both ML (+14%) and MSC (+27%) in comparison to CON and PRO ($p < 0.05$). Treatment tended to affect the concentration of SCFA ($p = 0.06$).

According to Tukey's test, MSC resulted in 10.5% higher SCFA concentration compared to PRO ($p < 0.10$), while CON and ML showed intermediate values. MSC showed the strongest shift in the SCFA composition compared to CON (Table 3). Specifically, MSC increased propionate at the expense of the acetate ($p < 0.05$), thereby significantly reducing the acetate to a propionate ratio in comparison to the other treatments. The relative proportions of butyrate and caproate were significantly uplifted exclusively with PRO in comparison to that of CON ($p < 0.05$). The percentage of isobutyrate ($p = 0.01$) and heptanoate ($p < 0.001$) were also affected by the treatment, while valerate and isovalerate were unaffected.

Table 3. Ruminal fermentation parameters as affected by dietary treatments *.

Item	CON	PRO	ML	MSC	SEM [1]	*p*-Value
pH	6.81	6.82	6.81	6.8	0.01	0.368
Redox potential (mV)	−197 [a]	−196 [a]	−254 [b]	−263 [b]	6.00	<0.001
Ammonia (mmol/L)	3.75 [c]	3.73 [c]	4.26 [b]	4.75 [a]	0.12	<0.001
Total SCFAs [2] (mmol/L)	69.6 [xy]	67.2 [y]	72.4 [xy]	74.3 [x]	5.79	0.064
SCFAs profile (mol/100 mol)						
Acetate	49.5 [a]	49.8 [a]	49.7 [a]	48.1 [b]	0.40	<0.001
Propionate	22.9 [b]	22.1 [b]	22.3 [b]	25.3 [a]	0.50	<0.001
Butyrate	7.20 [b]	7.66 [a]	7.32 [ab]	7.05 [b]	0.19	0.008
Isobutyrate	0.66 [ab]	0.67 [a]	0.64 [b]	0.64 [b]	0.01	0.018
Valerate	9.0	8.90	9.28	9.34	0.20	0.184
Isovalerate	4.27	4.03	4.08	4.37	0.36	0.103
Caproate	4.34 [b]	4.78 [a]	4.64 [ab]	3.63 [b]	0.25	<0.001
Heptanoate	2.29 [ab]	2.36 [b]	2.53 [a]	2.01 [b]	0.13	<0.001
Acetate to propionate	2.20 [a]	2.29 [a]	2.25 [a]	1.94 [b]	0.05	<0.001
Fermentation gases						
Total fermentation gas (mL/d)	386	427	403	387	38	0.685
Carbon dioxide (mL/d)	312.0	347.0	331.0	318.0	33.00	0.738
Methane (mL/d)	44.0	47.0	42.7	39.4	3.8	0.225
[3] MCR (% Gross energy intake)	0.84 [xy]	0.90 [x]	0.75 [xy]	0.72 [y]	0.07	0.020
Gross energy intake (MJ/d)	0.207 [a]	0.207 [a]	0.226 [b]	0.218 [c]	0.0001	<0.001
Antioxidant capacity (μg Trolox/mL) [4]	2.58	2.77	2.33	2.35	0.31	0.183

* A diet containing concentrate at 500 g/kg of diet dry matter without supplementation (CON) or top-dressed in propolis (PRO), dried moringa leaves (ML) or moringa seed cake (MSC). [abc] Least square means sharing no common superscripts differ significantly ($p \leq 0.05$) according to Tukey's method. [xy] Least square means sharing no common superscripts tend to differ ($0.05 < p \leq 0.10$) according to Tukey's method. [1] SEM: standard error of the mean. [2] Short chain fatty acids. [3] Methane conversion rate was estimated as follows: gross energy (Mcal/kg DM) according to Weiss and Tebbe [27], subsequently, gross energy intake (MJ/d) was quantified, and finally, methane production was adjusted to methane in MJ/100 MJ of gross energy intake or %. [4] Using ferric reducing antioxidant power (FRAP) assay

Treatment did not affect the absolute production (mL/d) of the total fermentation gases or individual CH_4 and CO_2 but affected the relative proportion (% of total gas) of both CO_2 ($p = 0.02$) and CH_4 ($p < 0.001$) (Table 3, Figure 1). The inclusion of the moringa by-products increased the estimated gross energy intake ($p < 0.001$) but decreased the methane conversion rate (MCR) relative to the gross energy intake ($p = 0.020$).

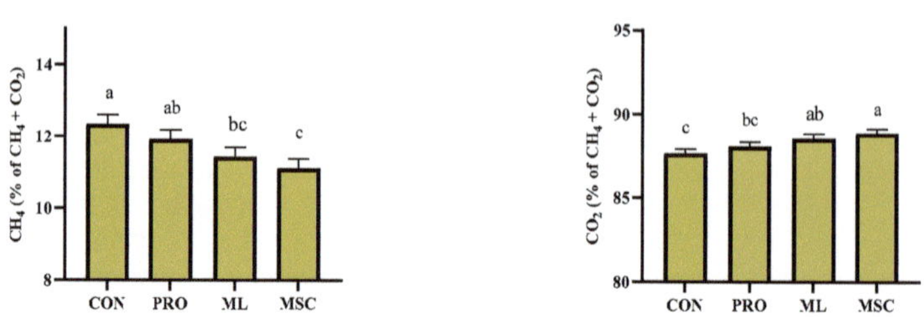

Figure 1. Relative proportion of carbon dioxide (CO_2) and methane (CH_4) as affected by PRO, ML, and MSC. Experimental diets included a diet with concentrate at 500 g kg^{-1} of diet dry matter without supplementation (CON) or top-dressed with propolis (PRO), dried moringa leaves (ML), or moringa seed cake (MSC). [abc] Least square means sharing no common superscripts differed significantly ($p \leq 0.05$) according to Tukey's method.

Accordingly, both ML and MSC resulted in 7% and 10% lower CH_4 percentages at the expense of CO_2 compared to CON ($p < 0.05$), respectively, while PRO did not show any difference from CON. As analyzed using the FRAP assay, the antioxidant capacity of the moringa seed cake, moringa leaf powder, and propolis was 1.14, 0.56, and 8.56 mg Trolox/g DM. However, the treatment did not affect the antioxidant capacity of the incubation fluid (Table 3).

4. Discussion

Moringa seed cake and moringa leaves are protein-rich ingredients that can be successfully used in ruminant diets. Moringa seed cake is a by-product obtained during the acquisition of oil from moringa seeds. Moringa seed cake and moringa leaves have been used as a substitute for good quality protein sources such as soybean meal [5,28]. The current data indicate that both moringa by-products promote the rumen microbial fermentation of proteins and carbohydrates, albeit the effect of MSC was often more prominent than that of ML. The most apparent effect of MSC and ML was the increased NH_3 concentrations (+27 and +14% of that of CON, respectively). Since the ML and MSC did not affect the CP degradation, the boosting effect of the NH_3 concentration was, therefore, associated with the increased substrate (CP) in the diet. Notably, the moringa seed cake and moringa leaves contained similar proportions of rumen degradable protein (58.4% and 66.8% of total CP, respectively) [29], which may explain the similar CP disappearances between ML and MSC observed in the present study. A higher inclusion rate (40% of diet DM) of the moringa seeds led to a greater increase in the ruminal NH_3 concentration [30] compared to the present findings. Karim et al. [12] documented the positive impact of moringa leaves on the characteristics of protein fermentation. The moringa treatments numerically increased the total SCFA concentration, which was possibly due to the extra dietary nutrients as well as an improvement in the protein and energy balance of the diet, which is important for microbial growth and activity [31]. The MSC diet also profoundly shifted the SCFA proportion to more propionate and less acetate. Aboamer et al. [32] evaluated the effect of moringa seed cake as a substitution for cottonseed meal on nutrient digestibility and milk production in Ossimi ewes. They found that the inclusion level of 2.5% of the diet DM increased gas production in vitro and increased milk lactose concentration in the ewes. Their findings were in line with the propionate-boosting effect observed in the present study. Propionate is a glucogenic precursor that is required for lactose synthesis [33]. On the contrary, the ML diet did not alter the SCFA profile despite the higher inclusion rate of moringa leaf powder compared with moringa seed cake. In agreement, Soliva et al. [34] observed no change in the SCFA production and composition even with the inclusion level as high as 30% of the diet DM. The different findings between MSC and ML hinted that the alteration of SCFA pathways might result from some secondary compounds unique to moringa seed cake, for instance, moringyne and vanillin [9]. Interestingly, ML decreased with the ash disappearance. The ruminal disappearance of minerals depended on the solubility of the minerals [35]. Our data may indicate a lower solubility of the minerals in the moringa leaf powder compared to the seed cake, which is likely due to the higher oxalate contents in moringa leaves than in the seed (1050 vs. 2.9 mg/100 g, respectively) [36].

We revealed that moringa by-products, especially seed cake, could have an added function in mitigating CH_4. It must be pointed out that the treatment difference in the absolute yield of CH_4 (mL/d) did not reach statistical significance. However, the absolute yield was also confounded with the top-dressing strategy that increased the amounts of substrates in moringa treatments. When the methane production was standardized by the diet, i.e., MCR (% of gross energy intake) or calculated as a proportion relative to the total gas production, the effect of moringa by-products became evident; therewith, MSC showed the strongest decrease in both variables. The CH_4-lowering effect observed with MSC could be explained by the increase in propionate, which is a metabolic hydrogen sink in the rumen [37], therefore, reducing the availability of metabolic hydrogen for methanogenesis. In line with our findings, the in sacco study revealed that out of seven different seed cakes from moringa, castor, cotton, palm kernel, radish, soybean, and sunflower, only moringa seed cake expressed a CH_4 mitigating property [13]. Their inclusion level was 40% of DM, which is about 10 times higher than the inclusion level used in the present study. This hints that low dosages of moringa seed cake can be effective in mitigating CH_4, which in turn can be attributed to the presence of oil in moringa seed cake, as oils likely reduce methanogenesis [38]. Surplus lipids could be detrimental to ruminal microbes especially those degrading fiber [39]. However, this was not the main explanation for the effect of MSC observed in the present study because the EE levels of all the diets (approx. 2% of diet DM) were still within the range deemed suitable for rumen microbial fermentation [40]. Moringa contains various functional compounds, including myricetin, quercetin, moringyne, vanillin, rutins, tannins, gallic acid, and kaempferol [9], which may contribute to the CH_4-lowering effect of moringa by-products tested in the present study. Some of these compounds in the extracts of other plants have been shown to mitigate enteric CH_4 production [41–43]. The weaker effect of ML on rumen fermentation variables, despite its higher inclusion level than MSC, might be related to the presence of different functional compounds. The contents and profiles of secondary compounds in moringa leaves vary from those of the moringa seed [9,11,12]. Additionally, we showed that moringa seed cake had twice the antioxidant power, based on the FRAP assay, compared to moringa leaf powder. Some plant secondary compounds, such as tannins, are known to reduce CH_4 synthesis via multiple routes, some of which do not involve an association with propionate production [44].

Despite the higher antioxidant value than those of moringa products, the addition of raw propolis did not drastically modulate the ruminal fermentation and the gas production parameters in the present experiment. Some researchers explored the effects of the propolis extract [14,21,45,46] or propolis phenols [42] and consistently documented higher butyrate production. Data from earlier studies suggest that the polyphenolic compounds of propolis, for instance, caffeic acid [47], might have played a role because it supports the growth of gut butyrate-producing bacteria [48], which might have replaced the population of other bacteria, such as propolis which also possess microbial inhibition properties [49]. Notably, we observed the highest proportion of butyrate with the raw propolis, but the change did not reach significance. This may be related to the low dosage as well as the form, i.e., raw propolis. To our knowledge, there is no comparative study using raw propolis on ruminal fermentation characteristics. In addition, our test product was from a local supplier, and thus, variation in the product quality and the effect on ruminal fermentation must be considered as well. Still, given the role of butyrate as a promoter of gut epithelial integrity [50], future research may invest in finding the effective (higher) dosages of propolis that could express a benefit on ruminal fermentation and gut health.

5. Conclusions

The inclusion of moringa seed cake at 3.8% of the diet DM modulated ruminal fermentation characteristics, leading to greater NH_3, favoring propionate production, and mitigating CH_4 without any negative effect on nutrient disappearance and physicochemical parameters in vitro. At a higher inclusion rate (7.4% of diet DM), the moringa leaf powder showed a similar direction, albeit a weaker effect, on NH_3 and CH_4 variables. Despite having a stronger antioxidant power compared to the moringa by-products, raw propolis supplemented at 0.1% of the diet DM did not affect the fermentation variables in vitro. Our data suggest that feeding moringa seed cake desirably modulates the rumen microbial metabolic activities of proteins and carbohydrates. Of note, in vitro studies do not account for host-dependent influences. In vivo studies are needed to identify the effective dosages that facilitate the health and production of ruminants.

Author Contributions: Conceptualization, Q.Z., R.K.-a. and H.E.S.; methodology, Q.Z and R.K.-a.; validation, Q.Z., R.K.-a. and M.M.; formal analysis, R.K.-a. and S.S.; investigation, M.M., H.E.S. and A.S.-A.; resources, Q.Z. and H.E.S.; data curation, R.K.-a., M.M. and H.E.S.; writing—original draft preparation, M.M.; writing—review and editing, H.E.S., S.S., A.S.-A., R.K.-a. and Q.Z.; visualization,

M.M.; supervision, Q.Z. and R.K.-a.; project administration, Q.Z. and R.K.-a.; funding acquisition, Q.Z. and H.E.S. All authors have read and agreed to the published version of the manuscript.

Funding: This research received no external funding.

Institutional Review Board Statement: Not applicable.

Informed Consent Statement: Not applicable.

Data Availability Statement: The data presented in this study are available in the main article.

Acknowledgments: We appreciate the technical assistance from A. Dockner, M. Hollmann, and S. Leiner (Institute of Animal Nutrition and Functional Plant Compounds, University of Veterinary Medicine, Vienna) for the sample analysis. We thank the staff of Ruminant Clinic (University of Veterinary Medicine, Vienna) for the care of the donor cows. We thank R. Schuhmacher (Horgen, Switzerland) for the kind support of the moringa products.

Conflicts of Interest: The authors declare no conflict of interest.

References

1. Dodd, D.; Mackie, R.I.; Cann, I.K. Xylan degradation, a metabolic property shared by rumen and human colonic *Bacteroidetes*. *Mol. Microbiol.* **2011**, *79*, 292–304. [CrossRef] [PubMed]
2. Jayasuriya, M.C.N. Use of crop residues and agro-industrial by-products in ruminant production systems in developing countries. *BSAP Occas. Publ.* **1993**, *16*, 47–55. [CrossRef]
3. Xia, C.; Rahman, M.A.U.; Yang, H.; Shao, T.; Qiu, Q.; Su, H.; Cao, B. Effect of increased dietary crude protein levels on production performance, nitrogen utilisation, blood metabolites and ruminal fermentation of Holstein bulls. *Asian-Australas. J. Anim. Sci.* **2018**, *31*, 1643–1653. [CrossRef] [PubMed]
4. Islam, Z.; Islam, S.M.; Hossen, F.; Mahtab-ul-Islam, K.; Hasan, M.; Karim, R. *Moringa oleifera* is a prominent source of nutrients with potential health benefits. *Int. J. Food Sci.* **2021**, *2021*, 6627265. [CrossRef] [PubMed]
5. Su, B.; Chen, X. Current status and potential of *Moringa oleifera* leaf as an alternative protein source for animal feeds. *Front. Vet. Sci.* **2020**, *7*, 53–66. [CrossRef]
6. Godino, M.; Arias, C.; Izquierdo, M.I. *Moringa oleifera: Potential* areas of cultivation on the Iberian Peninsula. *Acta Hortic.* **2017**, *1158*, 405–412. [CrossRef]
7. Gopalakrishnan, L.; Doriya, K.; Kumar, D.S. *Moringa oleifera*: A review on nutritive importance and its medicinal application. *Food Sci. Hum. Wellness* **2016**, *5*, 49–56. [CrossRef]
8. Patil, D.; Vaknin, Y.; Rytwo, G.; Lakemond, C.; Benjamin, O. Characterization of *Moringa oleifera* leaf and seed protein extract functionality in emulsion model system. *Innov. Food Sci. Emerg. Technol.* **2022**, *75*, 102903. [CrossRef]
9. Kumar, N.; Pareek, S. Bioactive compounds of moringa (Moringa species). In *Bioactive Compounds in Underutilized Vegetables and Legumes*, 1st ed.; Murthy, H.N., Paek, K.Y., Eds.; Springer: Cham, Switzerland, 2021; pp. 1–22. [CrossRef]
10. Soltan, Y.A.; Morsy, A.S.; Hashem, N.M.; Sallam, S.M. Utilization of *Moringa oleifera* in ruminant nutrition. In Proceedings of the Sustainable Development of Livestock's Production Systems "(SDLPS)", Alexandra University, Bab Sharqi, Egypt, 7–9 November 2017; pp. 7–9.
11. Mohammed, S.; Manan, F.A. Analysis of total phenolics, tannins and flavonoids from *Moringa oleifera* seed extract. *J. Chem. Pharm. Res.* **2015**, *7*, 132–135.
12. Karim, R.A.; Amin, M.R.; Moniruzzaman, M.; Sarker, M.B.; Kabir, A.K.M.A. Effect of *Moringa oleifera* leaf on the efficiency to increase protein supply to ruminants. *BJAS* **2015**, *44*, 46–51. [CrossRef]
13. Olivares-Palma, S.M.; Meale, S.J.; Pereira, L.G.R.; Machado, F.S.; Carneiro, H.; Lopes, F.C.F.; Maurício, R.M.; Chaves, A.V. In vitro fermentation, digestion kinetics and methane production of oilseed press cakes from biodiesel production. *Asian-Australas. J. Anim. Sci.* **2013**, *26*, 1102–1112. [CrossRef] [PubMed]
14. Morsy, T.A.; Gouda, G.A.; Kholif, A.E. In vitro fermentation and production of methane and carbon dioxide from rations containing *Moringa oleifera* leave silage as a replacement of soybean meal: In vitro assessment. *Environ. Sci. Pollut. Res.* **2022**, *29*, 69743–69752. [CrossRef] [PubMed]
15. ÖZTURK, H.; Pekcan, M.; Sireli, M.; Fidanci, U.R. Effects of propolis on in vitro rumen microbial fermentation. *Ankara Univ. Vet. Fak.* **2010**, *57*, 217–221.
16. Piccinelli, A.L.; Mencherini, T.; Celano, R.; Mouhoubi, Z.; Tamendjari, A.; Aquino, R.P.; Rastrelli, L. Chemical composition and antioxidant activity of Algerian propolis. *J. Agric. Food Chem.* **2013**, *61*, 5080–5088. [CrossRef] [PubMed]
17. Ahangari, Z.; Naseri, M.; Vatandoost, F. Propolis: Chemical composition and its applications in endodontics. *Iran. Endod. J.* **2018**, *13*, 285–292. [CrossRef]
18. Kumazawa, S. February. Bioactive compounds in bee propolis for drug discovery. *AIP Conf. Proc.* **2018**, *1933*, 030001. [CrossRef]
19. Mahmood, M.; Petri, R.M.; Gavrău, A.; Zebeli, Q.; Khiaosa-ard, R. Betaine addition as a potent ruminal fermentation modulator under hyperthermal and hyperosmotic conditions in vitro. *J. Sci. Food Agric.* **2020**, *100*, 2261–2271. [CrossRef]
20. Thomas, J.W. Protein requirements of milking cows. *J. Dairy Sci.* **1971**, *54*, 1629–1636. [CrossRef]

21. Santos, N.W.; Yoshimura, E.H.; Machado, E.; Matumoto-Pintro, P.T.; Montanher, P.F.; Visentainer, J.V.; dos Santos, G.T.; Zeoula, L.M. Antioxidant effects of a propolis extract and vitamin E in blood and milk of dairy cows fed diet containing flaxseed oil. *Livest. Sci.* **2016**, *191*, 132–138. [CrossRef]
22. BMSGPK. *Verordnung der Bundesministerin für Gesundheit und Frauen uber die Mindestanforderungen fur die Haltung von Pferden und Pferdeartigen, Schweinen, Rindern, Schafen, Ziegen, Schalenwild, Lamas, Kaninchen, Hausgeflugel, Strausen und Nutzfischen (1. Tierhaltungsverordnung) StF: BGBl. II Nr. 485/2004*; Bundeskanzleramt Osterreich: Vienna, Austria, 2004.
23. VDLUFA. Die Chemische Untersuchung von Futtermitteln. In *Handbuch der Landwirtschaftlichen Versuchs-und Untersuchungsmethodik (VDLUFA-Methodenbuch), Bd., Vol. III*; VDLUFA-Verlag: Darmstadt, Germany, 2012.
24. Weatherburn, M.W. Phenol-hypochlorite reaction for determination of ammonia. *Anal. Chem.* **1967**, *39*, 971–974. [CrossRef]
25. Soliva, C.R.; Hess, H.D. Measuring methane emission of ruminants by in vitro and in vivo techniques. In *Measuring Methane Production from Ruminants*, 1st ed.; Makkar, H.P., Vercoe, P.E., Eds.; Springer: Dordrecht, Holland, 2007; pp. 15–31. [CrossRef]
26. Benzie, I.F.; Strain, J.J. The ferric reducing ability of plasma (FRAP) as a measure of "antioxidant power": The FRAP assay. *Anal. Biochem.* **1996**, *239*, 70–76. [CrossRef] [PubMed]
27. Weiss, W.P.; Tebbe, A.W. Estimating digestible energy values of feeds and diets and integrating those values into net energy systems. *Transl. Anim. Sci.* **2019**, *3*, 953–961. [CrossRef] [PubMed]
28. El-Naggar, S.I. Impact of incorporating *Moringa oleifera* seed cake as protein source in growing lambs ration. *CIGR J.* **2017**, *2017*, 289–292.
29. Kakengi, A.M.V.; Shem, M.N.; Sarwatt, S.V.; Fujihara, T. Can *Moringa oleifera* be used as a protein supplement for ruminants? *Asian-Australas. J. Anim. Sci.* **2005**, *18*, 42–47. [CrossRef]
30. Lins, T.D.A.; Terry, S.A.; Silva, R.R.; Pereira, L.G.R.; Jancewicz, L.J.; He, M.L.; Wang, Y.; McAllister, T.A.; Chaves, A.V. Effects of the inclusion of *Moringa oleifera* seed on rumen fermentation and methane production in a beef cattle diet using the rumen simulation technique (Rusitec). *Animal* **2019**, *13*, 283–291. [CrossRef]
31. Nocek, J.E.; Russell, J. Protein and energy as an integrated system. Relationship of ruminal protein and carbohydrate availability to microbial synthesis and milk production. *J. Dairy Sci.* **1988**, *71*, 2070–2107. [CrossRef]
32. Aboamer, A.A.; Ebeid, H.M.; Shaaban, M.M.; Gaward, R.M.A.; Mostafa, M.M.; Abdalla, A.M. Effect of feeding moringa seed cake as an alternative protein source in lactating ewes rations. *Int. J. Dairy Sci.* **2020**, *15*, 80–87. [CrossRef]
33. Aschenbach, J.R.; Kristensen, N.B.; Donkin, S.S.; Hammon, H.M.; Penner, G.B. Gluconeogenesis in dairy cows: The secret of making sweet milk from sour dough. *IUBMB Life* **2010**, *62*, 869–877. [CrossRef]
34. Soliva, C.R.; Kreuzer, M.; Foidl, N.; Foidl, G.; Machmüller, A.; Hess, H.D. Feeding value of whole and extracted *Moringa oleifera* leaves for ruminants and their effects on ruminal fermentation in vitro. *Anim. Feed Sci. Technol.* **2005**, *118*, 47–62. [CrossRef]
35. Etcheverry, P.; Grusak, M.A.; Fleige, L.E. Application of in vitro bioaccessibility and bioavailability methods for calcium, carotenoids, folate, iron, magnesium, polyphenols, zinc, and vitamins B6, B12, D, and E. *Front. Physiol.* **2012**, *3*, 317. [CrossRef]
36. Grosshagauer, S.; Pirkwieser, P.; Kraemer, K.; Somoza, V. The future of Moringa foods: A food chemistry perspective. *Front. Nutr.* **2021**, *8*, 1–9. [CrossRef] [PubMed]
37. Guyader, J.; Ungerfeld, E.M.; Beauchemin, K.A. Redirection of metabolic hydrogen by inhibiting methanogenesis in the rumen simulation technique (RUSITEC). *Front. Microbiol.* **2017**, *8*, 393–409. [CrossRef]
38. Vargas, J.E.; Andrés, S.; López-Ferreras, L.; Snelling, T.J.; Yáñez-Ruíz, D.R.; García-Estrada, C.; López, S. Dietary supplemental plant oils reduce methanogenesis from anaerobic microbial fermentation in the rumen. *Sci. Rep.* **2020**, *10*, 1613. [CrossRef] [PubMed]
39. Ibrahim, N.A.; Alimon, A.R.; Yaakub, H.; Samsudin, A.A.; Candyrine, S.C.L.; Wan Mohamed, W.N.; Mookiah, S. Effects of vegetable oil supplementation on rumen fermentation and microbial population in ruminant: A review. *Trop. Anim. Health Prod.* **2021**, *53*, 422. [CrossRef] [PubMed]
40. Messana, J.D.; Berchielli, T.T.; Arcuri, P.B.; Reis, R.A.; Canesin, R.C.; Ribeiro, A.F.; Fernandes, J.J.D.R. Rumen fermentation and rumen microbes in Nellore steers receiving diets with different lipid contents. *Rev. Bras. Zootec.* **2013**, *42*, 204–212. [CrossRef]
41. Vergara-Jimenez, M.; Almatrafi, M.M.; Fernandez, M.L. Bioactive components in *Moringa oleifera* leaves protect against chronic disease. *Antioxidants* **2017**, *6*, 91. [CrossRef]
42. Oskoueian, E.; Abdullah, N.; Oskoueian, A. Effects of flavonoids on rumen fermentation activity, methane production, and microbial population. *BioMed Res. Int.* **2013**, *2013*, 349129. [CrossRef]
43. Aboagye, I.A.; Oba, M.; Koenig, K.M.; Zhao, G.Y.; Beauchemin, K.A. Use of gallic acid and hydrolyzable tannins to reduce methane emission and nitrogen excretion in beef cattle fed a diet containing alfalfa silage. *J. Anim. Sci.* **2019**, *97*, 2230–2244. [CrossRef]
44. Ku-Vera, J.C.; Jiménez-Ocampo, R.; Valencia-Salazar, S.S.; Montoya-Flores, M.D.; Molina-Botero, I.C.; Arango, J.; Gómez-Bravo, C.A.; Aguilar-Pérez, C.F.; Solorio-Sánchez, F.J. Role of secondary plant metabolites on enteric methane mitigation in ruminants. *Front. Vet. Sci.* **2020**, *7*, 584. [CrossRef]
45. Morsy, A.S.; Soltan, Y.A.; El-Zaiat, H.M.; Alencar, S.M.D.; Abdalla, A.L. Bee propolis extract as a phytogenic feed additive to enhance diet digestibility, rumen microbial biosynthesis, mitigating methane formation and health status of late pregnant ewes. *Anim. Feed Sci. Technol.* **2021**, *273*, 114834. [CrossRef]

46. Costa, J.B.G., Jr.; Zeoula, L.M.; Franco, S.L.; de Moura, L.P.; Valero, M.V.; Simioni, F.L.; da Paula, E.M.; Samensari, R.B. Effect of propolis product on digestibility and ruminal parameters in buffaloes consuming a forage-based diet. *Ital. J. Anim. Sci.* **2012**, *11*, e78. [CrossRef]
47. Celińska-Janowicz, K.; Zaręba, I.; Lazarek, U.; Teul, J.; Tomczyk, M.; Pałka, J.; Miltyk, W. Constituents of propolis: Chrysin, caffeic acid, p-coumaric acid, and ferulic acid induce PRODH/POX-dependent apoptosis in human tongue squamous cell carcinoma cell (CAL-27). *Front. Pharmacol.* **2018**, *9*, 336. [CrossRef]
48. Xu, J.; Ge, J.; He, X.; Sheng, Y.; Zheng, S.; Zhang, C.; Xu, W.; Huang, K. Caffeic acid reduces body weight by regulating gut microbiota in diet-induced-obese mice. *J. Funct. Foods.* **2020**, *74*, 104061. [CrossRef]
49. Kačániová, M.; Rovná, K.; Arpášová, H.; Čuboň, J.; Hleba, L.; Pochop, J.; Kunová, S.; Haščík, P. In vitro and in vivo antimicrobial activity of propolis on the microbiota from gastrointestinal tract of chickens. *J. Environ.* **2012**, *47*, 1665–1671. [CrossRef]
50. Penner, G.B.; Steele, M.A.; Aschenbach, J.R.; McBride, B.W. Ruminant Nutrition Symposium: Molecular adaptation of ruminal epithelia to highly fermentable diets. *J. Anim. Sci.* **2011**, *89*, 1108–1119. [CrossRef] [PubMed]

metabolites

Article

Enhancement of Cadmium Phytoremediation Potential of *Helianthus annuus* L. with Application of EDTA and IAA

Naila Shah [1,2], Muhammad Qadir [1], Muhammad Irshad [1,*], Anwar Hussain [1], Muhammad Hamayun [1], Waheed Murad [1], Ajmal Khan [3,*] and Ahmed Al-Harrasi [3,*]

[1] Department of Botany, Garden Campus, Abdul Wali Khan University Mardan, Mardan 23200, Pakistan
[2] Department of Botany, Government Girls College, Mardan 23200, Pakistan
[3] Natural and Medical Sciences Research Center, University of Nizwa, P.O. Box 33, Birkat Al-Mauz, Nizwa 616, Oman
* Correspondence: muhammad.irshad@awkum.edu.pk (M.I.); ajmalkhan@unizwa.edu.om (A.K.); aharrasi@unizwa.edu.om (A.A.-H.)

Abstract: The aim of the current study was to assess the cadmium (Cd) phytoremediation potential of *Helianthus annuus* L. that was exposed to 50, 100, and 150 mg/kg of cadmium for 15, 30, and 60 days with application of EDTA (Ethylenediaminetetraacetic acid) in the soil and IAA (indole acetic acid) as a foliar spray. The results indicated that the concentration, duration of exposure, and amount of Cd affect the phytoremediation potential. The maximum Cd was observed at 60 days (32.05, 16.86, and 10.63%) of Cd application, compared to 15 (2.04, 0.60, and 1.17%) or 30 days (8.41, 3.93, and 4.20%, respectively), in a dose-dependent manner. The application of EDTA in the soil and foliar IAA enhanced the Cd accumulation in the plants at 15, 30, and 60 days of exposure, with maximum accumulation at 60 days. Exposed plants with foliar IAA application showed 64.82%, 33.77%, and 25.84% absorption at 50, 100, and 150 mg/kg, respectively. Apart from higher absorption, the cadmium translocation to the edible part of the plants ceased, i.e., the seeds had 0% accumulation. The interesting fact was recorded that efficient phytoremediation was recorded at 15 days of exposure, whereas maximum phytoremediation was recorded at 60 days of exposure. To minimize the stress, the host also produced stress-related metabolites (i.e., flavonoids, phenolics, proline, and sugar) and antioxidants (i.e., catalases and ascorbate peroxidases). From the current evidence, it could be assumed that the use of EDTA and IAA, along with hyperaccumulating plants, could be a possible green method to remediate Cd-contaminated soil efficiently in a short period of time.

Keywords: sunflower; EDTA; IAA; short exposure duration; efficient cadmium remediation

Citation: Shah, N.; Qadir, M.; Irshad, M.; Hussain, A.; Hamayun, M.; Murad, W.; Khan, A.; Al-Harrasi, A. Enhancement of Cadmium Phytoremediation Potential of *Helianthus annuus* L. with Application of EDTA and IAA. *Metabolites* **2022**, *12*, 1049. https://doi.org/10.3390/metabo12111049

Academic Editor: Hirokazu Kawagishi

Received: 12 October 2022
Accepted: 28 October 2022
Published: 31 October 2022

1. Introduction

The release of heavy metals to the environment is very common in the current world. They mainly contaminate the soil and water, which plays havoc with the lives of both flora and fauna [1]. Soil reclamation through physical means is a cost-effective, labor-consuming, and time-efficient process with many side effects in the form of environmental imbalance in nutrients uptake, disturbed food chains, and effects on various nutrient-cycling phenomenon [2]. Moreover, increasing demand for food and forage crops for the world's population is the need of the hour. Striving for the fulfillment of the need for food for human beings and livestock results in the contamination of the environment by various agencies [3]. Anthropogenic activities are the main forces responsible for the contamination of soil by heavy metals [4]. The toxicity levels of heavy metals increase with increases in their sources and decrease in sinks or with no sinks. In such a scenario, even an essential heavy metal becomes toxic [5].

Cadmium is one of the heavy metals that is nonessential, and beyond the threshold levels is highly toxic and lethal to micro- and macro-flora and -fauna. The total Cd levels present in the soil do not inevitably reflect the Cd bioavailability to the host plants [6]

because Cd's varied distributions (adsorbed vs. free) and various chemical configurations (chemical speciation) impact its phytoavailability. Because it is primarily present in soil attached to the exchangeable solid phases, cadmium is comparatively accessible for plant absorption [7] and thus readily released into the soil solution. Cd is mostly found in the soil solution as the Cd2+ ion or as inorganic or organic compounds. Cd may reversibly bind to soil particles such as organic matter or Fe and Mn oxides in the solid phase. Cd is mostly absorbed by plants when they come into contact with pore water, which is the end result of Cd being divided between the liquid and solid phases of soil [8]. Damage to agricultural crops is a serious matter for plant physiologists and ecologists as a result of Cd availability and uptake [9]. It results in a reduction in the yield of the crop and, on the other hand, acts as a serious pollutant for soil and water. Apart from that, the cadmium is considered a group A contaminant and has carcinogenic and mutagenic properties in plants, animals, and particularly humans [10].

Due to ever-growing populations and increasing food demands, these contaminated lands can be left barren. In developing countries such as Pakistan, the cadmium concentration ranges from 0.03 to 0.07 mg/L [11]. To decrease the gap between production and demands, these agricultural lands must be made cultivable and contaminant-free. To remove the contaminants, particularly Cd, from the environment, several techniques are in practice to identify a cost-effective and viable method among them. Phytoremediation is considered a green way to remove the metal from the environment. It is a "plant-based method", in which elemental contaminants are extracted and removed from the environment or have their soil bioavailability reduced [12,13]. Even at low quantities, ionic substances in the soil can be absorbed by plants through their root systems. In order to absorb heavy metals and control their bioavailability, plants stretch their root systems into the soil matrix and create rhizosphere ecosystems, recovering the contaminated soil and maintaining soil fertility [14]. The differential adsorption–desorption properties of heavy metals alone and in combinations have differential effects on their transport in the soil, which may be utilized efficiently in phytoremediation practices. For instance, the recovery ratio of Cd ions from porous matter was higher than Pb ions, which was attributed to the lower adsorption of Cd ions on the solid matrix [15]. Similarly, the transport of Pb was found to be influenced by silica powders in porous media [16].

The use of phytoremediation has benefits. In addition to being easy to manage and inexpensive to install and maintain, phytoremediation has the following advantages: (i) it can reduce the introduction of pollutants to the environment and ecosystem; (ii) it can be used on a large-scale field; (iii) it can be easily disposed of; (iv) it prevents erosion and metal leaching by stabilizing heavy metals, reducing the risk of contaminants spreading; and (v) it stabilizes heavy metals [17]. In the current study, sunflowers were used as remedial plants for cadmium remediation, whereas EDTA was supplemented in the soil and IAA was applied as foliar spray to enhance their remediation potential and efficiently remove the cadmium from the contaminated site.

2. Materials and Methods

2.1. Soil Preparation

Plant growth medium was prepared by mixing sand (~0.5 mm), clay (~0.002 mm), and manure in a 2:1:1 ratio to make a sandy loam for better sunflower growth. The pots were filled with 5 kg of soil and were kept in the screen house at Abdul Wali Khan University Mardan.

2.2. Experimental Design

For the pot experiment, healthy and viable seeds of *H. annuus* L. (Hysun-33) were obtained from the agriculture research center in Mardan. The seeds were surface-sterilized with 70% ethanol, followed by rinsing with sterilized distilled water to remove the ethanol. The experiment was a factorial combination of three factors, i.e., Cd treatments (0, 50, 100, and 150 mg/kg of soil), EDTA supplemented in the soil in the form of an EDTA solution (5 mM (500 mL of solution in each pot)), the foliar application of IAA using a portable spray machine (2.5 µM solution sprayed at 5-day intervals until final harvest), and the combination of EDTA and IAA with selected metal levels. All the treatments had three replicates with four plants per replicate, which were allowed to grow in greenhouse conditions. The pots were properly irrigated with common tap water systematically in the morning and evening. For plants harvested at 15 and 30 days of exposure, only the root shoot lengths, fresh and dry weights, and metal contents were recorded. However, for plants harvested after 60 days of exposure, the effects of cadmium, EDTA, and IAA were recorded on different parameters, including growth and physiological attributes and cadmium accumulation in the presence and absence of EDTA and IAA.

2.3. *Estimation of Indole Acetic Acid and Salicylic Acid*

The indole acetic acid was determined by the protocol used by Hussain and Hasnain [18], whereas the salicylic acid was estimated using the protocol of Warrier et al. [19].

2.4. *Quantification of Metabolites*

The total flavonoid contents were extracted by macerating 0.5 g of fresh leaves of the host in 5 mL of 80% ethanol and incubating for 24 h in a shaker. Following incubation, centrifugation was performed at 10,000 rpm at 25 °C for 15 min. Pellets were removed, and supernatants were used for flavonoid determination using the $AlCl_3$ method, as mentioned earlier [20].

To extract the phenolics, 1 g of plant leaves were crushed in 16 mL of ethanol and incubated at 30 °C for 3 h. Following incubation, centrifugation was performed for 10 min at 10,000 rpm in normal conditions. The supernatants were filtered via Whatman No. 42 filter paper, and the volume was reduced to 1 mL using a rotary evaporator at 40 °C. The concentrated filtrate was resuspended in 10 mL of dH_2O, and phenolics were determined using the method of El Far [20].

The extraction of proline was performed from the leaves of the host using 0.2 g of fresh leaves (macerated to fine powder in liquid nitrogen) with 1 mL of 60% ethanol. The resultant mixture was kept for incubation at 4 °C for about 24 h. The reaction mixtures were centrifuged for 5 min at 10,000 rpm. To remove nearly all of the proline from the leaves, the procedure was repeated. Proline was estimated using the technique of Bates et al. [21].

2.5. *Determination of Antioxidant Response*

To assess the CAT activity, the cleavage of H_2O_2 was assessed [22]. A mixture of 3% H_2O_2 (0.4 mL) and 0.1 mM EDTA in 2.6 mL of 50 mM PBS (pH 7) was added to 0.1 mL of supernatant. The decrease in H_2O_2 was noted by the decline in optical density at 240 nm, which was considered degradation by µM H_2O_2 min^{-1}.

The protocol of Asada [23] was used for the estimation of APX in the leaves of the host. For the reaction to start, approximately 0.2 mL of leaf extracts were mixed with 0.1 mL of 0.5 mM ascorbic acid, 0.6 mL of 50 mM PBS (pH 7.0), and 0.1 mL of 0.1 mM H_2O_2. The decline in O.D was noted at 290 nm and expressed as U mg^{-1} protein (U = change of 0.1 absorbance min^{-1} mg^{-1} of protein).

The protein contents were estimated for each extract according to Bradford [24]. The chlorophyll and carotenoid pigments were quantified according to the method reported by Schoefs [25].

2.6. *Estimation of the Metal in Plant Biomass*

For the estimation of the metals in the plant parts treated with the aforementioned levels of Cd, 0.5 g of oven-dried samples were weighted and subjected to acid digestion. The process of acid digestion was started by adding 1 mL of perchloric acid ($HClO_4$) and 4 mL of nitric acid (HNO_3) to the oven-dried plant samples. The mixture was filtered using Whatman 42 filter paper after cooling at 30 °C. With distilled water, the mixture's final volume was changed to 25 mL. As a positive control solution, control plant samples underwent the same processing as the experimental samples. Similar steps were taken to produce the blank solution but without including the sample. The cadmium was quantified through an atomic absorption spectrophotometer (Perkin–Elmer model 700, MA, USA) following the protocol of Amin et al. [26].

2.7. Data Analysis

The experiments were repeated three times, and the data obtained from the factorial experiments were grouped into cadmium, cadmium/EDTA, cadmium/IAA, and cadmium/EDTA/IAA treatment conditions. An analysis of variance and Duncan's multiple range test were performed using SPSS Statistical Package v. 21 (IBM, Armonk, NY, USA) to determine the significance at $p \leq 0.05$.

3. Results

3.1. Effects of Cadmium, EDTA, and IAA on the Agronomic Attributes of *H. annuus* L.

When exposing sunflowers to the aforementioned supplementation of Cd and the application of the EDTA and IAA, the growth attributes were influenced significantly (Figure 1a). Dose-dependent decreases of 15, 22, and 25% were recorded at 50 mg/kg to 150 mg/kg, respectively, in the shoot and root length of the host plants. The application of EDTA in the presence of Cd significantly improved the shoot and root length as the level of Cd increased. However, the length was lower compared to the untreated control. On the other hand, a similar enhancement in the shoot and root length was also recorded upon the foliar application of IAA, reducing the effects of Cd by 8, 9, and 7% at 50, 100, and 150 mg/kg, respectively. Similar declines were also noted in the case of the total chlorophyll contents of the host plants, showing dose-dependent declines as the metal concentration increased from 0 mg to 150 mg/kg (Figure 1b). An improvement was recorded with the application of EDTA with the application of 50 mg/kg of Cd. However, no further improvement was recorded with the application of EDTA or IAA, separately or in combination, at all concentrations of the metal in the soil.

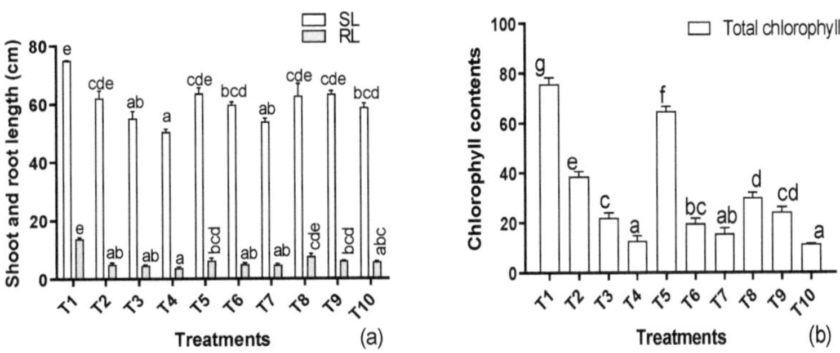

Figure 1. Effects of different levels of Cd under EDTA and IAA application on root and shoot length (**a**) and chlorophyll contents (**b**) of *H. annuus*. Data are the means of three replicates ± standard errors, where the different letters represent significant differences at $p \leq 0.05$. T1 = control, T2 = 50 mg/kg Cd, T3 = 100 mg/kg Cd, T4 = 150 mg/kg Cd, T5 = 50 mg/kg Cd + EDTA, T6 = 100 mg/kg Cd + EDTA, T7 = 150 mg/kg Cd + EDTA, T8 = 50 mg/kg Cd + EDTA + IAA, T9 = 100 mg/kg Cd + EDTA + IAA, T10 = 150 mg/kg Cd + EDTA + IAA.

3.2. Determination of Phytohormones

During the exposure of the host plant to the aforementioned levels of cadmium, a concentration-based significant decline was noted in the endogenous IAA synthesis in the sunflower as the metal level increased in the soil (Figure 2a). The supplementation of EDTA in the soil and IAA as a foliar spray improved the IAA production. However, the amount of endogenous IAA was lower than that of untreated plants. A contrary tendency was noted in the case of SA production, showing an opposite trend to IAA production

(Figure 2b). A direct relation of SA with the metal concentration was recorded, where an increase in the metal concentration increased the SA production. However, the application of EDTA and IAA inversely regulated the SA production, i.e., with the application of EDTA and IAA and an increasing metal concentration, a decline was recorded in the salicylic acid production.

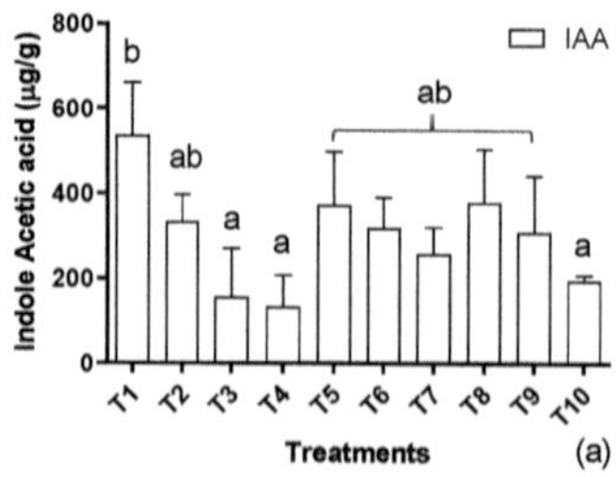

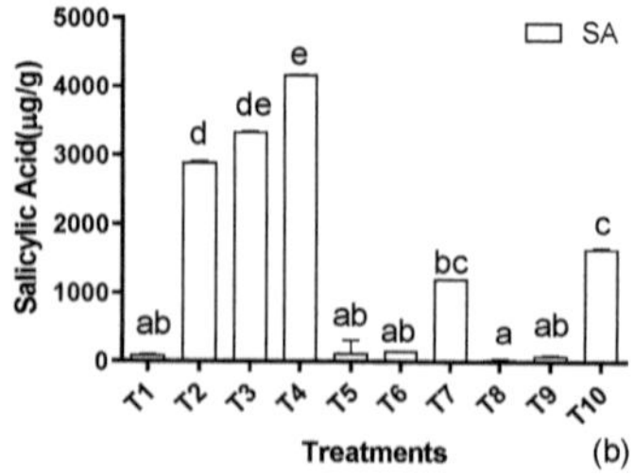

Figure 2. Effects of Cd levels, EDTA, and IAA on (**a**) indole acetic acid and (**b**) salicylic acid of *H. annuus*. Data are the means of three replicates ± standard errors, and the letters represent significant differences ($p < 0.05$).

3.3. Determination of Metabolites

Treating plants with the selected levels of cadmium negatively impacted the endogenous flavonoid production in the host plants. A concentration-based reduction in the accumulated flavonoid content was recorded (Figure 3a). The treatments of EDTA in the soil and IAA as a foliar spray significantly improved the flavonoid production of the host. However, the amounts were lower in comparison to the untreated control plants. On the other hand, endogenous phenolics were boosted when the soil was supplemented with the mentioned concentration of cadmium and showed a positive correlation with cadmium levels (Figure 3b). Similarly, the application of EDTA and IAA further significantly improved the phenolic production in comparison to the untreated hosts. A similar tendency was also noted in the case of endogenous proline production, i.e., increases in the cadmium levels were associated with accumulated proline contents, and they showed a direct proportion (Figure 3c). Improvements were also recorded with the application of EDTA and IAA in the presence of the mentioned supplements of cadmium in the soil.

Significant declines in the total protein and lipid levels were recorded in the host plants when the aforementioned levels of cadmium were supplemented in the soil (Figure 3d). EDTA-amended soil and foliar IAA enhanced the protein and lipid contents. However, the quantity was lowest compared to the untreated plants (Figure 3e). Similar tendencies were also noted in the accumulation of the total sugar contents of the sunflowers, which showed a decline with the increase in cadmium concentration (Figure 3f). Significant improvements were recorded with EDTA and IAA application. However, the sugar contents remained lower than in the control plants.

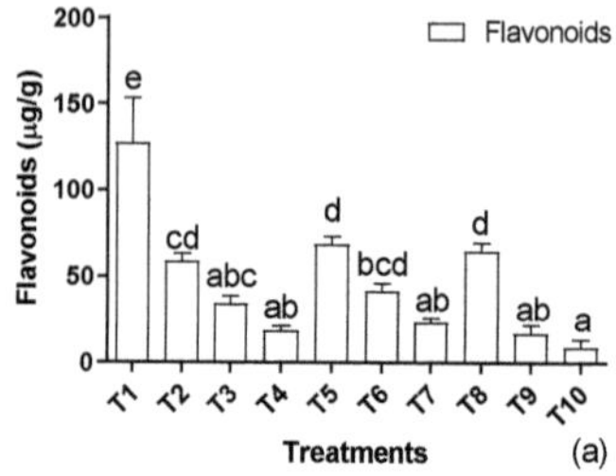

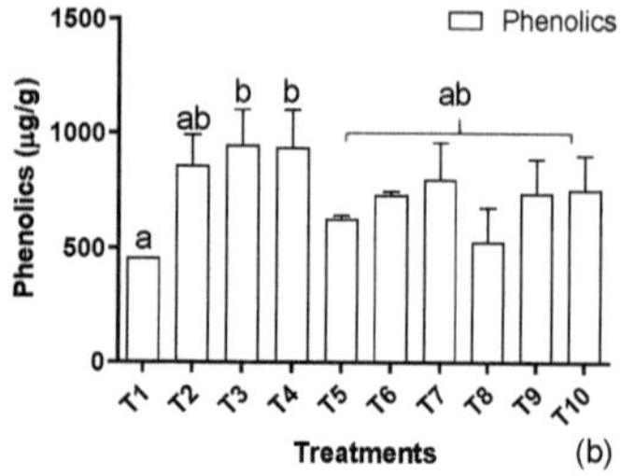

Figure 3. *Cont.*

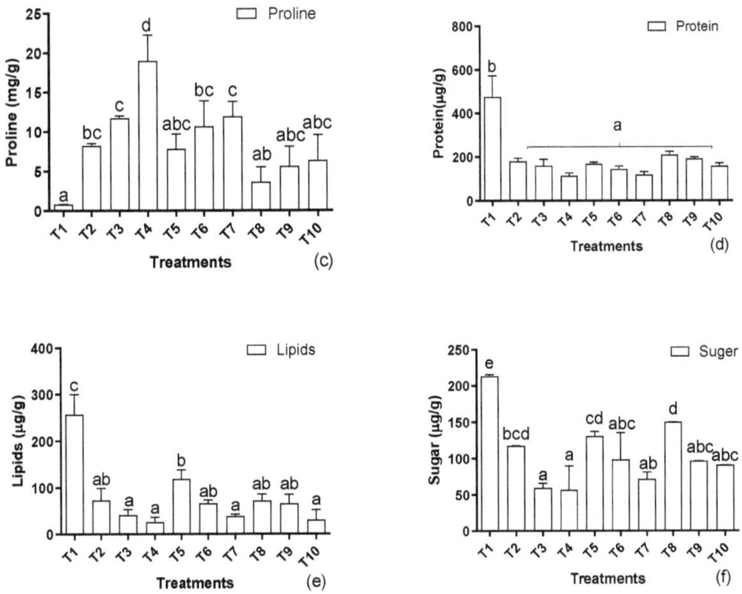

Figure 3. Effects of Cd levels, EDTA, and IAA on (**a**) flavonoids, (**b**) phenolics, (**c**) proline, (**d**) protein, (**e**) lipids, and (**f**) sugar contents of *H. annuus*. Data are the means of three replicates ± standard errors, and the letters represent significant differences ($p < 0.05$).

3.4. Antioxidant Response

Plants in stress boost their antioxidant system to cope with the primary and secondary stressful conditions. In the current scenario, treating plants with cadmium decreased the production of catalases as the metal level increased in the soil (Figure 4a). No further improvements were recorded with the application of EDTA in the soil or the foliar application of IAA at all supplemented levels of cadmium. A contrasting tendency was noted in the case of ascorbate peroxidases, indicating a multifold increase with the increase in metal in the soil up to 150 mg/kg (Figure 4b). Similar improvements were also recorded with the application of EDTA and IAA. Nonetheless, the enzyme units were lower than in cadmium-stressed plants.

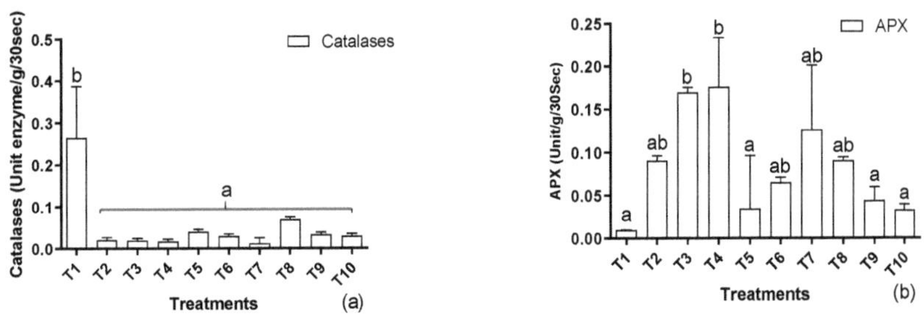

Figure 4. Effects of Cd levels, EDTA, and IAA on (**a**) catalases and (**b**) ascorbate peroxidase activity of *H. annuus*. Data are the means of three replicates ± standard errors, and the letters represent significant differences ($p < 0.05$).

3.5. Metal Determination

The total cadmium contents in the plant parts increased with the increase in cadmium supplementation in the soil from 0 to 150 mg/kg (Figure 5a). A similar tendency was also noted in the case of the application of EDTA, showing increased accumulation with the increase in metal in the soil. In the case of the foliar application of IAA, an even higher accumulation was recorded as the cadmium supplementation was elevated in the soil. The highest accumulation was noted in the plants treated with the foliar application of IAA, followed by EDTA application and only cadmium-stressed plants. Similarly, the accumulation was enhanced by an increase in exposure time. A lower accumulation was recorded in the plants exposed for 15 days to cadmium supplements. An interesting result was found in case of 30 days of exposure, which showed a decline in accumulation compared to 15 days of exposure. However, after 60 days of exposure, the maximum increase was recorded in the accumulation of cadmium from the soil, showing a direct relation to the duration of exposure to metals supplemented in the soil.

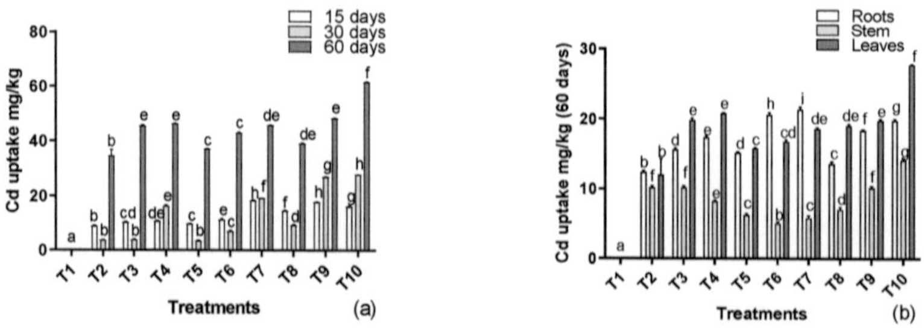

Figure 5. Effects of Cd levels, EDTA, and IAA on (**a**) cadmium uptake and (**b**) translocation to the aerial parts of *H. annuus*. Data are the means of three replicates ± standard errors, and the letters represent significant differences ($p < 0.05$).

Similar patterns were also recorded in the case of translocation of the metal to the aerial parts of the plants (Figure 5b). A higher amount was recorded in the roots of the plants, followed by the leaves of the host. The lowest accumulation was recorded in the stem, and no accumulation was recorded in the seeds of the sunflowers. When treating plants with cadmium levels, an increase in the accumulation in the roots was recorded. However, the application of EDTA in the soil and foliar IAA further enhanced the accumulation in the roots, thereby enhancing phytoremediation potential. A contrasting tendency was noted in case of the stems, i.e., with the inclination of metal level, the accumulation decreased in a concentration-based manner. The application of EDTA in the soil induced a significant improvement in the translocation to the stem. Interesting results were recorded when foliar IAA was applied, showing the highest accumulation in the stems at higher levels of cadmium supplementation. In the case of the leaves, a dose-dependent increase was recorded with the cadmium levels. Nonetheless, EDTA application lowered the translocation of cadmium to the leaves in a concentration-based manner in comparison to the stressed control plants. Similarly, a significant increase was recorded with IAA application, showing higher translocation and the accumulation of cadmium in the leaves.

4. Discussion

The increase in global population requires higher yields and production, which could be possible with better soil health and balanced nutrient conditions. Soil reclamation and the removal of these toxic metals are challenging tasks for agriculturists and other plants scientists [27,28]. The removal of toxic heavy metals from the agroecological zones in a sustainable way is the call of the day [29–31]. Among the techniques, phytoremediation with hyperaccumulating plants with short lives could be a possible solution to maintain better health in soils and increase their productivity to meet the food demands for the ever-growing human populations [32]. In the current scenario, sunflowers supplemented with the aforementioned levels of the cadmium were severely impaired in terms of plant height, root length, and chlorophyll contents [33]. The chlorophyll content decreased in a dose-dependent manner in response to cadmium. A similar pattern of chlorophyll reduction

has been reported in previous studies [34–39]. The excessive ROS production leads to the oxidation of the chlorophyll contents, leading to their degeneration and degradations as a result of secondary oxidative stress. The inhibition of chlorophyll may be attributed to the inhibition of enzymes involving pigment biosynthesis in response to cadmium, as has been previously demonstrated by Qian et al. (2009). Moreover, cadmium stress has shown to induce an imbalance in cell redox homeostasis, which leads to oxidative damage in plants (Hendrix et al., 2020).

The application of EDTA in the soil and foliar IAA reduced the effects of Cd by 8, 9, and 7% at 50, 100, and 150 mg/kg, respectively. To cope with stressful environmental constrains, plants produce a range of substances, including phytohormones, such as IAA and SA, and other metabolites, such as flavonoids, phenolics, proteins, proline, and sugar, to maintain homeostatic conditions inside the cell and maintain cell viability under environmental constrains [40,41]. Among the phytohormones, IAA not only acts as a plant growth promotor, but recently it was also revealed that it has a role in stress mitigation [42]. In the current scenario, a decline in the total IAA content was recorded with the increase in metal concentration in the medium. This was probably due to the fact that beyond a threshold level the metal causes the breaking down of IAA, which results in lower synthesis and the accumulation of IAA by the host, resulting in lower growth, development, and hence lower biomass production, which was recorded in the current study [43]. On the other hand, salicylic acid is a stress hormone that showed an increasing pattern with higher metal levels. Additionally, SA is known to reduce stress in individuals from several kingdoms, including humans, plants, and animals. Along these lines, it stands to reason that perhaps the host could be employing increased production as a stress management technique. In order to help their host to survive the harsh situation, chaperones, heat shock proteins, antioxidants, and genes involved in the manufacture of secondary metabolites such as sinapyl alcohol dehydrogenase, cinnamyl alcohol dehydrogenase, and cytochrome P450 are all activated by SA [44,45]. In the current condition, SA production was enhanced with increases in the metal concentration in a dose-dependent manner, which is an effective stress-mitigating strategy for ameliorating host stress tolerance [46]. Phytohormones and other stress-related metabolites relieve cadmium, and the resultant ROS accumulation also plays a key role in the mitigation of cadmium stress in sunflowers by improving growth and strengthening the antioxidant and metabolic systems [47]. To cope with metal stress, plants produce a range of secondary metabolites, including flavonoids, phenolics, low-molecular-weight protein, and sugar and proline contents [48]. In the current condition, lower flavonoid contents were recorded upon exposure to the mentioned levels of cadmium [49]. These lower flavonoids were perhaps because cadmium-induced stress results in the disturbance of the phenylalanine pathway, resulting in lower flavonoid production [50,51]. On the contrary, higher levels of phenolics were recorded at all mentioned levels of the metal. The phenolics were acting as effective ROS quenchers to relieve the stressful metal conditions [52]. Similarly, higher proline contents were recorded with the exposure of aforementioned concentrations of the metal. The proline contents act as an osmolyte, maintaining the host osmotic adjustment [53–55]. On the other hand, the protein, lipid, and sugar contents showed declines with the increase in metal [56]. This was probably due to the fact that a higher metal level caused toxicity to the cell, leading to systematic cell death. In some cases, the metal acts as a competing inhibitor and binds with the active sites of the enzymes, thus making them malfunction, resulting in lower production of protein, lipid, and sugar contents [57].

Plants produce ample amounts of antioxidants in response to environmental constrains, which aid the host to cope with the ROS produced as a result of the environmental stressors. In the current findings, lower catalase activity was recorded in all treatments compared to the control plants [58]. On the contrary, higher ascorbate peroxidase levels were recorded for each increase in metal in the soil. Higher ascorbate peroxidase contents showed efficient ROS scavenging and enhanced the stress tolerance of the host plants [59]. In order to control the ROS levels and preserve cellular homeostasis under stress, plants have a battery of antioxidant molecules. One essential antioxidant enzyme of such scavenging systems is ascorbate peroxidase (APX). It utilizes ascorbate as an electron donor to catalyze the transformation of H_2O_2 into H_2O and O_2. In response to environmental challenges and during typical plant growth and development, APX expression is variably regulated. Depending on their subcellular location and the presence of certain regulatory elements in the upstream regions of the corresponding genes, various isoforms of APX exhibit distinct responses to environmental stressors [60]. In a previous study, the supplementation of thiourea as a scavenger of ROS changed the expression of various arsenic (As)-related transporter genes in flag leaves and developing grains (inflorescence) of rice [61]. The antioxidant enzymes were also altered, and as accumulation was significantly reduced in rice. Moreover, vermicomposting in conjunction with phytoremediation can be an efficient strategy to restrict the bioavailability of soil pollutants [61].

Different factors affect the bioavailability of cadmium to the host plants, including pH, moisture, and the temperature of the surroundings. The bioavailability of cadmium increased with an increase in the pH based on pore water concentrations, explaining the reduced competition of H+ ions making cadmium more bioavailable in pore water at a high pH [62]. Cadmium sorption to the soils, estimated from water-soluble concentrations, was not significantly affected by the soil moisture content [63]. In a study with ryegrass, the uptake of both 109Cd and 65 Zn and their stable isotopes was higher in ryegrass grown at 21 °C than that grown at 9 °C. Results from a fractionation and speciation analysis of soil cadmium and zinc were correlated with plant uptake, and there was a good consistency between the observed plant uptake, the physicochemical forms of cadmium and zinc in the soil, and the soil solution presumed to be available to the plants [64]. However, in a study with metal and silicon, the bioavailability decreased with an increase in temperature [15]. The absorption of the metal by the host root showed a multifold dose-dependent increase when sunflower seedlings were grown in a soil condition spiked with the mentioned levels of the selected metal [45,65]. The higher uptake leads to higher bioremediation of the environment [48,66,67]. On the contrary, an increase was recorded in the Cd accumulation with the application of EDTA [68]. Similarly, the application of IAA tends to increase the accumulation of cadmium in the plant parts [69,70]. The IAA helps the plant to adjust to abiotic stresses. Moreover, IAA is an acidic hormone that helps to make the cell wall more flexible to expansion and cell division [71,72]. As a result of cell expansion and cell division, more compartments are available for metal compartmentalization, thereby dividing the stress to minimize its effects and allow the plant to grow normally, even at higher levels of metal in the medium [72]. In the current scenario, the accumulation of the metal increased with an increase in the exposure duration. Lower accumulation was noted in the sunflower with a duration of exposure to the metal of 15 days, followed by 30 days of exposure, and higher accumulation was recorded at 60 days of cadmium exposure. Similarly, higher translocation to the upper parts was also recorded. Higher absorption and accumulation were recorded by the roots and were subsequently translocated to the aerial parts of the plants. The accumulation of cadmium increased in the roots with the increase in metal supplements in the soil. The translocation and subsequent accumulation showed a decline in the case of the stem as the metal supplements increased in the soil [73]. The EDTA and IAA supplements increased the metal accumulation in the stem [68,74]. The highest translocation was recorded to the leaves of the sunflower, which showed a higher accumulation of cadmium. A positive correlation was recorded with the metal levels. Similarly, the rate of translocation and accumulation was higher with the application of both EDTA in the soil and foliar IAA [75]. For instance, no accumulation of cadmium was recorded in the seeds of the host plants.

5. Conclusions

The current evidence shows that Cd at higher concentrations actively accumulated in the hyperaccumulating plants, and beyond the threshold level (the WHO permissible level of 0.003 mg/kg), Cd exerted toxic effects in its host. The order of *H. annuus* plant parts based on the Cd concentration was stem > leaves > root and shoot > root. However, no accumulation or translocation were recorded in the seeds at any level of the metal and exposure time, ensuring food safety. The Cd accumulation in the leaves, stems, and roots increased in combination with EDTA and IAA compared to Cd applications alone in the control. The concentrations of Cd after 60 days in *H. annuus* subjected to Cd150 + EDTA + IAA exhibited a maximum accumulation of Cd of 64.80 mg/kg. The application of EDTA in soil and foliar IAA further improved the phytoremediation potential of the host and reduced the metal contaminant efficiently at the site. The application of these chemicals could be the possible solution for the rapid and enhanced bioremediation of sites contaminated with higher levels of cadmium.

Author Contributions: N.S.: Data curation, Formal analysis, Investigation, Methodology, Writing—original draft. M.Q.: Methodology, Writing—original draft. M.I.: Supervision. A.H.: Supervision. M.H.: Lab facility, resources. W.M.: Investigation, Writing—original draft. A.K.: Supervision, review and fund acquisition. A.A.-H.: supervision, review and fund acquisition. All authors have read and agreed to the published version of the manuscript.

Funding: The project was funded by The Oman Research Council (TRC) through the funded project (BFP/RGP/CBS/21/006).

Institutional Review Board Statement: Not applicable.

Informed Consent Statement: Not applicable.

Data Availability Statement: All the data are included in the manuscript.

Acknowledgments: The authors would like to thank the University of Nizwa for the generous support of this project. We thank technical staff for assistance.

Conflicts of Interest: The authors declare no conflict of interest.

References

1. Okereafor, U.; Makhatha, M.; Mekuto, L.; Uche-Okereafor, N.; Sebola, T.; Mavumengwana, V. Toxic metal implications on agricultural soils, plants, animals, aquatic life and human health. *Int. J. Environ. Res. Public Health* **2020**, *17*, 2204. [CrossRef] [PubMed]
2. Rajendran, S.; Priya, T.; Khoo, K.S.; Hoang, T.K.; Ng, H.-S.; Munawaroh, H.S.H.; Karaman, C.; Orooji, Y.; Show, P. A critical review on various remediation approaches for heavy metal contaminants removal from contaminated soils. *Chemosphere* **2022**, *287*, 132369. [CrossRef]
3. Mottet, A.; Tempio, G. Global poultry production: Current state and future outlook and challenges. *World's Poult. Sci. J.* **2017**, *73*, 245–256. [CrossRef]
4. Li, C.; Zhou, K.; Qin, W.; Tian, C.; Qi, M.; Yan, X.; Han, W.J.S. A review on heavy metals contamination in soil: Effects, sources, and remediation techniques. *Soil Sediment Contam. Int. J.* **2019**, *28*, 380–394. [CrossRef]
5. Mohammed, A.S.; Kapri, A.; Goel, R. Heavy metal pollution: Source, impact, and remedies. In *Biomanagement of Metal-Contaminated Soils*; Springer: Berlin/Heidelberg, Germany, 2011; pp. 1–28.
6. Duplay, J.; Semhi, K.; Errais, E.; Imfeld, G.; Babcsanyi, I.; Perrone, T. Copper, zinc, lead and cadmium bioavailability and retention in vineyard soils (Rouffach, France): The impact of cultural practices. *Geoderma* **2014**, *230*, 318–328. [CrossRef]
7. Degryse, F.; Shahbazi, A.; Verheyen, L.; Smolders, E. Diffusion limitations in root uptake of cadmium and zinc, but not nickel, and resulting bias in the Michaelis constant. *Plant Physiol.* **2012**, *160*, 1097–1109. [CrossRef]
8. Shahid, M.; Dumat, C.; Khalid, S.; Niazi, N.K.; Antunes, P. Cadmium bioavailability, uptake, toxicity and detoxification in soil-plant system. *Rev. Environ. Contam. Toxicol.* **2016**, *241*, 73–137.
9. Jadia, C.D.; Fulekar, M.H. Phytoremediation of heavy metals: Recent techniques. *Afr. J. Biotechnol.* **2009**, *8*, 921–928.
10. Kalay, M.; Canli, M. Elimination of essential (Cu, Zn) and non-essential (Cd, Pb) metals from tissues of a freshwater fish Tilapia zilli. *Turk. J. Zool.* **2000**, *24*, 429–436.
11. Waseem, A.; Arshad, J.; Iqbal, F.; Sajjad, A.; Mehmood, Z.; Murtaza, G. Pollution status of Pakistan: A retrospective review on heavy metal contamination of water, soil, and vegetables. *BioMed Res. Int.* **2014**, *2014*, 813206. [CrossRef]
12. Garbisu, C.; Alkorta, I. Phytoextraction: A cost-effective plant-based technology for the removal of metals from the environment. *Bioresour. Technol.* **2001**, *77*, 229–236. [CrossRef]
13. Ashraf, S.; Ali, Q.; Zahir, Z.A.; Ashraf, S.; Asghar, H.N. Phytoremediation: Environmentally sustainable way for reclamation of heavy metal polluted soils. *Ecotoxicol. Environ. Saf.* **2019**, *174*, 714–727. [CrossRef]
14. Yan, A.; Wang, Y.; Tan, S.N.; Mohd Yusof, M.L.; Ghosh, S.; Chen, Z. Phytoremediation: A promising approach for revegetation of heavy metal-polluted land. *Front. Plant Sci.* **2020**, *11*, 359. [CrossRef] [PubMed]

15. Bai, B.; Nie, Q.; Zhang, Y.; Wang, X.; Hu, W. Cotransport of heavy metals and SiO_2 particles at different temperatures by seepage. *J. Hydrol.* **2021**, *597*, 125771. [CrossRef]
16. Bai, B.; Jiang, S.; Liu, L.; Li, X.; Wu, H. The transport of silica powders and lead ions under unsteady flow and variable injection concentrations. *Powder Technol.* **2021**, *387*, 22–30. [CrossRef]
17. Singh, T.B.; Ali, A.; Prasad, M.; Yadav, A.; Shrivastav, P.; Goyal, D.; Dantu, P.K. Role of organic fertilizers in improving soil fertility. In *Contaminants in Agriculture*; Springer: Berlin/Heidelberg, Germany, 2020; pp. 61–77.
18. Hussain, A.; Hasnain, S. Interactions of bacterial cytokinins and IAA in the rhizosphere may alter phytostimulatory efficiency of rhizobacteria. *World J. Microbiol. Biotechnol.* **2011**, *27*, 2645–2654. [CrossRef]
19. Warrier, R.; Paul, M.; Vineetha, M. Estimation of salicylic acid in Eucalyptus leaves using spectrophotometric methods. *Genet. Plant Physiol.* **2013**, *3*, 90–97.
20. El Far, M.; Taie, H.A. Antioxidant activities, total anthocyanins, phenolics and flavonoids contents of some sweetpotato genotypes under stress of different concentrations of sucrose and sorbitol. *Aust. J. Basic Appl. Sci.* **2009**, *3*, 3609–3616.
21. Bates, L.S.; Waldren, R.P.; Teare, I. Rapid determination of free proline for water-stress studies. *Plant Soil* **1973**, *39*, 205–207. [CrossRef]
22. Radhakrishnan, R.; Lee, I.-J. Ameliorative effects of spermine against osmotic stress through antioxidants and abscisic acid changes in soybean pods and seeds. *Acta Physiol. Plant.* **2013**, *35*, 263–269. [CrossRef]
23. Asada, K. Ascorbate peroxidase—A hydrogen peroxide-scavenging enzyme in plants. *Physiol. Plant.* **1992**, *85*, 235–241. [CrossRef]
24. Bradford, M.M. A rapid and sensitive method for the quantitation of microgram quantities of protein utilizing the principle of protein-dye binding. *Anal. Biochem.* **1976**, *72*, 248–254. [CrossRef]
25. Schoefs, B.t. Chlorophyll and carotenoid analysis in food products. Properties of the pigments and methods of analysis. *Trends Food Sci. Technol.* **2002**, *13*, 361–371. [CrossRef]
26. Hussain, A.; Alamzeb, S.; Begum, S. Accumulation of heavy metals in edible parts of vegetables irrigated with waste water and their daily intake to adults and children, District Mardan, Pakistan. *Food Chem.* **2013**, *136*, 1515–1523.
27. Kloppenburg, J. Agriculturalists' Reclamation. Ph.D. Thesis, Stanford University, Stanford, CA, USA, June 2018; p. 302.
28. Lyons, G.; Stangoulis, J.; Graham, R. High-selenium wheat: Biofortification for better health. *Nutr. Res. Rev.* **2003**, *16*, 45–60. [CrossRef]
29. Bhargava, A.; Carmona, F.F.; Bhargava, M.; Srivastava, S. Approaches for enhanced phytoextraction of heavy metals. *J. Environ. Manag.* **2012**, *105*, 103–120. [CrossRef]
30. Rahman, Z.; Singh, V.P. The relative impact of toxic heavy metals (THMs)(arsenic (As), cadmium (Cd), chromium (Cr)(VI), mercury (Hg), and lead (Pb)) on the total environment: An overview. *Environ. Monit. Assess.* **2019**, *191*, 419. [CrossRef]
31. Kinuthia, G.K.; Ngure, V.; Beti, D.; Lugalia, R.; Wangila, A.; Kamau, L. Levels of heavy metals in wastewater and soil samples from open drainage channels in Nairobi, Kenya: Community health implication. *Sci. Rep.* **2020**, *10*, 8434. [CrossRef]
32. Liu, L.; Li, W.; Song, W.; Guo, M. Remediation techniques for heavy metal-contaminated soils: Principles and applicability. *Sci. Total Environ.* **2018**, *633*, 206–219. [CrossRef]
33. Zafar-ul-Hye, M.; Naeem, M.; Danish, S.; Khan, M.J.; Fahad, S.; Datta, R.; Brtnicky, M.; Kintl, A.; Hussain, G.S.; El-Esawi, M.A. Effect of cadmium-tolerant rhizobacteria on growth attributes and chlorophyll contents of bitter gourd under cadmium toxicity. *Plants* **2020**, *9*, 1386. [CrossRef]
34. Yang, H.; Shi, G.; Xu, Q.; Wang, H. Cadmium effects on mineral nutrition and stress in Potamogeton crispus. *Russ. J. Plant Physiol.* **2011**, *58*, 253–260. [CrossRef]
35. Gill, S.S.; Khan, N.A.; Tuteja, N. Cadmium at high dose perturbs growth, photosynthesis and nitrogen metabolism while at low dose it up regulates sulfur assimilation and antioxidant machinery in garden cress (*Lepidium sativum* L.). *Plant Sci.* **2012**, *182*, 112–120. [CrossRef] [PubMed]
36. Nada, E.; Ferjani, B.A.; Ali, R.; Bechir, B.R.; Imed, M.; Makki, B. Cadmium-induced growth inhibition and alteration of biochemical parameters in almond seedlings grown in solution culture. *Acta Physiol. Plant.* **2007**, *29*, 57–62. [CrossRef]
37. Nabi, A.; Aftab, T.; Masroor, M.; Khan, A.; Naeem, M. Exogenous triacontanol provides tolerance against arsenic-induced toxicity by scavenging ROS and improving morphology and physiological activities of *Mentha arvensis* L. *Environ. Pollut.* **2022**, *295*, 118609. [CrossRef] [PubMed]
38. Ma, J.; Saleem, M.H.; Alsafran, M.; Al Jabri, H.; Rizwan, M.; Nawaz, M.; Ali, S.; Usman, K. Response of cauliflower (*Brassica oleracea* L.) to nitric oxide application under cadmium stress. *Ecotoxicol. Environ. Saf.* **2022**, *243*, 113969. [CrossRef] [PubMed]
39. Muradoglu, F.; Gundogdu, M.; Ercisli, S.; Encu, T.; Balta, F.; Jaafar, H.Z.; Zia-Ul-Haq, M. Cadmium toxicity affects chlorophyll a and b content, antioxidant enzyme activities and mineral nutrient accumulation in strawberry. *Biol. Res.* **2015**, *48*, 11. [CrossRef]
40. Govind, G.; Kulkarni, J.; Shinde, H.; Dudhate, A.; Srivastava, A.; Suprasanna, P. Plant abiotic stress tolerance on the transcriptomics atlas. In *Advancements in Developing Abiotic Stress-Resilient Plants*; CRC Press: Boca Raton, FL, USA, 2022; pp. 193–236.
41. Husna; Hussain, A.; Shah, M.; Hamayun, M.; Iqbal, A.; Qadir, M.; Alataway, A.; Dewidar, A.Z.; Elansary, H.O.; Lee, I.-J. Phytohormones producing rhizobacteria alleviate heavy metals stress in soybean through multilayered response. *Microbiol. Res.* **2022**, *266*, 127237. [CrossRef]
42. Salvi, P.; Manna, M.; Kaur, H.; Thakur, T.; Gandass, N.; Bhatt, D.; Muthamilarasan, M. Phytohormone signaling and crosstalk in regulating drought stress response in plants. *Plant Cell Rep.* **2021**, *40*, 1305–1329. [CrossRef]

43. Ma, X.; Zhao, X.; Zhang, Q.; Zhou, Z.; Dou, Y.; Ji, W.; Li, J. Comparative transcriptome analysis of broccoli seedlings under different Cd exposure levels revealed possible pathways involved in hormesis. *Sci. Hortic.* **2022**, *304*, 111330. [CrossRef]
44. Fahad, S.; Hussain, S.; Bano, A.; Saud, S.; Hassan, S.; Shan, D.; Khan, F.A.; Khan, F.; Chen, Y.; Wu, C.; et al. Potential role of phytohormones and plant growth-promoting rhizobacteria in abiotic stresses: Consequences for changing environment. *Environ. Sci. Pollut. Res.* **2015**, *22*, 4907–4921. [CrossRef]
45. Qadir, M.; Hussain, A.; Shah, M.; Lee, I.J.; Iqbal, A.; Irshad, M.; Sayyed, A.; Ahmad, A.; Hamayun, M. Comparative assessment of chromate bioremediation potential of Pantoea conspicua and *Aspergillus niger*. *J. Hazard. Mater.* **2022**, *424*, 127314. [CrossRef] [PubMed]
46. Yan, Z.; Tam, N.F.Y. Effects of lead stress on anti-oxidative enzymes and stress-related hormones in seedlings of Excoecaria agallocha Linn. *Plant Soil* **2013**, *367*, 327–338. [CrossRef]
47. Husna; Hussain, A.; Shah, M.; Hamayun, M.; Iqbal, A.; Murad, W.; Irshad, M.; Qadir, M.; Kim, H.-Y. Pseudocitrobacter anthropi reduces heavy metal uptake and improves phytohormones and antioxidant system in *Glycine max* L. *World J. Microbiol. Biotechnol.* **2021**, *37*, 195. [CrossRef]
48. Qadir, M.; Hussain, A.; Hamayun, M.; Shah, M.; Iqbal, A.; Irshad, M.; Ahmad, A.; Lodhi, M.A.; Lee, I.-J. Phytohormones Producing Acinetobacter bouvetii P1 Mitigates Chromate Stress in Sunflower by Provoking Host Antioxidant Response. *Antioxidants* **2021**, *10*, 1868. [CrossRef]
49. Ibrahim, M.H.; Chee Kong, Y.; Mohd Zain, N.A. Effect of Cadmium and Copper Exposure on Growth, Secondary Metabolites and Antioxidant Activity in the Medicinal Plant Sambung Nyawa (*Gynura procumbens* (Lour.) Merr). *Molecules* **2017**, *22*, 1623. [CrossRef]
50. Chen, Q.Y.; Murphy, A.; Sun, H.; Costa, M. Molecular and epigenetic mechanisms of Cr (VI)-induced carcinogenesis. *Toxicol. Appl. Pharmacol.* **2019**, *377*, 114636. [CrossRef]
51. Hamayun, M.; Khan, N.; Khan, M.N.; Qadir, M.; Hussain, A.; Iqbal, A.; Khan, S.A.; Rehman, K.U.; Lee, I.-J. Antimicrobial and plant growth-promoting activities of bacterial endophytes isolated from Calotropis procera (Ait.) WT Aiton. *Biocell* **2021**, *45*, 363–369. [CrossRef]
52. Waśkiewicz, A.; Muzolf-Panek, M.; Goliński, P. Phenolic content changes in plants under salt stress. In *Ecophysiology and Responses of Plants under Salt Stress*; Ahmad, P., Azooz, M.M., Prasad, M.N.V., Eds.; Springer: New York, NY, USA, 2013; pp. 283–314.
53. Mitra, S.; Pramanik, K.; Sarkar, A.; Ghosh, P.K.; Soren, T.; Maiti, T.K. Bioaccumulation of cadmium by Enterobacter sp. and enhancement of rice seedling growth under cadmium stress. *Ecotoxicol. Environ. Saf.* **2018**, *156*, 183–196. [CrossRef]
54. Hamayun, M.; Hussain, A.; Iqbal, A.; Khan, S.A.; Lee, I.-J. Endophytic fungus Aspergillus japonicus mediates host plant growth under normal and heat stress conditions. *BioMed Res. Int.* **2018**, *2018*, 7696831.
55. Ismail, A.H.; Mehmood, A.; Qadir, M.; Husna, A.I.; Hamayun, M.; Khan, N. Thermal stress alleviating potential of endophytic fungus rhizopus oryzae inoculated to sunflower (*Helianthus annuus* L.) and soybean (*Glycine max* L.). *Pak. J. Bot.* **2020**, *52*, 1857–1865. [CrossRef]
56. Zhang, S.; Zhang, H.; Qin, R.; Jiang, W.; Liu, D. Cadmium induction of lipid peroxidation and effects on root tip cells and antioxidant enzyme activities in *Vicia faba* L. *Ecotoxicology* **2009**, *18*, 814–823. [CrossRef] [PubMed]
57. Zulfiqar, U.; Ayub, A.; Hussain, S.; Waraich, E.A.; El-Esawi, M.A.; Ishfaq, M.; Ahmad, M.; Ali, N.; Maqsood, M.F. Cadmium Toxicity in Plants: Recent Progress on Morpho-physiological Effects and Remediation Strategies. *J. Soil Sci. Plant Nutr.* **2022**, *22*, 212–269. [CrossRef]
58. Trchounian, A.; Petrosyan, M.; Sahakyan, N. Plant cell redox homeostasis and reactive oxygen species. In *Redox State as a Central Regulator of Plant-Cell Stress Responses*; Gupta, D.K., Palma, J.M., Corpas, F.J., Eds.; Springer International Publishing: Cham, Switzerland, 2016; pp. 25–50.
59. Saxena, S.C.; Salvi, P.; Kamble, N.U.; Joshi, P.K.; Majee, M.; Arora, S. Ectopic overexpression of cytosolic ascorbate peroxidase gene (*Apx1*) improves salinity stress tolerance in *Brassica juncea* by strengthening antioxidative defense mechanism. *Acta Physiol. Plant.* **2020**, *42*, 45. [CrossRef]
60. Verma, D.; Upadhyay, S.K.; Singh, K. Characterization of APX and APX-R gene family in Brassica juncea and B. rapa for tolerance against abiotic stresses. *Plant Cell Rep.* **2022**, *41*, 571–592. [CrossRef]
61. Upadhyay, M.K.; Majumdar, A.; Srivastava, A.K.; Bose, S.; Suprasanna, P.; Srivastava, S. Antioxidant enzymes and transporter genes mediate arsenic stress reduction in rice (*Oryza sativa* L.) upon thiourea supplementation. *Chemosphere* **2022**, *292*, 133482. [CrossRef] [PubMed]
62. Ardestani, M.M.; van Gestel, C.A.M. Using a toxicokinetics approach to explain the effect of soil pH on cadmium bioavailability to Folsomia candida. *Environ. Pollut.* **2013**, *180*, 122–130. [CrossRef]
63. Van Gestel, C.A.M.; van Diepen, A.M.F. The Influence of Soil Moisture Content on the Bioavailability and Toxicity of Cadmium forFolsomia candidaWillem (Collembola: Isotomidae). *Ecotoxicol. Environ. Saf.* **1997**, *36*, 123–132. [CrossRef]
64. Almås, Å.; Singh, B. Plant uptake of cadmium-109 and zinc-65 at different temperature and organic matter levels. *J. Environ. Qual.* **2001**, *30*, 869–877. [CrossRef]
65. Qadir, M.; Hussain, A.; Hamayun, M.; Shah, M.; Iqbal, A.; Husna; Murad, W. Phytohormones producing rhizobacterium alleviates chromium toxicity in *Helianthus annuus* L. by reducing chromate uptake and strengthening antioxidant system. *Chemosphere* **2020**, *258*, 127386. [CrossRef]

66. Guo, J.; Tang, S.; Ju, X.; Ding, Y.; Liao, S.; Song, N. Effects of inoculation of a plant growth promoting rhizobacterium *Burkholderia* sp. D54 on plant growth and metal uptake by a hyperaccumulator *Sedum alfredii* Hance grown on multiple metal contaminated soil. *World J. Microbiol. Biotechnol.* **2011**, *27*, 2835–2844. [CrossRef]
67. Zahoor, M.; Irshad, M.; Rahman, H.; Qasim, M.; Afridi, S.G.; Qadir, M.; Hussain, A. Alleviation of heavy metal toxicity and phytostimulation of *Brassica campestris* L. by endophytic *Mucor* sp. MHR-7. *Ecotoxicol. Environ. Saf.* **2017**, *142*, 139–149. [CrossRef] [PubMed]
68. Seth, C.S.; Misra, V.; Singh, R.R.; Zolla, L. EDTA-enhanced lead phytoremediation in sunflower (*Helianthus annuus* L.) hydroponic culture. *Plant Soil* **2011**, *347*, 231. [CrossRef]
69. Sun, S.; Zhou, X.; Cui, X.; Liu, C.; Fan, Y.; McBride, M.B.; Li, Y.; Li, Z.; Zhuang, P. Exogenous plant growth regulators improved phytoextraction efficiency by *Amaranths hypochondriacus* L. in cadmium contaminated soil. *Plant Growth Regul.* **2020**, *90*, 29–40. [CrossRef]
70. Husna; Hussain, A.; Shah, M.; Hamayun, M.; Qadir, M.; Iqbal, A. Heavy metal tolerant endophytic fungi Aspergillus welwitschiae improves growth, ceasing metal uptake and strengthening antioxidant system in Glycine maxL. *Environ. Sci. Pollut. Res.* **2022**, *29*, 15501–15515. [CrossRef] [PubMed]
71. Zhang, A.; Yang, X.; Lu, J.; Song, F.; Sun, J.; Wang, C.; Lian, J.; Zhao, L.; Zhao, B. OsIAA20, an Aux/IAA protein, mediates abiotic stress tolerance in rice through an ABA pathway. *Plant Sci.* **2021**, *308*, 110903. [CrossRef] [PubMed]
72. Sánchez-Rodríguez, C.; Rubio-Somoza, I.; Sibout, R.; Persson, S. Phytohormones and the cell wall in Arabidopsis during seedling growth. *Trends Plant Sci.* **2010**, *15*, 291–301. [CrossRef]
73. Fırat, Ö.; Çogun, H.Y.; Aslanyavrusu, S.; Kargın, F. Antioxidant responses and metal accumulation in tissues of Nile tilapia Oreochromis niloticus under Zn, Cd and Zn + Cd exposures. *J. Appl. Toxicol.* **2009**, *29*, 295–301. [CrossRef]
74. Qadir, M.; Hussain, A.; Shah, M.; Hamayun, M.; Iqbal, A. Enhancement of chromate phytoremediation and soil reclamation potential of *Brassica campestris* L. by *Aspergillus niger*. *Environ. Sci. Pollut. Res.* **2022**, *29*, 1–12. [CrossRef]
75. Hadi, F.; Bano, A.; Fuller, M.P. The improved phytoextraction of lead (Pb) and the growth of maize (*Zea mays* L.): The role of plant growth regulators (GA3 and IAA) and EDTA alone and in combinations. *Chemosphere* **2010**, *80*, 457–462. [CrossRef]

MDPI
St. Alban-Anlage 66
4052 Basel
Switzerland
www.mdpi.com

Metabolites Editorial Office
E-mail: metabolites@mdpi.com
www.mdpi.com/journal/metabolites

www.ingramcontent.com/pod-product-compliance
Lightning Source LLC
Chambersburg PA
CBHW041406210926
51850CB00013B/3192
* 9 7 8 3 7 2 5 8 3 5 1 8 8 *